Developments in Environmental Modelling, 14

Mathematical Submodels in Water Quality Systems

1. ENERGY AND ECOLOGICAL MODELLING
edited by W.J. Mitsch, R.W. Bossermann and J.M. Klopatek
1981 839 pp.

2. WATER MANAGEMENT MODELS IN PRACTICE:
A CASE STUDY OF THE ASWAN HIGH DAM
by D. Whittington and G. Guariso
1983 xxii+246 pp.

3. NUMERICAL ECOLOGY
by L. Legendre and P. Legendre
1983 xvi+419 pp.

4A. APPLICATION OF ECOLOGICAL MODELLING IN ENVIRONMENTAL MANAGEMENT
PART A
edited by S.E. Jørgensen
1983 viii+735 pp.

4.B APPLICATION OF ECOLOGICAL MODELLING IN ENVIRONMENTAL MANAGEMENT
PART B
edited by S.E. Jørgensen and W.J. Mitsch
1983 viii+438 pp.

5. ANALYSIS OF ECOLOGICAL SYSTEMS: STATE-OF-THE-ART IN ECOLOGICAL MODELLING
edited by W.K. Lauenroth, G.V. Skogerboe and M. Flug
1983 992 pp.

6. MODELLING THE FATE AND EFFECT OF TOXIC SUBSTANCES IN THE ENVIRONMENT
edited by S.E. Jørgensen
1984 viii+342 pp.

7. MATHEMATICAL MODELS IN BIOLOGICAL WASTE WATER TREATMENT
edited by S.E. Jørgensen and M.J. Gromiec
1985 vi+802 pp.

8. FRESHWATER ECOSYSTEMS: MODELLING AND SIMULATION
by M. Straškaba and A.H. Gnauck
1985 309 pp.

9. FUNDAMENTALS OF ECOLOGICAL MODELLING
by S.E. Jørgensen
1986 389 pp.

10. AGRICULTURAL NONPOINT SOURCE POLLUTION:
MODEL SELECTION AND APPLICATION
edited by A. Giorgini and F. Zingales
1986 409 pp.

11. MATHEMATICAL MODELLING OF ENVIRONMENTAL AND ECOLOGICAL SYSTENS
edited by J.B. Shukla, T.G. Hallam and V. Capasso
1987 xii+254pp.

12. WETLAND MODELLING
edited by W.J. Mitsch, M. Straskraba and S.E. Jørgensen
1988 x+228 p.

13. ADVANCES IN ENVIRONMENTAL MODELLING
edited by A. Marani
1988 691 pp.

Developments in Environmental Modelling, 14

Mathematical Submodels in Water Quality Systems

Edited by

S.E. Jørgensen
Langkaer Vaenge 9, 3500 Vaerløse, Copenhagen, Denmark

and

M.J. Gromiec
Instytut Meteorologii i Gospodarki Wodnej, 01-673 Warszawa u. Podlesna 61, Poland

ELSEVIER
Amsterdam — Oxford — New York — Tokyo 1989

ELSEVIER SCIENCE PUBLISHERS B.V.
Sara Burgerhartstraat 25
P.O. Box 211, 1000 AE Amsterdam, The Netherlands

Distributors for the United States and Canada:

ELSEVIER SCIENCE PUBLISHING COMPANY INC.
655, Avenue of the Americas
New York, NY 10010, U.S.A.

Library of Congress Cataloging-in-Publication Data

Mathematical submodels in water quality systems / edited by S.E.
Jørgensen and M.J. Gromiec.
p. cm. -- (Developments in environmental modelling ; 14)
Bibliography: p.
Includes index.
ISBN 0-444-88030-5
1. Water quality--Mathematical models. 2. Aquatic ecology--Mathematical models. 3. Hydrology--Mathematical models.
I. Jørgensen, Sven Erik, 1934- . II. Gromiec, Marek Jerzy.
III. Series.
TD370.M43 1989
628.1'61--dc20 89-16789
CIP

ISBN 0-444-88030-5 (Vol. 14)
ISBN 0-444-41948-9 (Series)

Printed in The Netherlands

CONTENTS

LIST OF TABLES

CHAPTER 1

INTRODUCTION

by
Sven Erik Jørgensen

1.1 THE APPLICATION OF SUBMODELS

Environmental models may be divided into empirical and mechanistic models. The latter type is based on quantitative descriptions of physical, chemical and biological processes in the environment. The description attempts to describe the mechanism behind the process on a sound scientifical basis.

Comprehensive environmental models have been used since the early seventies. Mechanistic models are used in this context more and more, probably because of attempts to put more ecology into the models. Experience has shown that it is absolutely necessary to develop models on a sound ecological basis, to find a balanced complexity and to build reality into the models, see for instance Jørgensen (1988). One may go as far as to state that 15-20 years ago the development of models was 75% mathematics and 25% ecology in its widest sense. Today it is the opposite: 75% ecology and 25% mathematics.

The result has been an increasing demand for the development of good quantitative and ecological descriptions of physical, chemical and biological processes in ecosystems. Such descriptions may be called 'submodels'.

For each process there are several and, to a certain extent, equally valid descriptions, which vary in complexity and the number of side reactions and details, that they include. As in the development of total models, the selection of the right submodels is a matter for the ecosystem and its problems (e.g. which side reactions and/or details are significant in the considered case?), and the required accuracy, including the amount and quality of the available data. Therefore it must not be expected that it is possible to give a general answer to the question as to which submodel to use for each of the relevant processes? Several submodels are presented in this book together with the considerations needed in individual cases for making the right selection of the submodel, best fitted to the ecosystem, the problem and the data.

Submodels are developed similarly to total models. It means that the Figure 1.1. is also valid for the development of a submodel, although the complexity and the needed data of course in most cases are less. It is recommended to make all the steps presented in Figure 1.1 also in development of a submodel. Some of the steps may be considered redundant, but it is still recommended that the procedures are used to avoid exclusion of important steps and to assure that all the required considerations are included.

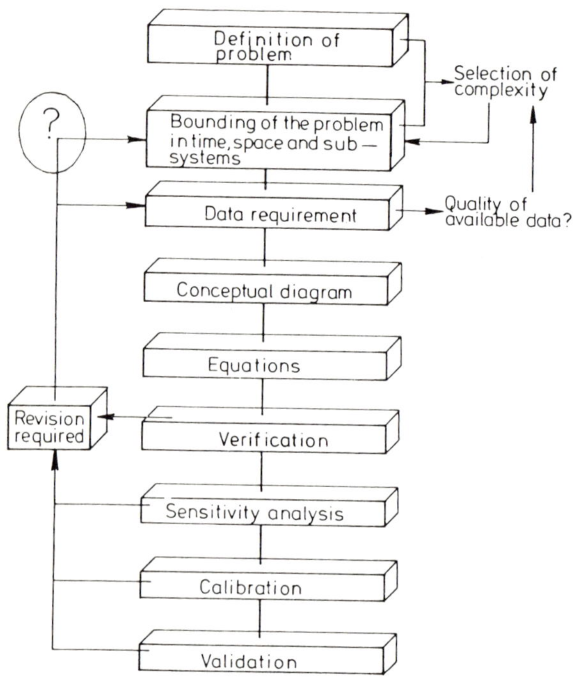

Fig. 1.1: A tentative modelling procedure.

The development or selection of a submodel is often an important part of a total model development. New requirements often emerge in the verification phase, as a result of sensitivity analysis or calibration. The verification may show that something is missing in the internal logic of the total model which of course may call for another selection of submodels or for the development of an additional submodel. The sensitivity analysis may reveal that the model results are particularly sensitive to the right description of one or more processes, which implies that more detailed submodels for these processes should be considered. If the calibration is not giving acceptable results, a change to some of the applied submodels should be considered. To a certain extent the same is also true after the validation. In the tentative procedure Figure 1.1 also shows how the development of a model is an iterative procedure. After the verification, sensitivity analysis, calibration and validation, revisions may be required. These revisions may all include development or selection of submodels.

1.2 OVERVIEW OF THE PRESENTED SUBMODELS

This book presents the most important, but not all, submodels applied in water quality modelling. Attempts are made to include physical as well as chemical and biological submodels to demonstrate the different classes available.

The diffusion and advection processes have not been included among the physical processes, because these processes are widely covered in books on hydrodynamics.

Each of the following chapters are devoted to one process and its related submodels. The importance of the process is discussed in the introduction of the chapter, followed by a presentation of the most applied submodels. A comprehensive list of applicable submodels may be presented but emphasis is laid on the question as to which submodel to select in which case?

The chapter discusses furthermore the parameter values and how to determine them. For some of the submodels examples of application be considered and, finally, the state of the art of the submodel: where are we today? Which research is needed to fill the gaps in our knowledge?

This book should not be considered a handbook in aquatic submodels. Jørgensen (1988) and Jørgensen et al (1979) already give tables of some important submodels. The present book gives an overview of some of the most important submodels in water quality modelling, and its scope is rather to discuss the advantages and disadvantages of various submodels to be applied to various case studies and total models.

Chapter two covers **Volatilization** i.e. the tranfer of a component from an aquatic ecosystem to the atmosphere. It is an important process in toxic substance models, as a substantial part of toxic organics may be lost from aquatic ecosystems to the atmosphere by volatilization. Furthermore, the nitrogen balance of aquatic ecosystems is often significantly influenced by the volatilization of ammonia.

Reaeration is mentioned in chapter three and may be considered the tranfer of gasses from the atmosphere to aquatic ecosystems. The process is of particular importance for the oxygen balances of aquatic ecosystems, for this process is the major source of oxygen in polluted streams.

Adsorption and Ion Exchange are presented in the following chapter. These processes explain the removal of many components, including ammonia, phosphate, toxic organics and metal ions, from water to a solid phase. While volatilization, reaeration and adsorption are considered physical processes, pure ion exchange is a chemical process, as the amount of uptaken and released ions are equivalent, in accordance with the definition of ion exchange. However, pure ion exchange or adsorption rarely occur independently in nature, for field processes are often mixtures of the two. For this reason they have been treated together in one chapter.

Chapter five treats **Heat Exchange.** All processes presented in the other chapters are dependent on temperature. Therefore it is considered extremely important to get an accurate prediction of the temperature, which again requires an accurate description of heat exchange.

Chapter six is devoted to **Sedimentation,** which is the process that describes the removal of suspended matter from the water phase to the bottom, i.e. from the hydrosphere to the litosphere. The process is often significant for the mass balances of aquatic ecosystems. Many pollutants are involved in this pathway of removal from aquatic ecosystems. It is a physical process, which, however, is related to **Coagulation,** mentioned in chapter seven and to **Precipitation**, mentioned in chapter eight. These two processes describe two physical-chemical processes, which consider the formation of suspended matter from dissolved components.

The two following chapters consider two chemical processes of great importance for the effect and removal of pollutants from aquatic ecosystems, including heavy metal and nutrients. Chapter nine covers **Complex Formation** and chapter ten treats **Hydrolysis and Chemical Redox Processes.** Furthermore, the latter process is of importance for the oxygen balance of aquatic ecosystems and for the oxidation state of two important nutrients - nitrogen and sulphur. Nitrification, denitrification and formation of hydrogen sulphide are all processes, that are dependent on the redox potential of the system.

Chapter eleven presents submodels for the description of **Photo-chemical Reactions.** These processes are of particular interest for the decomposition of toxic organics.

Chapter twelve starts the presentation of biological processes of importance for water quality modelling. This is **Microbiological Decom-position,** which is the major process for consumption of oxygen in most aquatic ecosystem.

Nitrification is mentioned in chapter thirteen. It is one of the important processes in the biogeochemical cycle of nitrogen in aquatic ecosystems. It consumes oxygen and its therefore not only of importance for the nitrogen balance but also for the oxygen balance of the system.

Predator-Prey submodels are presented in chapter fourteen. This is the only submodel of population dynamics included in this survey of important submodels. It is an important submodel in eutrophication models as it describes grazing as well as fish predation on zooplankton - both processes determining indirectly the phytoplankton concentration. Furthermore this submodel is an integrated part of a fishery model for a large lake or a sea ecosystem.

Primary Productivity is the major process in eutrophication models as it determines directly the phytoplankton concentration in aquatic ecosystems and it is of particular interest for lake ecosystems. This important submodel is covered in chapter fifteen.

Fish Growth is another important submodel in fishery models. It is treated in chapter sixteen.

The last chapter is devoted to **The Exchange between Sediment and Water.** Sediment accumulates nutrients, biodegradable organic matter and toxic substances. The amounts of these components in the sediment are often significantly larger than the amounts in the water column, particularly when shallow waters are considered. Therefore it is often very important for the mass balances to consider the reaction rates for the release of these components from the sediment to the water phase.

CHAPTER 2

VOLATILIZATION

by
Sven Erik Jørgensen

2.1 THE IMPORTANCE OF VOLATILIZATION IN ENVIRONMENTAL CONTEXT

Volatilization is the transfer of components from the liquid phase to the atmosphere. And has an environmental role in the following contexts:

1. When *the oxygen concentration is in water higher than* equilibrium concentration between water and the atmosphere, oxygen is tranferred from water to air. This may be true for eutrophic lakes during the summer when photosynthesis is dominant. In such cases when the oxygen profile is modelled, this process may be of significance. In the next chapter submodels of reaeration will be presented and the equations presented here can also be applied to cover the opposite process, since the reaeration expression will be negative.

2. When *the carbon dioxide concentration in water is higher than* the equilibrium concentration between water and the atmosphere, surplus carbon dioxide will be transferred from the hydrosphere to the atmosphere. The transfer of carbon dioxide from the atmosphere to the sea is, in general, more important than the opposite process. Mankind produces carbon dioxide by combustion of fossil fuel and about 60% of this is dissolved in the sea, a feature of great importance for the so called 'greenhouse effect' which would otherwise be considerably more pronounced. However, the transfer of carbon dioxide from water to air is environmentally significant for lakes with low pH. The acidification of lakes means that hydrogen carbonate is transformed into carbon dioxide, which then ecapes from the water. This implies that carbon may become a limiting nutrient and the lakes turn oligotrophic. This is a regional problem of great consequence in Scandinativa and Northeast United States. Removal of sulphur from fossil fuels seems to be the only acceptable solution to the problem.

3) Ammonia is very soluble in water, but since the concentration of ammonia in the atmosphere is small, a certain volatilization of ammonia

from water to air will always take place. However at neutral pH most of (ammonia + ammonium) is in the form of ammonium, whereas during summer in eutrophic lakes a pH of 8,5, or even more, may be observed, implying that the ratio of ammonia to ammonium is larger. This ratio can be found from the mass equation:

$$[NH_3]^*[H^+] / [NH_4^+] = K_a \tag{2.1}$$

where [] indicates molar concentrations, and K_a the dissociation constant for the ammonium ion.

This equation may be expressed in logarithmic form as follows:

$$\log([NH_3]/[NH_4^+]) = pH - pK_a \tag{2.2}$$

where p indicates - log. The pK_a for the ammonium ion is 9.3, but is dependent on the salinity and the temperature (see Table 2.1).

TABLE 2.1
Concentration of ammonia (NH_3 + NH_4^+) which contains a non-ionized ammonia concentration of 0.025 mg NH_3 l^{-1}

C°	pH = 7.0	pH = 7.5	pH = 8.0	pH = 8.5	pH = 9.0	pH = 9.5
5	19.6	6.3	2	0.65	0.22	0.088
10	12.4	4.3	1.37	0.45	0.16	0.068
15	9.4	5.9	0.93	0.31	0.12	0.054
20	6.3	2	0.65	0.22	0.088	0.045
25	4.4	1.43	0.47	0.17	0.069	0.039
30	3.1	1	0.33	0.12	0.056	0.035

The influence of the salinity on the ammonia/ammonium ratio may be considered by use of the concept of ionic strength, I:

$$I = \sum 1/2\, C * Z^2 \tag{2.3}$$

where C = molar concentration of ions and Z = the change.

All concentrations are multiplied by P, the activity coefficient, in the mass equation:

$$\frac{[NH_3]\,[H^+]}{[NH_4^+]} = 10^{-9.3} \tag{2.4}$$

It is found from I by application of Table 3 in Appendix 1.

At pH= 9.3 about half of the (ammonia + ammonium) is ammonia. *Therefore the volatilization of ammonia may be significant in very eutrophic lakes* during the summer months. A pH more than 9.0 has been observed in very eutrophic lakes and even pH = 11.0 has been measured in the shallow Søllerød Lake of Denmark.

4) Volatilization is an important process for *many toxic substances in the environment.* Many organic toxic substances, discharged into the environment in small concentrations, have sufficient vapour pressure to give rise to significant volatilization. In many cases volatilization may the most important removal process for the toxic substance from the aquatic ecosystem, as other processes inCluding biodegradation may be very slow.

It should be Clear from this survey of the more important volatilization processes, that it is very useful to have a good quantative description as a submodel for this process in environmental management models.

2.2 MODELS OF THE VOLATILIZATION PROCESS

Transfer processes, describing the flow from one phase to another are of major environmental interest. Volatilization can be described by use of the so-called two-film theory.

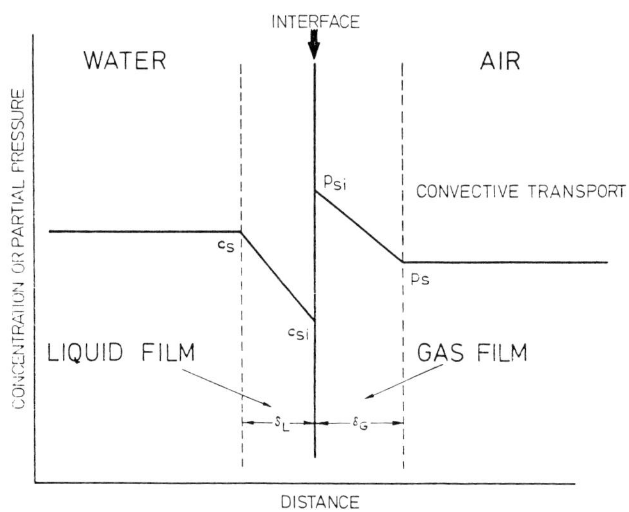

Fig. 2.1: The two-film model of volatilization from the surface of water bodies.

Figure 2.1 illustrates the major features of a *two-film model* of mass transfer, which is generally applied in chemical engineering. The water phase is assumed to be well mixed, so that any volatile compound is at a uniform concentration C_s except in the vicinity of the interface. A stagnant liquid film of thickness ∂L separates the bulk of the water phase from the interphase. Movement of a volatile component through this film is due to diffusion. The concentration decreases across the film from C_s to C_{si}, and the rate at which the Si component is transported across the film, N_s, is

$$N_s = K_l^s (C_s - C_{si}) \tag{2.5}$$

where K_l^s is the liquid film mass transfer coefficient (m h^{-1}).

A stagnant gas film having a thickness of ∂_G is on the air side. The partial pressure, P_{Si}, on the air side is related to C_{Si} (molar concentration on the water side of the interface) in accordance with Henry´s law: (x_{Si} mole fraction)

$$P_{Si} = H_c C_{Si} = H * x_{Si} \tag{2.6}$$

where H_c and H are Henry´s law´s constants expressed in M or mole fraction unit. The relation between H_c and H is:

$$H_c = H * 18/100 = 1.8 * 10^{-3} H \tag{2.7}$$

The transport across the gas film, N_s, is

$$N_S = \frac{K_G{}^S}{R\,T} (P_{Si} - P_s) \tag{2.8}$$

where $K_G{}^S$ is the gas film mass transfer coefficient (m h^{-1}).

By combining these equations, we obtain

$$K_V{}^S = \frac{A}{V} \left(\frac{1}{K_l{}^S} + \frac{R\,T}{H_C{}^S K_G{}^S} \right)^{-1} \tag{2.9}$$

where $K_V{}^S$ is the overall transfer coefficient (h^{-1})
A is the interfacial area (m^2)
V is the liquid volume

T is the absolute temperature

We assume that

$$K_l = D / \partial_L \tag{2.20}$$

where D is the molecular diffusion coefficient, and similarly:

$$K_G = D / \partial_G \tag{2.11}$$

It has been shown that if molecules are spherical, molecular diffusion coefficients in solution are inversely proportional to molecular diameters, d, so that

$$\frac{K_V{}^S}{K_G{}^O} = \frac{D^S}{D^O} = \frac{d^O}{d^S} \tag{2.12}$$

where S indicates the considered component and O is oxygen. d^O is 2.98 A.

If data on the diffusion coefficients or molecular diameter for the component are not obtainable, the molecular diameters can be estimated from the critical volume, V_c, since

$$\frac{\pi d^3}{6} = \frac{V_c}{2N} \text{ or } \frac{V_c}{3N} \tag{2.13}$$

where N is Avogadro´s number.

$H_C{}^S$ can be estimated from solubility and vapor pressure:

$$H_C{}^S = \frac{P^S}{S_{wo}} \tag{2.14}$$

where P^S is the vapour pressure of S in pure form and S_{wo} is its solubility in water. When data for the considered component are not available, data for a related component can be used.

The theory of this process has been developed by several authors (Liss et al., 1974; Mackay et al., 1976 and Smith et al., 1977).

The application of "fugacity" in environmental modelling has been discussed by Mackay and Paterson (1981 and 1982). Fugacity has dimensions of pressure and may be considered a measure of " escaping tendency" from the phase. When phases are at equlibrium their fugacities are equal.

Fugacity, f, is related to concentration by the following equation:

$$f = C / Z, \tag{2.15}$$

where C is the concentration (mol/m^3) and Z is the fugacity capacity (mol/m^3 Pa). As seen from this equation f has the unit of pressure or in this case, of Pa.

Equilibrium involves:

$$C_a / C_w = Z_a / Z_w = K_p \tag{2.16}$$

where the indices a and w are used to indicate air and water respectively as examples for the phases. K_p is a partition coefficient.

Transfer rates between two phases by diffusion are expressed by the following equation (models per unit of area and time) :

$$N = D * \Delta f, \tag{2.17}$$

where N is the rate of tranfer, D is the diffusion coefficient and Δf is the difference in fugacity. D is the total resistence for the tranfer consisting of the resistances of the two phases in series. Notice that D may be found as K*Z, where K is the transfer coeffcient and Z is the fugacity capacity defined below.

The so-called unit world model is used to find the distribution of particular organic chemicals in the various spheres. It consists of six compartments: air, water, soil, sediment, suspended sediment and biota. This simplified model aims toward identification of the partition among these six compartment of toxic substances emitted to the environment. The conceptual diagram of the unit world model is shown in Figure 2.2.

The volumes and densities of the unit world and the definition of fugacity capacities are given in Table 1 Appendix 1.

It is also possible to use this approach regionally or for particular ecosystems. The volumes used in this latter case are listed in Table 2 Appendix 2.

The fugacity model may be used in four levels.

The first level calculates the equilibrium distribution of a chemical between phases. It assumes that each compartment is well mixed and there is no reaction or advection into or out of the system.

If M is the total amount in moles, m_i and V_i are the amounts and volumes of each compartment, then the following equation is valid:

$$M = \Sigma m_i = \Sigma C_i * V_i = \Sigma f_i * Z_i * V_i = f * \Sigma Z_i * V_i \qquad (2.18)$$

Since M is known, f may be calculated and the amount in each compartment is given by:

$$m_i = f * V_i * Z_i \qquad (2.19)$$

The percentage distribution is found as m/M and the concentrations are found by:

$$C_i = Z_i * f \qquad (2.20)$$

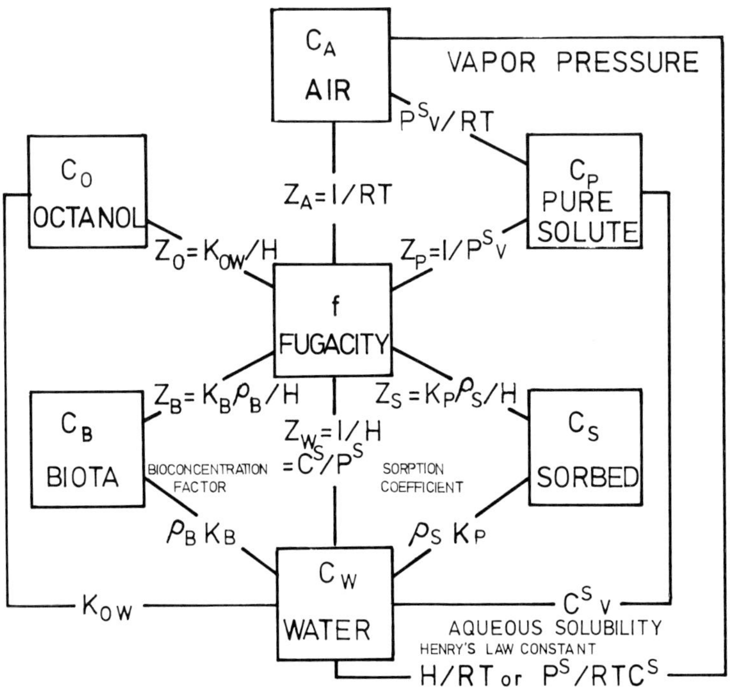

Fig. 2.2: Conceptual Diagram of the Fugacity Model.

Level two considers equilibrium but inCludes also reaction and advection. Reactions comprise photolysis, hydrolysis, biodegradation, oxidation etc. All these processes are assumed to be first order reactions. If the rate constants are not known, or cannot be estimated, they are set to zero. Since process rates are expressed in first-order form they are additive and the total removal rate from a compartment can be found as:

$$\Sigma V_i * C_i * \Sigma k_j = \Sigma V_i * C_i * k_i \quad (2.21)$$

As in level one a common fugacity,f, prevails and if we substitute C by f*Z, we get for the total rate, E:

$$E = f * \Sigma V_i * Z_i * k_i \quad (2.22)$$

The average residence time, tr, due to reactions may be found by use of the following equation:

$$tr = M / E \quad (2.23)$$

and the overall rate constant, K, is E / M, or 1 / tr.

The third level is devoted to a steady state, nonequilibrium situation, which implies that the fugacities are different in each phase. Equation (2.19) is used to account for the transfer.

The D values may be calculated from quantities such as interphase areas, mass transfer coefficients (as indicated above D is the product of the transfer coeffecient and the fugacity capacity: D = K * Z), release rate of chemicals into phases such as biota or sediment, and Z values, or by use of the estimation methods presented in Jørgensen (1989a).

Level four involves a dynamic version of level three, where emissions and thus concentrations, are varying with time. This implies that differential equations must be applied for each compartment to calculate the change in concentrations with time, for instance:

$$V_i * dC_i/ dt = E_i - V_i * C_i * k_i - \Sigma D_{ij} * \Delta f_{ij} \quad (2.24)$$

This model level is similar in concept to the EXAMS model, see Mackay et al 1983.

Levels one or two are sufficient in most cases, but if the environmental management problem requires the prediction of

1) the time taken for a substance to accumulate to a certain level in a phase after emission has started, or

2) the length of time for the system to recover after the emission has ceased,

the fourth level must be applied.

2.3 THE APPLICATION OF THE VOLATILIZATION SUBMODEL IN ENVIRONMENTAL MODELLING

Submodels of Volatilization are used in many toxic substance models; see, for instance, Mackay et al. (1982), Cohen (1983), Smith et al. (1982) and Jørgensen (1989a); and in eutrophication models to account for the escape of ammonia in very eutrophic lakes with high pH during the summer.

The release of carbon dioxide in acidic lakes is in most cases taken into account by computation of the equilibrium concentration, when the temperature and pH is given.

The fugacity approach has been widely used and a typical example is given by Mackay (1982). It concerns the distribution of PCB between air and water at the Great Lakes:

1) H is in this case 49.1 and
2) the distribution coeffcient for air water (= H/R*T) therefore 0.02.
3) the fugacity capacity for water = 1/H is found to be 0.0204 and
4) the fugacity capacity for air = 1/ R*T = 0.000404.
5) the distribution coeffecient between water and suspended matter in the water was estimated to be 100 000.
6) as the concentration of suspended matter in the Great Lakes has been found to be 2 * 10^{-6} on a volume basis (approximately 4 mg/l, the density being 2000 g/l), the fraction dissolved is 1 / (l+0.2) = 0.833

At the Great Lakes 2 ng/l of PCB is found in the water and 2 ng/m^3 in the air. The fugacity can be calculated in water and air as C/Z and it was found that the fugacity in water is 17 times higher than in air, which implies that volatilization will occur.

If it is assumed that the transfer coefficient in water is 10^{-5} m/s and in air 10^{-3} m/s, the volatilization rate can be calculated from the traditional two-resistance model, using the relation D = K *Z to find the overall diffusion coefficient, D:

$$1 / D = 1/ 10^{-5} * 0.0204 + 1/10^{-3} * 0.000404 \qquad (2.25)$$

D is found to be 1.36 * 10^{-7}. N is calculated by use of (2.19):

$$N = D (f_w - f_a) = D (2.8 * 10^{-7} - 1.53 * 10^{-8}) \quad (2.26)$$

N is found be to be $35.9 * 10^{-15}$ mol/m^2/s.

It can be shown that the transfer with precipitation is negligible compared with the volatilization rate, while the washout of partiCles and dry deposition are important processes. If these processes are considered the net flux to the atmosphere becomes about 75% of the flux found above.

2.4 PARAMETER ESTIMATION IN THE VOLATILIZATION SUBMODEL

k_a and k_l may be estimated from the following equations: (from Wolff and Van der Heidje 1982)

$$k_a = 1.0*10^{-3} + 46.2* 10^{-3}* U * Sca^{-0.67} \quad (2.27)$$

$$k_l = 1.0*10^{-3} + 34.1* 10^{-4}* U * Scw^{-0.50} \quad (2.28)$$

$$k_l = 1.0*10^{-3} + 144* 10^{-4}* U * Scw^{-0.50} \quad (2.29)$$

where U is the air-phase friction velocity, which is related to the actual wind velocity, measured at a height of 10 m in the unit m/s, U10:

$$U = 10^{-2} U10 (6.1 + 0.63 * U10)^{0.5} \quad (2.30)$$

(2.28) is valid for $U > 0.3$ m/s, while (2.29) is applicable for $U < 0.3$m/s.

Sca and Scw are the Schmidt's numbers for air and water respectively. They are defined as μ/d*D, where μ is the air or water viscosity in kg/m/s, d is the air or water density (kg/m^3) and D is the molecular diffusivity of the chemical in air or water (m^2/s). Sca is typically in the range 0.6 - 2.5, while Scw is the range 500-2000.

It was mentioned in section 2.2 how it is possible to estimate the molecular diffusivity in air.

For large unhydrated molecules with a molecular weight > 1000 the following simple expression at room temperature for estimation of diffusion coefficients in dilute solutions, Dl may be used:

$$Dl = 2.74 *10^{-9} * M^{-1/3} \quad (2.31)$$

where M is the molecular weight.

Dl may also be found from the following equation:

$$Dl = 25.2 * 10^{-5} * T/\mu * V_b^{1/3} \quad (2.32)$$

where T is the absolute temperature, μ is the viscosity of water at that temperature in centipoise and V_b is the molal volume of solute at normal boiling point in cm^3/mole. Notice that Dl is found in cm^2/s and the result must therefore be multiplied by 10^{-4} to obtain the molecular diffusivity in the right unit for application in the equations (2.29) and (2.30).

Only a few estimation methods for the parameters needed for this submodel are presented above. A wide spectrum of estimation methods are actually available. For those interested in a relatively simple, but complete, outline of estimation methods can refer to Jørgensen (1989a). The idea behind the construction of a outline of estimation methods is that of being able to estimate all parameters relevant for ecological and environmental modelling, when these parameters are not available in the literature. For many toxic substance of environmental interest the required parameters are unfortunately not available in the literature and it is therefore preferable to be able to estimate these parameters from a few physical-chemical data. The same physical-chemical parameters are, however, not available for all chemicals and therefore a methodological strategy is required for deriving parameters and physical-chemical properties from existing physical-chemical data. See Jørgensen (1989a).

Some useful parameters are shown in Table 2.2.

TABLE 2.2
Evaporation rate of chemical compounds (Dilling et al, 1975 and Hutzinger et al., 1974)

Item	Value	Conditions
CCl2=CCl2	25.7 min	50% evaporation, dilute water sol., initial=1 ppm, 298K
CCl3CCl3	45 min	50% evaporation, dilute water sol., initial=1 ppm, 298K
CCl4	29 min	50% evaporation, dilute water sol., initial=1 ppm, 298K
CHCl=CClCH2Cl	49 min	50% evaporation, dilute water sol., initial=1 ppm, 298K
CHCl=CCl2	21.3 min	50% evaporation, dilute water sol., initial=1 ppm, 298K
CHCl=CHCH2Cl	31 min	50% evaporation, dilute water sol., initial=1 ppm, 298K
CHCl=CHCH3	16 min	50% evaporation, dilute water sol., initial=1 ppm, 298K
CHCl=CHCl (cis)	18 min	50% evaporation, dilute water sol., initial=1 ppm, 298K

TABLE 2.2 (continued)

Item	Value	Conditions
CHClOCHCl (trans)	24 min	50% evaporation, dilute water sol., initial=1 ppm, 298K
CHCl2CCl3	45 min	50% evaporation, dilute water sol., initial=1 ppm, 298K
CHCl2CHCl2	56 min	50% evaporation, dilute water sol., initial=1 ppm, 298K
CHCl3	21 min	50% evaporation, dilute water sol., initial=1 ppm, 298K
CH2=CClCH2Cl	20 min	50% evaporation, dilute water sol., initial=1 ppm, 298K
CH2=CClCH3	29 min	50% evaporation, dilute water sol., initial=1 ppm, 298K
CH2=CCl2	22 min	50% evaporation, dilute water sol., initial=1 ppm, 298K
CH2=CHCH2Cl	27 min	50% evaporation, dilute water sol., initial=1 ppm, 298K
CH2=CHCl	26 min	50% evaporation, dilute water sol., initial=1 ppm, 298K
CH2ClCCl2CH2Cl	47 min	50% evaporation, dilute water sol., initial=1 ppm, 298K
CH2ClCCl3	43 min	50% evaporation, dilute water sol., initial=1 ppm, 298K
CH2ClCHClCHCl2	51 min	50% evaporation, dilute water sol., initial=1 ppm, 298K
CH2ClCHCl2	21 min	50% evaporation, dilute water sol., initial=1 ppm, 298K
CH2ClCH2Cl	29 min	50% evaporation, dilute water sol., initial=1 ppm, 298K
CH2Cl2	20.7 min	50% evaporation, dilute water sol., initial=1 ppm, 298K
CH3CCl3	20 min	50% evaporation, dilute water sol., initial=1 ppm, 298K
CH3CHCl2	22 min	50% evaporation, dilute water sol., initial=1 ppm, 298K
CH3CH2Cl	21 min	50% evaporation, dilute water sol., initial=1 ppm, 298K
CH3Cl	27 min	50% evaporation, dilute water sol., initial=1 ppm, 298K
PCB	1.74 mg/((cm^2)*hour)	Surface=12.28 cm^2, 373K, 24 H, AroClor 1221
PCB	0.874 mg/((cm^2)*hour)	Surface=12.28 cm^2, 373K, 24 H, AroClor 1232
PCB	0.338 mg/((cm^2)*hour)	Surface=12.28 cm^2, 373K, 24 H, AroClor 1242
PCB	0.053 mg/((cm^2)*hour)	Surface=12.28 cm^2, 373K, 24 H, AroClor 1254
PCB	0.009 mg/((cm^2)*hour)	Surface=12.28 cm^2, 373K, 24 H, AroClor 1260
PCB	0.013 mg/((cm^2)*hour)	Surface=12.28 cm^2, 373K, 24 H, AroClor 1262
PCB	0.152 mg/((cm^2)*hour)	Surface=12.28 cm^2, 373K, 24 H, AroClor 1448
PCB	0.007 mg/((cm^2)*hour)	Surface=12.28 cm^2, 373K, 72 H, AroClor 4465
PCB	0.004 mg/((cm^2)*hour)	Surface=12.28 cm^2, 373K, 72 H, AroClor 5442
PCB	0.004 mg/((cm^2)*hour)	Surface=12.28 cm^2, 373K, 72 H, AroClor 5460
PCB	1.74 mg/((cm^2)*hour)	Surface=12.28 cm^2, 373K, 24 H, AroClor 1221
PCB	0.874 mg/((cm^2)*hour)	Surface=12.28 cm^2, 373K, 24 H, AroClor 1232
PCB	0.338 mg/((cm^2)*hour)	Surface=12.28 cm^2, 373K, 24 H, AroClor 1242
PCB	0.053 mg/((cm^2)*hour)	Surface=12.28 cm^2, 373K, 24 H, AroClor 1254

TABLE 2.2 (continued)

Item	Value	Conditions
PCB	0.009 mg/((cm^2)*hour)	Surface=12.28 cm^2, 373K, 24 H, AroClor 1260
PCB	0.013 mg/((cm^2)*hour)	Surface=12.28 cm^2, 373K, 24 H, AroClor 1262
PCB	0.152 mg/((cm^2)*hour)	Surface=12.28 cm^2, 373K, 24 H, AroClor 1448
PCB	0.007 mg/((cm^2)*hour)	Surface=12.28 cm^2, 373K, 72 H, AroClor 4465
PCB	0.004 mg/((cm^2)*hour)	Surface=12.28 cm^2, 373K, 72 H, AroClor 5442
PCB	0.004 mg/((cm^2)*hour)	Surface=12.28 cm^2, 373K, 72 H, AroClor 5460

2.5 CONCLUSIONS AND FUTURE RESEARCH NEEDS

A reasonably accurate volatilization submodel can be constructed for the cases presented in section 2.1. In cases where estimation methods are used, i.e. for many toxic substances, the accuracy is limited by the accuracy of the estimation methods. Experience in the application of estimation methods is unfortunately rather limited to date, but much effort is being devoted to further development of these methods and we should have much more reliable methods in the nearest future. Much research is also being devoted to direct measurements of many model parameters, which will enable us to test the estimation methods on more data and thereby improve the methods further. The key phrase for further research needs is: "more experience" and this is valid not only for the application of parameter estimation methods, but also for the application of submodels and total models.

REFERENCES

Cohen, Y., 1983. Mass transfer across a sheared wavy air-water interface. Int. J. Heat Mass Trans., 26, 1289.

Dilling, W.L, Tefertiller, N.B., Kallos, G.J., 1975. Environ. Sci. Technol. 9(9), pp 833-837.

Hutzinger, O., Safe, S., Zitko, V., 1974. The Chemistry of PCBs, CRC-Press.

Jørgensen, S.E., 1989a. Modelling in Ecotoxicology. Elsevier, Amsterdam, New York, Oxford.

Liss, P.S. and Slater, P.G., 1974. Flux of gases across the air-sea interface. Nature, 247: 181-184.

Mackay, D., 1982. Basic properties of materials. In: Environmental Risk Analysis for Chemicals, (Conway, R.A., ed). Van Nostrand Reinhold, New York, chap. 2.

Mackay, D., 1979. Finding fugacity feasible. Environ. Sci. Technol., 13, 1218.

Mackay, D. and Cohen, Y., 1976. Prediction of Volatilization Rate of Pollutants in Aqueous Systems. Symposium on Non-biological Transport and Transformation of Pollutants on Land and Water, May 11-13. National Bureau of Standards, Gaithersburg, Maryland.

Mackay, D., Joy, M. and Paterson, S., 1983. A quantitative water, air, sediment interaction (QWASI) fugacity model for describing chemical fate in lakes and rivers. Chemosphere, 12, 827.

Mackay, D. and Paterson, S., 1982. Fugacity revisited. Environ. Sci. Technol., 16, 654A.

Mackay, D. and Paterson, S., 1981. Calculating fugacity. Environ. Sci. Technol., 15, 1006.

Mackay, D., Shiu, W.Y., Bobra, A., Billington, J., Chau, E., Yeun, A., Ng, C. and Szeto, F., 1982. Volatilization of Organic Pollutants from Water. EPA Rep. No. 600/63-82-019, NTIS No. PB 82-830-934, Athens, Georgia.

Smith, J.H. et al., 1977. Environmental Pathways of Selected Chemicals in Freshwater Systems. part I, EPA 600/7-7-113.

Smith, J.H., Mackay, D. and Ng, C., 1982. Volatilization of pesticides from water. Res. Rev., 85, 73.

Wolff, C.J.M. and Van der Heidje, H.B., 1982. A model to assess the rate of evaporation of chemical compounds from surface waters. Chemosphere, 11, 103.

CHAPTER 3

REAERATION

by
M.J. Gromiec

3.1 INTRODUCTION

Reaeration, the process of oxygen absorption from the atmosphere by a body of water, is one of the main sources of oxygen in aquatic systems. The reaeration process in an aquatic system is characterized by its surface reaeration coefficient.

The absorption of oxygen from the atmosphere by water is normally assumed to be a first-order process, i.e. the rate of absorption is directly proportional to the dissolved oxygen deficit (Adeney and Becker, 1919). This principle was first applied to water quality modelling by Streeter and Phelps (1925), but it should be noted that they considered the reaeration process as the only oxygen source in a river system.

The mass transfer rate of oxygen from air to water is usually presented as:

$$\frac{dC}{dt} = \frac{K_V A}{V} (C_s - C_L) \tag{3.1}$$

in which K_V is the mass transfer coefficient (LT^{-1}), A is the surface area through which the diffusion takes place (L^2), V is the volume of water (L^3), C_s is the oxygen saturation concentration (ML^{-3}), and C_L is the oxygen concentration in the water body (ML^{-3}).

Since for a stream, A/V = 1/H, where H is a stream depth (L), the term $(C_s - C_L)$ represents oxygen deficit, equation (3.1) can be written as:

$$\frac{dD}{dt} = -\frac{K_V}{H} D = -K_a D \tag{3.2}$$

in which K_a is the reaeration coefficient, base e, (T^{-1}).

Integrating equation (3.2), and solving for the reaeration coefficient between time limits t_1 and t_2, and deficit D_1 at time t_1 and deficit D_2 at time t_2, gives:

$$k_a = (\log D_1 - \log D_2)\, \Delta t^{-1} \tag{3.3}$$

where k_a is the reaeration coefficient, base 10, (T^{-1}), $k_a = K_a/2.303$, $\Delta t = t_2 - t_1$, and t(T) is a time of flow.

The most important factors that affect the value of k_a are temperature and turbulence. The effect of temperature on the reaeration coefficient is usually expressed as:

$$k_{a(T°C)} = k_{a(20°C)}\ \Omega^{T-20} \tag{3.4}$$

where Ω is the temperature coefficient, and the temperature, T, is in °C. The U.S. Committee on Sanitary Engineering Research (1961) concluded that the temperature coefficient remains constant over a wide range of turbulent conditions with a value of 1.0241. This value was first reported by Elmore and West (1961). Typically values of 1.016 to 1.047 are used in a number of water quality modelling applications, with the value of 1.024 being most common.

In addition, many other factors can have an influence on k_a, such as surface active agents, suspended solids, and winds. However, from the practical point of view the effects of these factors are seldom included in the reaeration prediction models. The surfactants depress k_a, and the degree of this effect is related to turbulence. Also, winds have an effect on the reaeration coefficient, and this effect appears to be particularly significant during high wind speeds. The influence of winds on reaeration is discussed by Eloubaidy and Plate (1972), Mattingly (1977), Gulliver and Stefan (1981) and Frexes et al. (1984). The influence of rainfall on reaeration is of rather theoretical significance, and it has been analyzed by Banks et al. (1984). Dams and other hydraulic structures can influence the reaeration process in rivers and streams (Butts and Evans, 1983).

3.2 MEASUREMENT TECHNIQUES

The methods for direct measurement of the atmospheric reaertion coefficient were classified by Zogorski and Faust (1973) as follows: the disturbed equilibrium method, developed by Gameson et al. (1955), with various modifications; the method of monitoring dissolved oxygen /DO/

deficit change between two sites, first applied by Churchill et al. (1962); the gasometric /dome/ method, proposed by Copeland and Duffer (1964).

The disturbed equilibrium methods are used for clean or polluted streams (in general for small streams with flows less than 1.4 m^3 sec^{-1}). According to Zogorski and Faust (1973) the following criteria must be met in order to apply these methods: the rates of these processes (except reaeration) which affect the dissolved oxygen concentration, are unchanged by the injection of a reducing chemical (usually sodium sulphite); the existing dissolved oxygen profile is such that the dissolved oxygen deficit can be increased by the injection of reducing chemicals; the reducing chemical is oxidized before the first sampling point is reached; complete mixing (vertical and horizontal) occurs before the first sampling point; and the same water body is measured at the two sampling points.

The method of measuring the change in dissolved oxygen deficit between two sites should be used when a stream is relatively free of organic pollution, and other sinks and sources of dissolved oxygen are small compared with the oxygen change due to the reaeration. The gasometric method is especially suitable for an estuarine environment.

The methods for indirect estimation of the atmospheric reaeration coefficient were divided by Zogorski and Faust (1973) into the following classes: the diurnal curve method, pioneered by Odum (1956), the tracer techniques, first described by Tsivoglou et al. (1965); the dissolved oxygen sag technique, developed by Streeter and Phelps (1925); and predictive models. However, the productivity methods, proposed by Hornberger and Kelly (1975), should also be added to this group. These methods are based on these assumptions: productivity is lineary related to incident light; night-time respiration is constant; and the reaeration coefficient is constant over a single day. The first productivity method uses the first two assumptions, and the second method uses the last two assumptions.

The diurnal curve analysis can be performed by taking measurements of dissolved oxygen at certain time intervals at one or two sites during a twenty-four hour period. The single curve obtained from one site is good only for a stream with the same upstream and downstream conditions. However, since normally the upstream and downstream conditions differ, it is necessary to take measurements at two sites, and to perform the double curve analysis. Assuming that accrual of ground water is absent, and that the respiration is constant at night, then the reaeration coefficient can be estimated using measured values of the oxygen content just after sunset and just before dawn. A summary of assumptions for the diurnal curve method was presented by Zogorski and Faust (1973) as: the reaeration coefficient and the respiration are constant at day and night; the diurnal variations in the dissolved oxygen content are a result of variations in photosynthetic production and changes in oxygen diffusion across the air-water interface resulting from changes in the oxygen saturation deficit; water passing a sampling site has the same physical, chemical and biological characteristics.

The radioactive tracer technique consists of simultaneously injecting to a stream three tracers, such as a fluorescent dye (rhodamine), a dispersion indicator (tritium), a gaseous tracer (krypton-85), and measuring their concentration at two sampling points. The average mass-transfer coefficient ratio for krypton and dissolved oxygen was established in the laboratory by Tsivoglou et al. (1965), Tsivoglou (1967) as 0.83. It was found that this ratio is not significantly influenced by pollutants, temperature, aorturbulence conditions. Thus this method can be used in polluted streams without interference, and it appears to produce reliable results (Tsivoglou and Wallance, 1972, and Tsivoglou and Neal, 1976). Due to certain hazards involved in applying this method, other modified tracer methods have been developed, such as the hydrocarbon tracer techniques (Rainwater and Holley, 1984). These techniques use a hydrocarbon gas tracer and a fluorescent dye. Various tracer gases (such as ethylene and propane) can be used to measure the reaeration coefficient. Wilcock (1984) proposed methyl chloride as a gas tracer.

The estimation of the reaeration coefficient by the dissolved oxygen balance method is subject to many sources of error, mainly associated with the accuracy of measurements of all parameters in a dissolved oxygen balance equation. Various theoretical, semi-empirical and empirical models for the mass transfer coefficient or the reaeration coefficient are discussed in the following section.

3.3 PREDICTIVE MODELS

Many excellent reviews on theoretical, mathematical, and semi-empirical models for the reaeration coefficient have been presented, such as by Bennett and Rathbun (1972), Lau (1972a), Wilson and Macleod (1974), Zison et al. (1978) and Bowie et al. (1985). A summary of these models is given below to indicate the variety of different approaches.

Theoretical models

Various theories and theoretical models have been presented for predicting the gas-liquid mass-transfer coefficient or the reaeration coefficient. The theoretical models can be divided into the following five groups (Bennett and Rathbun, 1972):

1. Film models, including a two-film model;
2. Renewal models, including penetration, film-penetration, and surface renewal-damped eddy-diffusivity models;
3. Kinetic-theory models;
4. Large-eddy models;
5. Turbulent-diffusion models.

These theoretical models (Table 3.1) are based on equations describing the gas transfer through a gas-liquid interface. The differential equations describing diffusion, as well as the boundary conditions chosen, are presented in Table 3.2.

Semi-empirical and empirical models for rivers

The first empirical model for the river reaeration coefficient was formulated by Streeter and Phelps (1925) as follows:

$$k_a = C_1 U^{n_1} H_1^{-2} \tag{3.14}$$

in which U is the mean velocity of flow (LT^{-1}), H_1 is the mean depth of water above extreme low water (L) and C_1 and n_1 are empirical constants. These constants were evaluated by an indirect method on a number of the Ohio River reaches. When U is expressed in ms^{-1}, and H_1 in m, the values of C_1 range from 13.06 to 23.96 and the values of n_1 varied from 0.57 to 5.40. The exponent n_1, was related to the function defined as the mean velocity increase per c. 1.5 m increase in the river stage, and the constant, C_1, was related to the slope of the stream and to a factor, which is a measure of the relative roughness of the stream bottom.

Table 3.1
Theoretical models for the reaeration coefficient

Model name	Equation	eq. no.	Reference
1. Two-film model	$K_a = D_m L_f^{-1} H^{-1}$	3.5	Lewis and Whitman, 1924
2. Renewal models:			
a. Penetration model	$K_a = 2(D_m/\Pi\ t_e)^{-0.5}\ H^{-1}$	3.6	Higbie, 1935
	$K_a = (D_m\ s)^{0.5}\ H^{-1}$	3.7	Danckwers, 1951
b. Film penetration model	$K_a = (D_m\ s)^{0.5}\ H^{-1}\ \coth(^a{}_s)^{0.5}$	3.8	Dobbins, 1956
	$K_a = 0.5(D_m\ s)^{0.5}\ H^{-1}\ [1+\cosh(^a{}_s)^{0.5}]\ /\ \sinh\ (^a{}_s)^{0.5}$	3.9	Marchello and Toor, 1963
c. Surface renewal - damped eddy diffusivity model	$K_a = n_o a^{1/n_o}\ \Pi^{-1}\ D_m^{1-1/n_o}\ H^{-1}\ \sin(\Pi/n_o)$	3.10	King, 1966
3. Kinetic theory model	$K_a = b\ n_s\ L_f\ H^{-1}$	3.11	Tsivoglou, 1967
4. Large-eddy model	$K_a = c(D_m\ u'/¶)^{0.5}\ H^{-1}$	3.12	Fortescue and Pearson, 1967
5. Turbulent diffusion model	$K_a = 4.62(\Pi)^{-0.5}\ (E_y\ s)^{0.5}\ H^{-1}$	3.13	Kishinevsky, 1955

The notations used in table 3.1 are:

D_m = the molecular-diffusion coefficient (L^2T^{-1})
L_f = the thickness of the film (L)
H = the depth of flow (L)
t_e = the age of surface at which elements are replaced (T)
$^a{}_s$ = $(s\ L_f^2\ /\ D_m)^{0.5}$
s = the rate of surface renewal (T)
E_y = the eddy-diffusion coefficient for mass in vertical direction (L^2T^{-1}), $E_y = ay^n$
y = the distance from the interface in vertical direction (L)
a, n_o = parameters independent of time
Π = the constant 3.1416
b = the proportion of available molecules in the surface layer that actually escape
n_s = the number of new surface layers exposed per unit of time (T^{-1})
u' = the longitudinal turbulent intensity (LT^{-1})
¶ = the turbulent macroscale of the flow (L)
c = the constant in the Fortescue and Pearson (1967) model

Table 3.2
Differential equations and boundary conditions used in the development of theoretical reaeration models

Reference	Diffusion equation	Boundary Conditions		
Higbie, 1935	$\frac{\partial C}{\partial t} = D_m \frac{\partial^2 C}{\partial y^2}$	$C = C_L$ $C = C_L$ $C = C_s$	$t = 0$ $t > 0$ $t > 0$	$y > 0$ $y - \infty$ $y = 0$
Dobbins, 1956	$\frac{\partial C}{\partial t} = D_m \frac{\partial^2 C}{\partial y^2}$	$C = C_L$ $C = C_s$ $C = C_L$	$t = 0$ $t \geq 0$ $t \geq 0$	$0 < y \leq L_f$ $y = 0$ $y = L_f$
Harriott, 1962	$\frac{\partial C}{\partial t} = D_m \frac{\partial^2 C}{\partial y^2}$	$C = C_s$ $C = C_L$ $C = f(y)$	$t \geq 0$ $t \geq 0$ $t = 0$	$y = 0$ $y - \infty$ $0 \leq y \leq 0$
King, 1966	$\frac{\partial C}{\partial t} = \frac{\partial}{\partial y} [(D_m + a y^{n_o}) \frac{\partial C}{\partial y}]$	$C = C_L$ $C = C_L$ $C = C_s$	$t = 0$ $t > 0$ $t > 0$	$y > 0$ $y - \infty$ $y = 0$
Fortescue and Pearson, 1967	$\frac{\partial C}{\partial x} u_e + \frac{\partial C}{\partial y} v_e = D_m (\frac{\partial^2 C}{\partial x^2} + \frac{\partial^2 C}{\partial y^2})$	$C = C_s$ $\partial C/\partial x = 0$ $C = C_L$	$y = 0$ $0 \leq y \leq ¶$ $y = ¶$	$0 \leq x \leq ¶$ $x = n¶$ $0 \leq x \leq ¶$

The notations used in table 3.2 are:
C_s = the dissolved oxygen concentration at saturation (ML^{-3})
C_L = the dissolved oxygen concentration in the bulk-liquid phase (ML^{-3})
L_f = the thickness of the film (L)
t = the time elapsed since the formulation of the interface (T)
y = the distance from the interface in vertical direction (L)
a, n_o = the parameters independent of time
u_e = the eddy velocity in the longitudinal direction (LT^{-1})
v_e = the eddy velocity in the vertical direction (LT^{-1})
x = longitudinal position (L)
¶ = the turbulent macroscale of the flow (L)

O'Connor and Dobbins (1956) presented an analysis of the reaeration mechanism based on certain concepts from the field of fluid turbulence and on relationships defining the rate of surface renewal. They developed two equations for the use under different conditions of turbulence. The equation for stream conditions of anisotropic turbulence was presented in the following form:

$$k_a = C_2 D_M^{0.5} S^{0.25} H^{-1.25} \tag{3.15}$$

in which D_M is the coefficient of molecular diffusivity (L^2T^{-1}), S is the slope of the river channel (LL^{-1}), H is the average depth (L), and C_2 is constant. For relatively deep streams, where conditions of isotropic turbulence might be approached (see Fig. 3.1) the following equation was presented:

$$k_a = C_3 D_M^{0.5} U^{0.5} H^{-1.5} = C_3' U^{0.5} H^{-1.5} \tag{3.16}$$

where U is the mean velocity of flow (LT^{-1}) and C_3 is constant.

The verification of these models was based on both field and laboratory data. If in equation (3.16) the reaeration coefficient at 20°C has units of day^{-1}, D_M is in $m^2\ day^{-1}$ ($D_M = 0.0001806\ m^2d^{-1}$), U in m sec^{-1}, H in m, then $C_3' = 1.707$.

O'Connor (1958) showed that the differences between the values predicted by equations (3.15) and (3.16) are insignificant and proposed that equation (3.16) should be generally applied.

Krenkel and Orlob (1962) conduced a series of experiments in a laboratory flume in order to examine the relationship between k_a (T^{-1}) and the longitudinal mixing coefficient, D_L (L^2T^{-1}). The basic equation was assumed to have the following form:

$$K_a = C_4 D_L^{n_1'} H^{m_1} \tag{3.17}$$

in which C_4, n_1', m_1 are constants, and k_a, D_L and H are variables. In addition, they hypothesized that

$$k_a = D_5 E_D^{b_1} H^{c_1} \tag{3.18}$$

where C_5, b_1 and c_1 are constants, and E_D is the energy dissipation per unit mass of flow (L^2T^{-3}), which was approximated for open channel flow as:

$E_D = U\ g\ S_e$ in which U is the mean stream velocity (LT^{-1}), g is the gravitational constant (LT^{-2}), and S_e is the slope of the energy gradient. The application of a multiple correlation analysis to the data from fifty-eight independent experiments, after correcting all test values to 20°C, produced the equation

$$k_{a\ (20°C)} = C_4' D_L^{1.321} H^{-2.32} \tag{3.19}$$

An alternative empirical relation was

$$k_{a\ (20°C)} = C_5'\ E_D^{0.408}\ H^{-0.66} \tag{3.20}$$

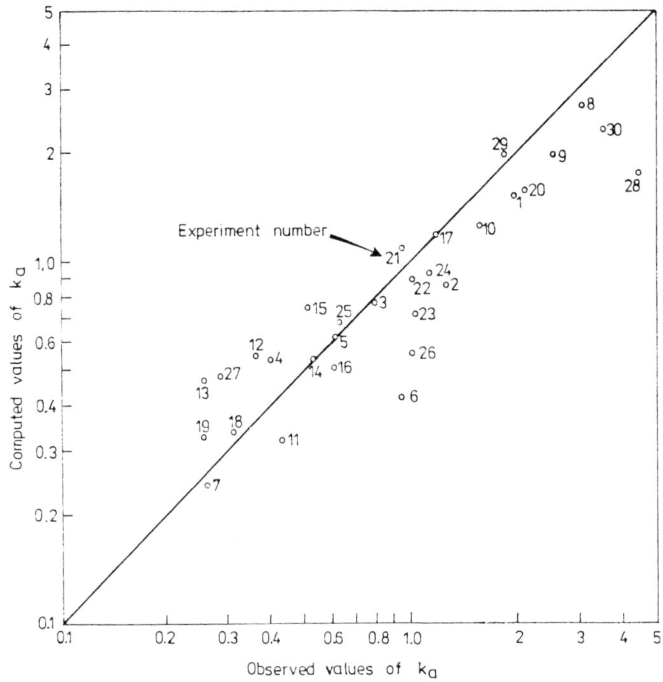

Fig. 3.1: Observed vs. computed values of k_a in O'Connor Dobbins model for isotropic conditions, after Churchill et al. (1962).

If, in equation (3.19), $k_{a\ (20°C)}$ has units of day^{-1}, D_L is in m^2/min, and H in m, then $C_4' = 2.4*10^{-2}$. Also, if in equation (3.20) $k_{a\ (20°C)}$ has units of day^{-1}, E_D is in $m^2\ min^{-3}$, and H in m, then $C_5' = 0.198$. Harleman and Holley (1962) analyzed the data of Krenkel and Orlob (1962) and have shown that a linear relationship exists between the coefficient of longitudinal dispersion, D_L (L^2T^{-1}) and the average vertical eddy diffusivity, !e_y (L^2T^{-1}). The latter parameter was used by Krenkel and Orlob (1962) to determine the reaeration coefficient. A regression analysis of their data indicated that

$$k_a = C_6\ !e_y^{1.237}\ H^{-2.087} \tag{3.21}$$

If k_a is expressed in day^{-1}, !e_y in m^2/sec, and H in m, then $C_6 = 2.60$, and the correlation coefficient for this equation was equal to 0.98.

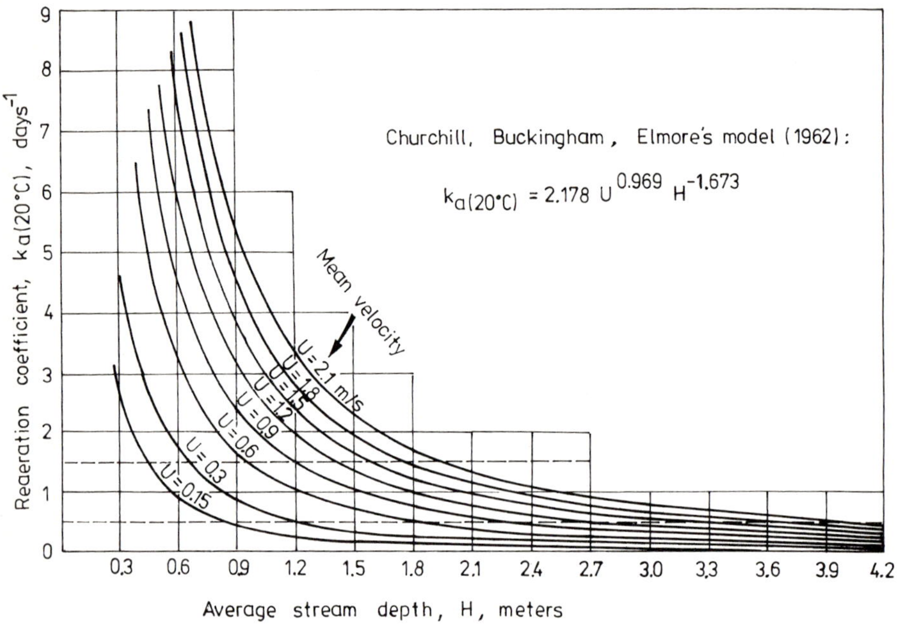

Fig. 3.2: Churchill model for atmospheric reaeration coefficient, after Churchill et al. (1962).

Churchill, Buckingham and Elmore (1962) performed in extensive empirical study on several tributary streams in the upper Tennessee basin, and applied dimensional analysis and multiple regression techniques to analyze the data. Nineteen equations for k_a resulting from various combinations of the following variables: mean velocity, mean depth, energy slope, resistance coefficient, fluid density, fluid viscosity, surface tension, molecular diffusion coefficient, and a vertical diffusion coefficient, were presented. The recommended equation had the form (see Fig. 3.2):

$$k_{a\ (20°C)} = C_7\ U^{0.969}\ H^{-1.673} \qquad (3.22)$$

If k_a has units of day^{-1}, when U is in m/sec and H in m, then C_7 = 2.178. The data were corrected to a constant temperature of 20°C using the temperature coefficient ø = 1.024. The multiple correlation coefficient for equation (3.22) was 0.822. It was suggested that for many applications equation (3.22) can be used in the following simplified form:

$$k_{a\,(20°C)} = C_7{}' U H^{-1.67} \qquad (3.23)$$

Again, when k_a has units of day^{-1}, and U is in m/sec and H in m, then $C_7{}' = 2.26$. Both equations were considered to be applicable for reaches with constant variables U and H, and to relatively unpolluted water. For polluted streams the authors indicated the necessity of making appropriate corrections.

Dobbins (1964) introduced a coefficient, C_A, to relate the actual interfacial area to the superficial area or area of the projection of the surface to the surface of a horizontal plane that assumed that C_A is equal to $1 + 0.3\,(N'_{Fr})^2$, in which N'_{Fr} is the Froude Number. Dobbins (1965) gave C_A as $1 + (N'_{Fr})^2$ and proposed the following model, which had originated from his earlier work:

$$k_a = C_8\, C_A\, A_D\, E^{0.375} \coth\left(\frac{BE^{0.125}}{C_D{}^{0.5}}\right) C_D{}^{-1.5}\, H^{-1} \qquad (3.24)$$

If, in this model k_a, is given as day^{-1}, and U in m/sec, H in m, s in $m/10^{3*}m$, then $C_8 = 0.12$, and the remaining parameters are given by the following expressions:

$$C_A = 1 + (N'_{Fr})^2$$
$$C_D = 0.9 + N'_{Fr}$$
$$A_D = 9.68 + 0.055\,(T-20)$$
$$B = 0.976 + 0.0137\,(30-T)^{1.5}$$
$$E = 30\,s\,U$$
$$N'_{Fr} = U\,(gH)^{-0.5}$$

where N'_{Fr} is the Froude Number and T is the temperature in °C.

Churchill's et al. (1962) data have been used by many investigators for the development of additional models of the reaeration coefficient.

Owens, Edwards and Gibbs (1964) combined their data with that collected by Churchill et al. (1962) and by Gameson et al. (1955), and from regression analysis obtained the following equation:

$$K_{a\,(20°C)} = C_9\, U^{0.67}\, H^{-1.85} \qquad (3.25)$$

For the restricted stream data set obtained by the British Water Pollution Research Laboratory (WPCL) the equation was in the form:

$$K_{a\,(20°C)} = C_{10}\, U^{0.73}\, H^{-1.75} \tag{3.26}$$

If K_a has units of day^{-1} and U, H are in m/sec and m, respectively, C_9 = 2.316 and C_{10} = 3.0.

Langbein and Durum (1967) presented graphically Churchill's et al. (1962) field data, together with the data reported by O'Connor and Dobbins (1956), plus the laboratory data of Streeter et al. (1936) and of Krenkel and Orlob (1962) and obtained an equation of the form:

$$K_{a\,(20°C)} = C_{11}\, U\, H^{-1.33} \tag{3.27}$$

in which C_{11} = 2.23 if k_a is in day^{-1}, U in m/sec, and H in m.

Isaacs and Gaudy (1968) applied a least sguares analysis to the field data of Churchill et al. (1962) and to the laboratory data of Krenkel (1960) and Krenkel and Orlob (1962), assuming that

$$K_{a\,(20°C)} = C_{12}\, U\, H^{-1.15} \tag{3.28}$$

For the data of Churchill et al. (1962) a value of C_{12} = 2.219, and the correlation coefficient was 0.898. When one of the experiments, No. 28, was omitted (see Fig. 3.1) C_{12} = 2.064, and the correlation coefficient increased to 0.948. For the data of Krenkel (1960) and Krenkel and Orlob (1962) a value of C_{12} = 1.347, and the correlation coefficient was 0.95. These values of C_{12} were obtained when U and H are in m/sec, and m, respectively. In addition, Isaacs and Gaudy (1968) have applied the Buckingham theorem along with dimensional analysis to develop a model for the reaeration coefficient, relating the following dimensionless parameters: the Reynolds Number N_{Re} = UHv^{-1}, the Froude Number $N_{Fr} = U^2/gH/^{-1}$, the Schmidt Number $N_{Sc} = vD_M^{-1}$, and $k_aH^2D_M^{-1}$.

The model relating these dimensionless parameters was presented as

$$k_a = C_{13}\, N_{Re}^{\;a}\, N_{Fr}^{\;b}\, N_{Sc}^{\;d}\, D_M\, H^{-2} \tag{3.29}$$

where k_a is in (T^{-1}), D_M in (L^2T^{-1}), H in (L); and other variables are: the kinematic viscosity of water v (L^2T^{-1}), the mean velocity U (LT^{-1}) and the gravitational constant g (LT^{-2}). Equation (3.29) can be rewritten as

$$k_a = C_{14}\ U^n\ H^m \tag{3.30}$$

in which n is used to replace the exponent (a+2b), and m to replace (a-b-2), and $C_{14} = C_{13}\ D_M{}^{1-d} * v^{d-a} * g^{-b}$.

Since a preliminary analysis performed on the data represented by fifty two reaeration experiments (conducted in an experimental apparatus) yielded values of n = 1.0027, and m = -1.4859, the value of n was assumed to be unity, and the value of m to approximate to -1.5. It was calculated that a = 2/3 and b = 1/6, and by assuming that the group $D_M{}^{1-d} * v^{d-2/3}$ reflects changes in k_a due to the temperature effects (when ø = 1.0241), the value of d was obtained as 1/2. When equation (3.30) was applied to the laboratory data of Isaacs and Gaudy (1968) a value of C_{14} = 1.685 was found for k_a with dimension in day^{-1}, U in m/sec and H in m. The difference between the values of constants (i.e. 1.685 and 2.064) were assumed to be attributed to differences in channel geometry.

Isaacs and Maag (1969) pointed out the failure of Langbein and Durum (1967) to include all of Churchill's data (1962), and, based on their use of these data they recommended a model in the form:

$$k_a = C_{15}\, v_1\, v_2\, UH^{-1.5} = C_{16}\, UH^{-1.5} \tag{3.31}$$

in which C_{15} and C_{16} are constants, v_1 is a non-dimensional variable which varies with the channel geometry, and v_2 is a non-dimensional variable which is a measure of the surface velocity. If k_a is in day^{-1}, U in m/sec, H in m, then the average value of v_1 was found to be 1.078, the average for v_2 was 1.16 and C_{15} = 1.645, C_{16} = 2.057.

Thackston and Krenken (1969) assumed that the reaeration coefficient is directly proportional to the vertical mass coefficient and inversely proportional to H^2, as was originally proposed by Krenkel (1960). In addition, assuming that the value of the vertical mass transfer coefficient is proportional to the value of the momentum transfer coefficient they have proposed the following model:

$$k_a = C_{17}\ U_*\ H^{-1} \tag{3.32}$$

where U_* is the shear velocity (LT^{-1}), usually computed from $U_* = (HSg)^{0.5}$. However, when the Froude number, given as $N'_{Fr} = U(gH)^{-0.5}$, was incorporated into equation (3.32) final form of this equation was presented as:

$$k_a = C_{17} [1 + (N'_{Fr})^{0.5}] U_* H^{-1} \tag{3.33}$$

The constant C_{17} was estimated from the laboratory flume data as 18.6 and then reevaluated and found to be $C_{17} = 10.8$.

Cadwallader and McDonnell (1969) analyzed a restricted data set, viz the data of Churchill et al. (1962), Owens et al. (1964) and the British Water Pollution Research Laboratory channel data, by a multivariate analysis, and proposed a model for the reaeration coefficient similar to the model expressed by equation (3.20):

$$k_a = C_{18} E_D^{0.5} H^{-1} \tag{3.34}$$

When k_a is in day^{-1} and E_D is the energy dissipation per unit mass of flow in $m^2 sec^{-3}$, then $C_{18} = 25.7$.

Negulescu and Rojanski (1969) performed a series of experiments in a channel with circulated water and by the application of a regression analysis to their data, they proposed the following equations:

$$k_a = C_{19} (U/H)^{0.85} \tag{3.35}$$

and

$$k_a = C_{20} D_L (U/H)^{1.63} \tag{3.36}$$

If, in the above equations, k_a is in day^{-1}, U in m/sec, H in m, and D_L in cm^2/sec, then $C_{19} = 4.74$ and $C_{20} = 0.0153$.

The authors proposed to use these equations for streams in which the velocity was not greater than 1.2 m/sec and the depth no more than 0.5 m.

Shastry et al. (1969) used Bard's absolute Gauss method for testing some of the models for the reaeration coefficient. The data of Churchill et al. (1962) and of Thackston and Krenkel (1969) were used in this analysis and k_a was related to the average stream velocity U (LT^{-1}), the stream depth H (L), the shear velocity U_* (LT^{-1}), the coefficient of longitudinal dispersion D_L (L^2T^{-1}), and the Froude Number, N_{Fr}, and the Peclet Numbers based on depth, N_{PeH} and N^*_{PeH}. The models corresponded to the following equation:

$$k_a = C_{21} A_1^{a_1} B_1^{b_1} \tag{3.37}$$

The values of the parameters estimated by Bard's nonlinear parameter estimation program are given in Table 3.3. It was concluded that models IV and VI in this table gave much a better fit for data used than other models.

Padden and Gloyna (1971) developed the following relationship for the reaeration coefficient in a research flume:

$$k_{a(20°C)} = C_{22} U^{0.703} H^{-1.0545} \tag{3.38}$$

If k_a is in day^{-1}, U in m/sec, and H in m then C_{22} = 1.963.
During this experimental study the velocities were ranging from 0.03 m/sec to 0.14 m/sec and depths from 0.035 m to 0.19 m.

Table 3.3
Values of parameters for models in studies of Shastry et al. (1969) as estimated by Bard's method

Model	A_1	B_1	C_{21}	a_1	b_1
I	U	H	$0.840*10^{-5}$	1.1631	-1.485
II	D_L	H	$0.238*10^{-4}$	1.3410	-2.049
III	U_*	H	$0.221*10^{-3}$	0.5833	-1.132
IV	N_{Fr}	U_*/U	$0.214*10^{-3}$	0.1080	-1.007
V	$1+N_{Fr}$	U_*/H	$0.202*10^{-3}$	1.4281	0.106
VI	N_{Fr}	N_{PeH}	$0.263*10^{-3}$	1.2933	1.137
VII	N_{Fr}	N^*_{PeH}	not quoted (model did not give a good fit)		

where $N_{PeH} = D_L (UH^{-1})$, $N^*_{PeH} = D_L (U_* H)^{-1}$

Bennett and Rathbun (1972) performed a statistical analysis on most of the available data in the literature. They recommended use of the equation which was developed using the data points of Churchill et al. (1962) and Owens et al. (1964):

$$k_a = C_{23}\, U^{0.674}\, H^{-1.865} \tag{3.39}$$

If in equation (3.39) k_a is in day^{-1} with U and H in m/sec and m, respectively, then $C_{23} = 2.33$.

Parkhust and Pomeroy (1972) hypothesized that the reaeration coefficient k_a (T^{-1}) is directly proportional to the oxygen exchange coefficient D_E (LT^{-1}) and inversely proportional to H(L):

$$k_a = D_E\, H^{-1} \tag{3.40}$$

where $D_E = C_{24}\, C_A\, ^a\, ß\, \Omega\, (SU)^{n_2}\, R^{m_2}$ in which $C_{24} = a_p C_E$, and a_p is a coefficient for the units of measurement (defined as unity if D_E is expressed m/hr and U is m/sec); C_A is an empirical factor relating the coefficient for a stream with disturbed surface to the coefficient calculated by extrapolation from quiet streams (term C_A was originally proposed by Dobbins); ^{a} is the ratio of the exchange coefficient in impure water to the coefficient in pure water (at the same temperature and flow conditions); ß is the ratio of the exchange coefficient in stream of irregular shape to the coefficient in stream of simple shape (rate of energy dissipation, temperature and purity of water being the same); Ω is the ratio of the exchange coefficient at temperature T to the coefficient at 20°C (other conditions being the same, since Ω is a function not only of temperature but also the intensity of turbulence); R is the hydraulic radius (L); S is the slope of energy line of stream; U is the velocity of flow (LT^{-1}); and n_2, m_2 are exponents. The exchange coefficient D_E was measured in 12 functioning sewers, and the equation for the sewer data was presented as:

$$k_a = C_{25}\, [\, (\, 1 + p(N'_{Fr})^2\,]\; \Omega\; (SU)^{0.375}\; H^{-1} \tag{3.41}$$

in which $N'_{Fr} = U\, (gH)^{-0.5}$ and the value p derived from these data was 0.17. If D_E is in m/hr, and U in m/sec, then $C_{25} = a_p\, C_E\, ^a\, ß$ is equal to 0.96. The above equation was assumed to be also applicable to streams, except those with low Reynolds number (up to 5000).

Tsivoglou and Wallace (1972) postulated that the reaeration coefficient is directly proportional to the rate of energy expenditure in nontidal fresh water streams, and proposed the following model

$$K_a = C_{26} (\Delta H/t_f) \tag{3.42}$$

where K_a is the reaeration coefficient (base e) per hour, ΔH is the water surface elevation change (L), t is the time of flow (T) and C_{26} is the constant of proportionality (L^{-1}), which was designated as the escape coefficient. The escape coefficient C_{26} was assumed to remain unchanged until the degree or type of pollution is changed, or the hydraulic mixing regime is altered significantly, and was related to the water temperature in the same way as K_a ($\varnothing$ = 1.022). For the length of stream between points 1 and 2 the following model was presented:

$$D_2/D_1 = \exp [-C_{26} \Delta H] \tag{3.43}$$

where D_1 is the DO deficit at the upstream end of a stream section (ML^{-3}), D_2 is the DO saturation deficit (ML^{-3}), that would occur at the downstream end if there were no concurrent sources of oxygen consumption in the stream section, and C_{26}, ΔH are the same as in equation (3.42). In reaeration tracer studies performed over a period of 1966-1970 on five American non-tidal fresh water streams (the Flint, South, Patuxent, Jackson and Chattahoochee Rivers) they have applied the field tracer procedure (developed by the U.S. FWPCA), in which a radioactive form of the noble gas krypton serves as a tracer for oxygen. From summary analysis of the observed results for all five rivers studied, the mean values of the oxygen escape coefficient, C_{26}, at the temperature of 25°C in equations (3.42) and (3.43) were 0.172 and 0.180 per metre, respectively. Using the mean of the two separately derived results the following model for the reaeration coefficient was presented:

$$K_{a\,(25°C)} = 0.176 \; (\Delta H/t_f) \tag{3.44}$$

The authors have stated that the above model may be used for moderately polluted streams (5-days BOD about 15 mg/l) and reasonably well mixed. The lowest values of C_{26} are associated with more pollution, and vice versa.

Lau (1972b) employed a dimensional analysis to the problem of reaeration and presented the following functional relationship

$$k_a RU^{-1} = f\,(UR\,f\,\mu^{-1},\, D_M\,f\,\mu^{-1},\, U_*U^{-1},\, wR^{-1}) \tag{3.45}$$

Since it was found that very little difference was made when the depth,

H, was used as a substitute for the hydraulic radius, R, therefore the depth, H, was used in this analysis, and equation (3.45) can be presented as

$$k_a\, HU^{-1} = f(N_{Re}, N'_{Sc}, U_*U^{-1}, wH^{-1}) \quad (3.45)$$

in which the Reynolds Number, $N_{Re} = UH\, f\, \mu^{-1}$, the Schmidt Number, $N'_{Sc} = D_M\, f\, \mu^{-1}$, and the velocity U ($LT^{-1}$), the friction velocity U_* (LT^{-1}), the width of the channel w (L), and the density and viscosity of the fluid (water), f and μ, are in (ML^{-3}) and ($ML^{-1}T^{-1}$), respectively. From the analysis of the effect of the independent parameters on k_a, using the data of Churchill et al. (1962), of Krenkel (1960), and of Thackston and Krenkel (1969), it was concluded that the parameter, $k_a\, HU^{-1}$, increased with U_*U^{-1} but was not affected by N_{Re} and $w\, H^{-1}$. Therefore, the following model was presented:

$$k_a = C_{27}\, (U_*/U)^3\, U\, H^{-1} \quad (3.47)$$

In this model the effect of the parameter, $D_M\, f\, \mu^{-1}$ has not been included since it was not investigated due to lack of suitable data. If k_a is in day^{-1}, and H/U in sec, then C_{27} = 1088.64. It was also pointed that this equation gave a reasonable fit to data with temperature at 20°C, and in which w/H varied from 6 to 290, U_*/U from 0.06 to 0.25 and with N_{Re} ranging from $2*10^3$ to $2*10^6$.

Bansal (1973) reanalyzed the reaeration data of Churchill et al. (1962), O'Connor and Dobbins (1956) and of Langbein and Durum (1967) on natural streams and obtained the following relationships:

$$\log P_R = -C_{28} + C'_{28} \log N_{Re} \quad (3.48)$$
$$\log P'_R = -C_{29} - C'_{29} \log N_{Re} \quad (3.49)$$
$$\log P''_R = C_{30} - C'_{30} \log N_{Re} \quad (3.50)$$
$$\log P_R = C_{31} + C'_{31} \log P_v \quad (3.51)$$
$$\log P_R = C_{32} + C'_{32} \log P_L \quad (3.52)$$

in which P_R is a dimensionless reaeration parameter, $P_R = k_a H^2 v^{-1}$; and $P'_R = k_a\, HU^{-1}$ and $P''_R = K_a H^2 D_z^{-1}$ are different forms of the reaeration parameter; $P_v = D_z v^{-1}$ is the dimensionless dispersion parameter, and $P_L = F_D\, (U_s/U)\, S_L\, (UH)^{-1}$ is the dimensionless vertical parameter; $N_{Re} = f\, UH\, \mu^{-1}$, F_D is a regional dispersion factor, U the average flow velocity for the reach (LT^{-1}),

U_s the mean flow velocity at the sampling station (LT^{-1}), μ the coefficient of viscosity of water ($ML^{-1}T^{-1}$), v the kinematic viscosity of water (L^2T^{-1}), f the mass density of water (ML^{-3}). The values of the longitudinal dispersion coefficient D_L (L^2T^{-1}), and of the vertical dispersion coefficient D_z (L^2T^{-1}) were calculated by using the following relationships, respectively:

$$\log P_L = C_{33} - C'_{33} \log N_{Re} \tag{3.53}$$

$$\log P_v = C_{34} - C'_{34} \log N_{Re} \tag{3.54}$$

in which $C_{33} = 4.66$. The correlation coefficients for equations (3.48)-(3.52) varied from 0.865 to 0.972.

Alonso, McHenry, Hong (1975) studied the atmospheric reaeration of uniform streams in a laboratory flume, with and without suspended sediments, and derived the following relationship, respectively:

$$k_{d(20°C)} = C_{36}\ U_*\ R^{0.956} \tag{3.56}$$

and

$$k_{a(20°C)} = C_{36}\ [1 + p_1 C_s^{1.9}]^{-1} * U_* R^{-1} \tag{3.57}$$

If k_a is in 10^{-4} sec^{-1}, and U_* in m/sec, R in m, then $C_{36} = 6.19 * 10^{-4}$, $p_1 = 9.6 * 10^{-8}$.

Foree (1976) used the radioactive technique developed by Tsivoglou (1967) to determine the reaeration coefficient on streams with varying characteristics distributed across the state of Kentucky. For a particular reach under study k_a (T^{-1}) was calculated from

$$k_a = (0.83)^{-1} \ln\left(\frac{R_u}{R_d}\right) t_f^{-1} \tag{3.58}$$

in which R_u, R_d are measured ratios of krypton-85 activity to tritium activity at the upstream and downstream end of the reach, respectively; t_f is the time of passage of the dye peak through the reach T; and 0.83 is a ratio of exchange coefficient for krypton to exchange coefficent for oxygen, as previously determined by Tsivoglou. From the observed data a very simple model was presented by regression analysis

$$k_{a\,(20°C)} = C_{37} + C_{38}\,S^{1.2} \tag{3.59}$$

If k_a in day^{-1} and the slope of reach, S in m/km, then $C_{37} = 0.116$, $C_{38} = 0.539$. The correlation coefficient was 0.94.

Long in 1984, as cited by Bowie et al. (1985), proposed the following model based on data collected on streams in Texas:

$$k_a = C_{39}\,U^{0.273}\,H^{-0.894} \tag{3.60}$$

in which $C_{39} = 1.923$ if k_a is in day^{-1}, U in m/sec, and H in m.

Reaeration Models for Estuaries and Lakes

Some of the predictive models for the river reaeration coefficient have been applied to estuaries. One of the more widely applied models is the model (equation 3.16) of O'Connor and Dobbins (1956) in which the mean tidal velocity over a complete cycle is used. This model has been subsequently modified by Thomann and Fitzpatric (1982) in order to include wind speed effects. Also, Ozturk (1979) proposed the formula as follows:

$$k_a = C_{40}\,U^{4/3}\,H^{-1} \tag{3.61}$$

in which U is the mean tidal velocity LT^{-1}, and H is the mean depth. If k_a has units of day^{-1} when U is in m/sec and H in m, then $C_{40} = 4.56$.

Wen et al. (1984) have shown that the reaeration coefficient of an estuary increases with increasing chloride concentration. The reaeration coefficient in saline water, $k'_{a\,20°C}$, at any chloride concentration Cl mg/l at 20°C in day^{-1}, was expressed as

$$k'_{a\,(20°C)} = k_{a\,(20°C)}\,\exp\,(0.0127\,Cl) \tag{3.62}$$

where $k_{a\,(20°C)}$ is the reaeration coefficient in fresh water at 20°C in day^{-1}. The correlation coefficient of $k'_{a\,(20°C)} / k_{a\,(20°C)}$ and Cl was 0.88.

All models for reaeration of lakes are usually for the surface transfer coefficient, $k_L = k_a * H$, in units of LT^{-1} (see Table 3.4). The effects of winds are included in reaeration predictions in lakes.

Table 3.4
Reaeration coefficients for lakes

Reference	Surface transfer coefficient, k_L md^{-1}
Downing and Truesdale, 1955	$k_L = 0.0276\, V_w^{2.0}$
Kanwisher, 1963	$k_L = 0.0432\, V_w^{2.0}$
Liss, 1973	$k_L = 0.0156\, V_w^{2.0}$ $(V_w \leq 4.1$ m/sec)
	$k_L = 0.0269\, V_w^{1.9}$ $(V_w > 4.1$ m/sec)
Broecker and Peng, 1974	$k_L = 0.0449\, V_w^{2.0}$
Banks, 1975	$k_L = 0.0362\, V_w^{0.5}$ $(0 \leq V_w \leq 5.5$ m/sec)
	$k_L = 0.0277\, V_w^{2.0}$ $(V_w > 5.5$ m/sec)
Weiler, 1974	$k_L = 0.398$ $(V_w < 1.6$ m/sec)
	$k_L = 0.0155\, V_w^{2.0}$ $(V_w \geq 1.6$ m/sec)
Baca and Arnett, 1976	$k_L = a + b\, V_w$
Chen et al., 1976	$k_L = (86400\, D_M) / (200\text{-}60\, V_w^{0.5})\, 10^{-6})$
Yu et al., 1977	$k_L = 0.319\, V_w$
Broecker et al., 1978	$k_L = 0.864\, V_w$
Smith, 1978	$k_L = a' + b'\, V_w$

where
D_M = the molecular diffusion coefficient of oxygen in water, m^2sec^{-1}
V_w = the wind speed, m/sec
a = 0.05-1.01 md^{-1}; b = 10^{-6} - 10^{-5} m^{-1}
a' = 0.64 md^{-1}; b' = 0.128 $sec^2\, m^{-1}\, d^{-1}$

3.4 ANALYSIS OF PREDICTIVE MODELS FOR RIVERS

Comparison of Behaviour of Predictive Models

The variation within reaeration coefficient values for rivers obtained by application of various models is extremely high. Tsivoglou and Wallance (1972) tested the five selected models for predicting the river reaeration coefficient from the hydraulic properties, by applying a correlation-regression analysis. The reaeration coefficients measured in their field tracer studies were compared with values predicted by these five models. The results of the statistical analysis are presented in Table 3.5. In this analysis a model capable of predicting values of the reaeration coefficient close to values observed in the field has the value of the intercept of the line near zero, and the value of the slope of the line near unity. Also, the value of the correlation coefficient close to unity indicates that the variations of the predicted values of the reaeration coefficient about the regression line is relatively small.

This analysis has indicated that none of the five selected models was able to predict accurately the values of the reaeration coefficient obtained from tracer studies. In most cases there was a tendency to overpredict observed values, and the differences between predicted and observed values of the reaeration coefficient were significant for highly turbulent stream reaches and for reaches with non-uniform flows.

The performance of the various prediction models was comprehensively evaluated by Bennett and Rathbun (1972), by comparing their standard error of estimate (SE), and the percent standard error of estimate (PE), defined as:

$$SE = [\sum_{i=1}^{N} (k_a^e - k_a^c)_i^2 / N]^{0.5} \qquad (3.63)$$

$$PE = 100\,(1 - 10^{-SLE}) \qquad (3.64)$$

where k_a^e is the estimated k_a from a prediction model, k_a^c is the calculated k_a from the results of an experimantal test, and N is number of k_a values used. The computed values of SE in day^{-1} and PE in percent for the various prediction models are presented in Table 3.6. These values are for the original data from which the models were derived, as well as for data available in the literature of field and flume observations. From a plot (Fig. 3.3) of the models given in Table 3.6, for an assumed slope of 0.0001 mm^{-1} and a mean velocity of about 0.3 m/sec, it is evident that models derived from field data have slopes of the order of -1.5, and those derived from flume data have slopes of the order of -0.5. This analysis, performed on field data from natural streams and data from laboratory channel flumes, has indicated a significant difference between the k_a regression models obtained from the two groups of data.

Table 3.5
Statistical test of selected predictive models (Tsivoglou and Wallance, 1972)

Predictive models for k_a (25°C)	Flint River (84 observations)			South River (96 observations)			Potuxent River (30 observations)		
	A_o	B_o	r_{xy}	A_o	B_o	r_{xy}	A_o	B_o	r_{xy}
O'Connor-Dobbins, 1958	-0.06	1.14	0.68	0.34	-0.03	0.00	0.08	0.19	0.50
Churchill et al., 1962	-0.02	1.53	0.75	0.04	-0.26	-0.06	0.09	0.25	0.53
Langbein-Durum, 1967	-0.09	2.90	0.74	0.42	-0.48	-0.07	0.09	0.46	0.56
Owens et al., 1964	-0.01	0.73	0.76	0.39	-0.13	-0.05	0.10	0.12	0.51
Thackston-Krenkel, 1969	0.06	0.72	0.34	0.44	3.03	-0.90	0.02	0.54	0.44

Table 3.5 (continued)

Predictive models for k_a (25°C)	Jackson River (80 observations)			Chattahoochee River (30 observations)		
	A_o	B_o	r_{xy}	A_o	B_o	r_{xy}
O'Connor-Dobbins, 1958	0.12	0.27	0.14	0.05	-0.05	-0.06
Churchill et al., 1962	0.12	0.52	0.21	0.05	-0.05	-0.06
Langbein-Durum, 1967	0.10	0.84	0.24	0.05	-0.04	-0.04
Owens et al., 1964	0.12	0.27	0.19	0.05	-0.05	-0.06
Thackston-Krenkel, 1969	0.11	0.18	0.20	0.03	0.25	0.10

Regression equations: $X = A_o + B_oX$

where: Y = the measured value of k_a

X = the value of k_a calculated from the predictive equations

B_o = the slope, and A_o = the intercept

r_{xy} = the correlation coefficient

Table 3.6
Error analysis of various predictive models for reaeration coefficient (Bennett and Rathbun, 1972)

Model represented by equation no.	Data from source publication N	SE, day^{-1}	PE, %	Total applicable data N'	SE, day^{-1}	PE, %
15	13	0.34	45	38	33.7	46
16	22	0.42	44	169	115.0	71
20	58	7.53	15	207	16.4	51
22	30	0.52	28	239	140.3	73
24	123	7.74	33	207	21.0	43
25	68	5.00	32	239	474.9	85
26	32	5.46	35	239	419.8	84
28 a)	30	0.55	30	239	55.3	63
28 b)	58	10.44	29	239	30.8	65
32	40	9.96	25	207	20.6	45
33	105	5.90	37	207	20.0	45
35	23	1.63	29	239	21.2	61

a) $C_{12} = 2.064$ b) $C_{12} = 1.347$

N = number of data points used in original publication
N' = number of data points which can be used

This conclusion was confirmed by Wilson and Macload (1974) in their statistical analysis of predictive performance of various models for the river reaeration coefficient. These models were categorized as those which are a function of stream depth and velocity and those which are concerned with energy dissipation or surface renewal. About five hundred data points obtained from the literature were used. As statistical measures were selected: the standard error of estimate (SE) in day^{-1} and a normalized mean error of prediction (NME) in % defined as:

$$NME = \left(\sum_{i=1}^{N} \frac{k_a^e - k_a^c}{k_a} * 100\% \right) / N \qquad (3.65)$$

where k_a^e is the calculated value from a prediction model, and k_a^c is the observed values. The results of this analysis are presented in Table 3.7.

Table 3.7
Predictive performance of various models for reaeration coefficient (Wilson and Macleod, 1974)

Model represented by equation no.	Number of data points	Standard error, SE (day^{-1})	Normalized mean error, NME, %
16	482	114.43	196.1
20	382	116.63	-34.8
22	482	104.31	194.9
24	382	73.43	34.5
28 a)	482	71.88	14.2
28 b)	482	75.91	42.9
28 c)	482	71.77	75.1
32 d)	382	112.36	-15.0
33 e)	382	115.62	-27.6
34	382	66.19	69.6
35	482	99.15	54.4
39	482	358.36	701.4
41	382	33.25	31.4
47	382	150.37	330.9

a) C_{12} = 1.347 b) C_{12} = 1.685 c) C_{12} = 2.064 d) C_{17} = 18.6 e) C_{17} = 10.8

The models incorporating only the simple hydraulic parameters showed a tendency to predict higher values of the reaeration coefficient than those observed. Also, models derived from the stream data generally over-predicted the values of the coefficient observed in flumes, and those derived from the flume data underpredicted the values of the reaeration coefficient observed in streams. Since none of the investigated models permit an accurate prediction over the all range of data tested it was assumed that the inadequacy of these models is attributable to the omission of some variables to the feature of the flow structure from the consideration. Covar (1976) postulated that models derived from the laboratory flume data should not be applied to natural streams with entirely different hydraulic conditions. Fig. 3.4a shows the data used by O'Connor and Dobbins (1956), Churchill et al. (1962) and Owens et al. (1964) and the regions selected to divide the applicable models, Fig. 3.4b gives the effect of such a selection method.

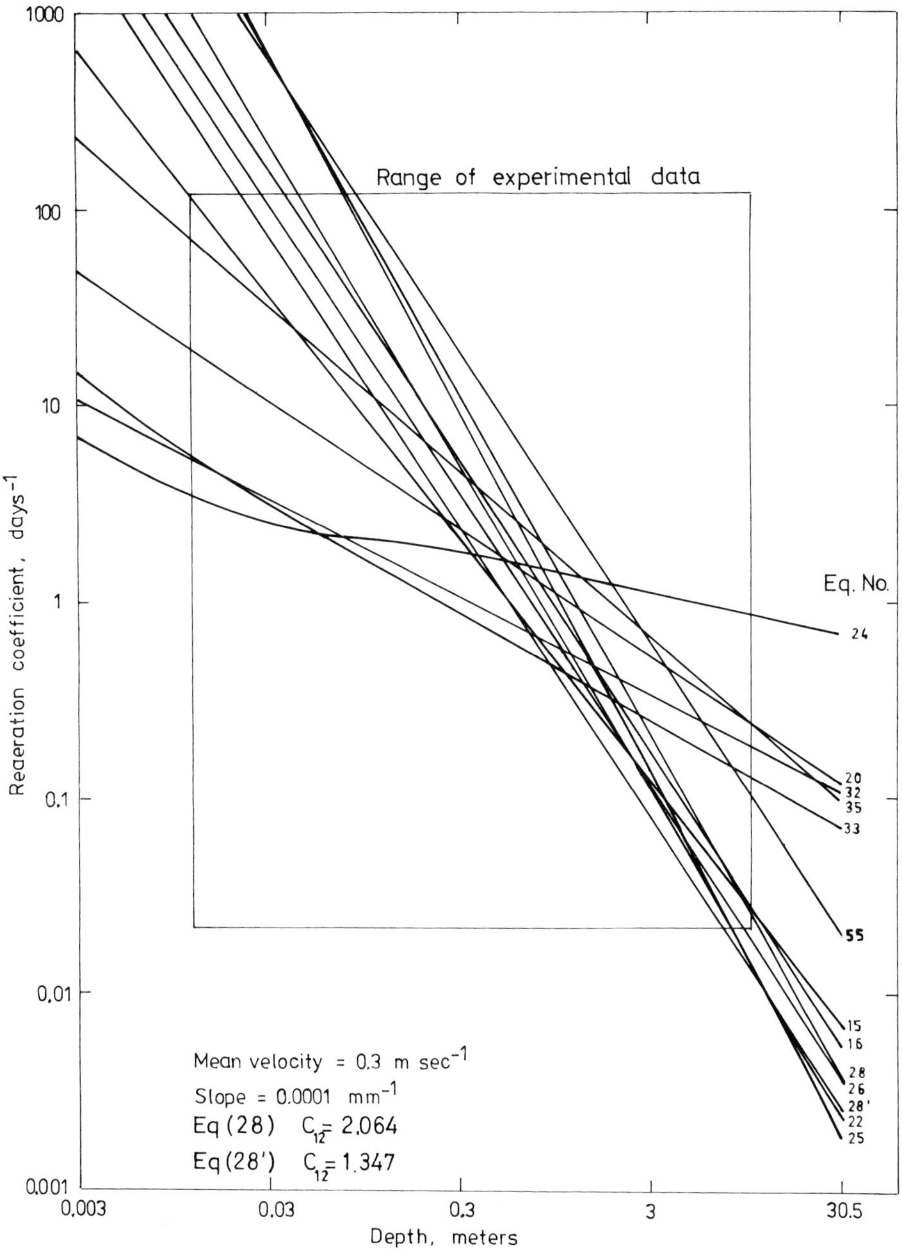

Fig. 3.3: Reaeration coefficient as a function of depth, from prediction equations, after Bennett and Rathbun (1972).

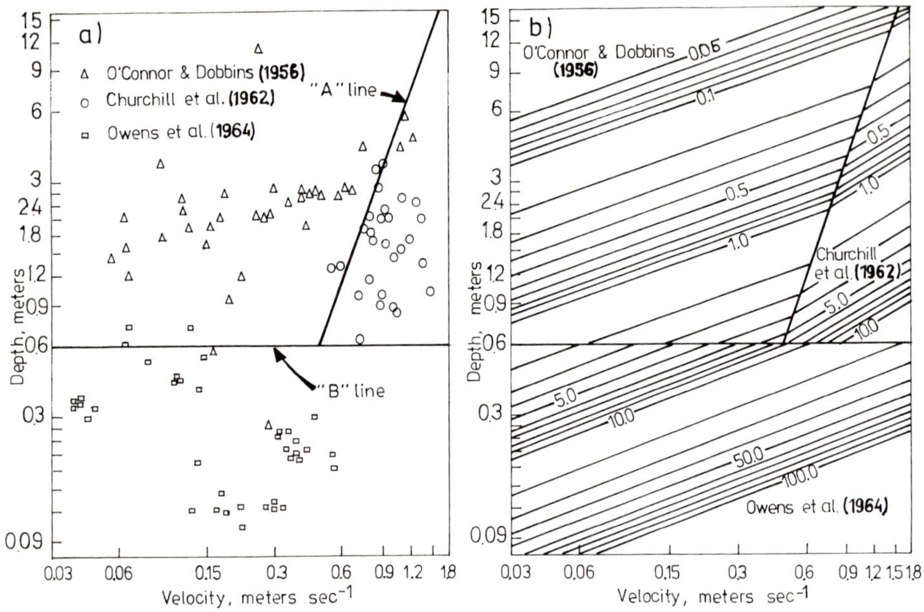

Fig. 3.4: Field data (a), and (b) K_a vs depth and velocity, after Covar (1976).

Effect of Predictive Models on the Dissolved Oxygen Balance

The results from the BOD-DO simulation in the San Antonio River Basin, using the QUAL-1 model, developed by Texas Water Development Board (2971), are shown in Fig. 3.5. Since the reaeration coefficient has been estimated by various predictive models the observed variation is quite high. Fig. 3.4 demonstrates clearly that the river water quality model can closely simulate observed conditions only if the appropriate method for estimation of the reaeration coefficient is used.

In many cases the main interest in the practical application of river water quality models is directed at the determination of the critical time, t_c for the critical deficit, D_c. At this critical point there is no change in the deficit since deoxygeneration and reaeration are balanced. A hypothetical example by Gromiec (1986), shown in Fig. 3.6, illustrates the possibility of a quite significant effect (in this case 236% difference), which an inappropriate selection of the reaeration coefficient predictive model may have on the computed critical time t_c. Large errors may occur when such values of computed critical times are applied to determine the degree of waste treatment required.

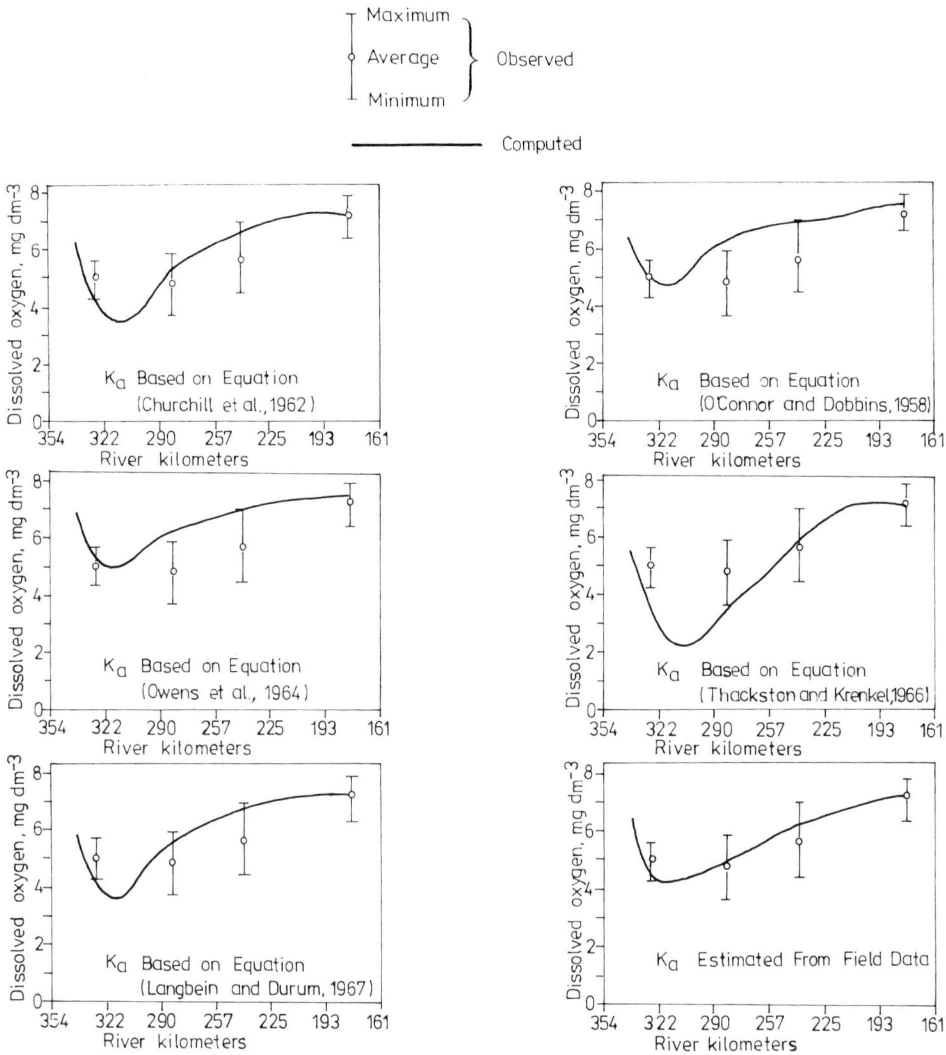

Fig. 3.5: DO simulation results obtained by application of QUAL-1 with various reaeration coefficients to the San Antonio-River Basin - after Frank D. Masch and Associates and the Texas Water Development Board (1971).

3.5 SUMMARY AND CONCLUSION

A theory of the atmospheric reaeration process has evolved from the film theory to energy dissipation or surface renewal approach. At present, estimation of the reaeration coefficient is still more an art than a science.

Many currently available models for the reaeration coefficient are modifications of the model proposed by Streeter and Phelps (1925) more than fifty years ago. As a result the reaeration coefficient is usually presented as a function of the simple hydraulic parameters, such as stream velocity and depth. The reaeration coefficient is also presented as being related to molecular diffusivity or to a turbulent dispersion coefficient. A number of other predictive relationships of the reaeration coefficient is also applied.

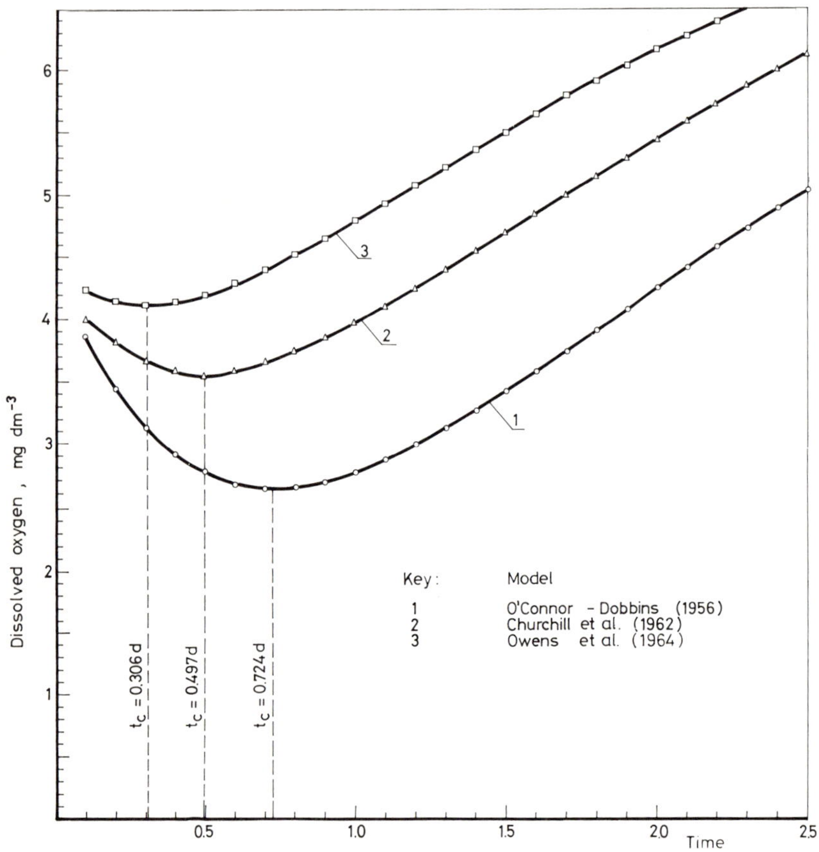

Fig. 3.6: Influence of the selected for k_a on the critical time t_c.

Since different approaches have been taken to the problem, the results provided by various models differ significantly. In addition, the arbitrary selection of a model for the reaeration coefficient can bias the results of an analysis. Therefore, the predictive models for the reaeration coefficient should be applied only when the conditions are the same as those for which the models were derived. The potential inaccuracies of prediction from such models should be recognized. The application of a given model outside the range of values of physical variables for which it has been developed can provide significant error. One of the main difficulties in developing an adequate model for the reaeration coefficient seems to be the lack of independently observed and accurate data for natural streams. The weakness in the prediction models for the reaeration coefficient is also related to the lack of well established and generally accepted means of their verification. Since atmospheric reaeration is one of the main sources of oxygen in streams the development of reliable methods for measuring and predicting the reaeration coefficient would contribute significantly to water quality modelling.

REFERENCES

Adeney, W.E. and H.G. Becker, 1919. The determination of the rate of solution of atmospheric nitrogen and oxygen in water. Philosophical Magazine 38, 317-338.

Alonso, C.V., J.R. McHenry and J.C.S. Hong, 1975. The influence of suspended sediment on the reaeration of uniform streams. Water Research 9, 8, 695-700.

Baca, R.G. and R.C. Arnett, 1976. A limnological model for eutrophic lakes and impoundments. Battelle Inc., Pasific Northwest Laboratories, Richard, Washington.

Banks, R.B., 1975. Some features of wind action on shallow lakes. Journal of the environmental engineering division ASCE, 101, EE5. 813-827.

Banks, R.B., G.B. Wickramanayake and B.N. Lohani, 1984. Effects of rain on surface reaeration. Journal of the environmental engineering division, ASCE, 110, EE1, 1-14.

Bansal, M.K., 1973. Atmospheric reaeration in natural streams. Water Research 7, 5, 769-782.

Bennett, J.P. and R.E. Rathbun 1972. Reaeration in open-channel flow. Geological survey professional paper no. 737. United States Government Printing Office, Washington, D.C.

Bowie, G.L. et al., 1985. Rates, constants and kinetics formulations in surface water qualityh modelling, second edition. U.S. Environmental Protection Agency, Athens, Georgia, EPA 600-3-85-040.

Butts, T.A. and R.L. Evans, 1983. Small stream channel dam aeration characteristics. Journal environmental engineering 109, 3, 555-573.

Broecker, H.C., J. Petermann and W. Siems, 1978. The influence of wind on CO_2 exchange in a wind-wave tunnel. Journal of Marine Research 36, 4, 595-610.

Cadwallader, T.E. and A.J. McDonnel, 1969. A multivariate analysis of reaeration data. Water Research 3, 731-742.

Chen, C.W., D.J. Smith and S. Lee, 1984. Documentation of water quality models for the helms pumped storage project. Tetra Tech, Incorporated.

Churchill, M.A., H.L. Elmore and R.A. Buckingham, 1962. The prediction of stream reaeration rates. Journal of the Sanitary Engineering Division, ASCE, 88, SA4, 1-46.

Commitee on Sanitary Engineering Research, 1961. Effects on water temperature on stream reaeration, 31st progress report. Journal of the Sanitary Engineering Division, ASCE, 87, SA6, 59-71.

Copeland, B.J. and W.R. Duffer, 1964. Use of a clear plastic dome to measure gaseous diffusion rates in natural waters. Limnology and Oceanography 9, 494.

Covar, A.P., 1976. Selecting the proper reaeration coefficient for use in water quality models. Proceedings of the conference on environmental modelling and simulation, EPA 600/9-76-016, U.S. Environmental Protection Agency. Cincinnati, Ohio.

Danckwerts, P.V., 1951. Significance of liquid-film coefficients in gas absorption. Industrial and Engineering Chemistry, 43, 6, 1460-1467.

Dobbins, W.E., 1964. BOD and oxygen relationship in streams. Journal of the Sanitary Engineering Division, ASCE 90, SA3, 53-78.

Dobbins, W.E., 1956. The nature of the oxygen transfer coefficient in aeration systems. Biological treatment of sewage and industrial wastes (eds.: J. McCabe, W.W. Eckenfelder, Jr.), vol. 1, Reinhold Publishing Co., Inc., New York, N.Y.

Downing, A.L. and G.A. Truesdale, 1955. Some factors affecting the rate of solution of oxygen in water. Journal of Applied Chemistry 5, 570-581.

Elmore, H.L. and W.F. West, 1961. Effect of water temperature on stream reaeration. Journal of the Sanitary Engineering Division, ASCE, 87, SA6, 59-71.

Eloubaidy, A.F. and E.J. Plate, 1972. Wind-shear turbulence and reaertion coefficient. Journal of the Hydraulics Division, ASCE, 98, HY1, 153-170.

Foree, E.G., 1976. Reaeration and velocity prediction for small streams. Journal of the Environmental Engineering Division, ASCE, 102, EE5, 937-952.

Fortescue, G.E. and J.R.A. Pearson, 1967. On gas absorption into a turbulent liquid. Chemical Engineering Science 22, 1163-1176.

Frexes, P., G.H. Jirka and W. Brutsaert, 1984. Examination of recent field data on stream reaeration. Journal of the Environmental Engineering Division, ASCE, 110, EE6, 1179-1183.

Gameson, A.L.H., G.A. Trusdale and A.L. Downing, 1955. Reaeration studies in a lakeland beck. Journal of the Institution of Water Engineers, 9, 7, 571-594.

Gromiec, M.J., 1986. Basic processes and parameters of Stream Water Quality Modelling. Proc. Estimation of Parameters in Mathematical Modelling of Water Quality (in Polish), Warsaw.

Gulliver, J.S. and H.G. Stefan, 1981. Air-Water Surface Exchange of Oxygen: Theory and Application to the USEPA Monticello Experimental Field Channels, St. Anthony Falls Hydraulic Laboratory External Momorandum No. 173, University of Minnesota, 1-108.

Harriott, P., 1962. A Random Eddy Modification of the Penetration Theory. Chemical Engineering Sci., 17, 149-154.

Higbie, R., 1935. The Rate of Absorption of Pure Gas into a Still Liquid During Short Periods of Exposure. Transactions, American Institute of Chemical Engineers 31, 365-390.

Hornberger, G.M. and M.G. Kelly, 1975f. Atmospheric Reaeration in a River Using Productivity Analysis. Journal of the Environmental Engineering Division, ASCE, 101, EE5, 729-739.

Isaacs, W.P. and A.F. Gaudy, 1968. Atmospheric Oxygenation in a Simulated Stream. Journal of the Sanitary Engineering Division, ASCE, 94, SA2, 319-344.
Isaacs, W.P. and J.S. Maag, 1969. Investigation of the Effect of Channel Geometry and Surface Velocity on the Reaeration Rate Coefficient. Engineering Bulletin, Purdue University, vol. 53.

Kanwisher, J., 1963. On Exchange of Gases Between the Atmosphere and the Sea. Deep-Sea Research 10, 195-207.
King, C.J., 1966. Turbulence Liquid Phase Mass Transfer at a Free Gas-Liquid Interface. Industrial and Engineering Chemistry Fundamentals, 5, 1, 1-8.
Kishinevsky, M.K., 1955. Two Approaches to the Theoretical Analysis of Absorption Processes. Journal Applied Chemistry, U.S.S.R. 28, 881-886.
Krenkel, P.A., 1960. Turbulent Diffusion and the Kinetics of Oxygen Absorption. Ph.D. dissertation. University of California, Department of Sanitary Engineering, Berkely.
Krenkel, P.A. and G.T. Orlob, 1962. Turbulent Diffusion and the Reaeration Coefficient. Journal of the Sanitary Engineering Division, ASCE, 88, SA2, 53-83.

Langbein, W.B. and W.H. Durum, 1967. The Reaeration Capacity of Streams. United States Geological Survey Circular No. 542, Reston, Va.
Lau, Y.L., 1972a. A Review of Conceptual Models and Prediction Equations for Reaeration in Open-Channel Flow. Technical Bulletin No. 61, Inland Water Branch, Department of the Environment, Ottawa, Canada.
Lau, Y.L., 1972b. Prediction Equation for Reaeration in Open-Channel Flow. Journal of the Sanitary Engineering Division, ASCE, 98, SA6, 1063-1068.
Lewis, W.K. and W.G. Whitman, 1924. Principles of Gas Absorption. Industrial and Engineering Chemistry 16, 12, 1215-1220.
Liss, P.S., 1973. Processes of Gas Exchange Across an Air-Water Interface. Deep Sea Research 20, 221-238.

Marchello, J.M. and H.L. Toor, 1963. A Mixing Model for Transfer Near a Boundary. Industrial and Engineering Chemistry Fundamentals 2, 8-12.
Mattingly, G.E., 1977. Experimental Study of Wind Effects on Reaeration. Journal of the Hydraulic Division, ASCE, 103, HY3, 311-323.

Negulescu, M. and V. Rojanski, 1969. Recent Research to Determine Reaeration Coefficient. Water Research 3, 3 189-202.

O'Connor, D.J. and W.E. Dobbins, 1956. Mechanism of Reaeration in Natural Streams. Journal of the Sanitary Engineering Division, ASCE, 82, SA6, 1-30.
O'Connor, D.J., 1958. Measurement and Calculation of Stream Reaeration Ratio. Seminar on Oxygen Relationships in Streams, U.S. Public Health Service, Robert A. Taft Sanitary Engineering Center, Report W 58-2, 35-45, Cincinnati.
Odum, H.T., 1956. Primary Production in Flowing Waters. Limnology and Oceanography 1, 102-117.
Owens, M., R.W. Edwards and J.W. Gibbs, 1964. Some Reaeration Studies in Streams. International Journal of Air and Water Pollution 8, 469-486.
Ozturk, Y.F., 1979. Mathematical Modelling of Dissolved Oxygen in Mixed Estuaries. Journal of the Environmental Engineering Division, ASCE, 105, EE5, 883-904.

Padden, T.J. and E.F. Gloyna, 1971. Simulation of Stream Processes in a Model River. Report No. EHE-70,23, CRWR-72, Center for Research in Water Resources, The University of Texas, Austin, Texas.
Parkhurst, J.D. and R.D. Pomeroy, 1972. Oxygen Absorption in Streams. Journal of the Sanitary Engineering Division, ASCE, 98, SA1, 101-124.

Rainwater, K.A. and E.R. Holley, 1984. Laboratory Studies on Hydrocarbon Tracer Gases. Journal of the Environmental Engineering Division, ASCE, 110, EE1, 27-41.

Shastry, J.S., L.T. Fan and L.E. Erickson, 1969. Discussion of "Reaeration Prediction in Natural Streams" by E.L. Thackston and P.A. Krenkel (1969), Journal of the Sanitary Engineering Division, ASCE, 95, SA6, 1186-1190.

Smith, D.J., 1978. WQRRS, Generalized Computer Program for River-Reservoir Systems. U.S. Army Corps of Engineers, Hydrologic Engineering Center (HEC), Davis, California. User's Manual 401-100, 100A, 210.

Streeter, H.W. and E.B. Phelps, 1925. A Study of the Pollution and Natural Purification of the Ohio River. Public Health Bulletin 146, U.S. Public Health Service, Washington D.C.

Streeter, H.W., C.T. Wrigh and R.W. Kehr, 1936. Measures of Natural Oxidation in Polluted Streams, Part III. An Experimental Study of Atmospheric Reaeration Under Stream-Flow Conditions, Sewage Works Journal 8, 2, 282-316.

Texas Water Development Board, 1971. Simulation of Water Quality in Streams and Canals. Report 128. Reproduced by NTIS, Springfield, Va.

Thackston, E.L. and P.A. Krenkel, 1969. Reaeration Prediction in Natural Streams. Journal of the Sanitary Engineering Division, ASCE, 95, SA1, 65-94.

Thomann, R.V. and J.J. Fitzpatrick, 1982. Calibration and Verification of a Mathematical Model of the Eutrophication of the Potomac Estuary. Prepared for Department of Environmental Services, Government of the District of Columbia.

Tsivoglou, E.C., 1967. Tracer Measurement of Stream Reaeration. Report to the Federal Water Pollution Control Administration, Department of Interior, Washington, D.C.

Tsivoglou, E.C., R.L. O'Connell, C.M. Walter, P.J. Godsil and G.S. Logsdon, 1965. Tracer Measurement of Atmospheric Reaeration, Part I, Laboratory Studies, Journal of Water Pollution Control Federation 37, 10, 1343-1362.

Tsivoglou, E.C. and J.R. Wallace, 1972. Characterization of Stream Reaeration Capacity. Report No. EPA-R3-72-012, U.S. Environmental Protection Agency, Washington, D.C.

Tsivoglou, E.C. and L.A. Neal, 1976. Tracer Measurement of Reaeration: III. Predicting the Reaeration Capacity of Inland Streams. Journal Water Pollution Control Federation 48, 12, 2669-2689.

Weiler, R.R., 1974. Carbon Dioxide Exchange Between Water and Atmosphere. Journal of Fisheries Research, Board Committee 31, 329-332.

Wen, C.G., J.F. Kao, L.K. Wang and C.C. Liaw, 1984. Effect of Salinity on Reaeration Coefficient of Receiving Waters. Water Science and Technology 16, 5-7, 139-154.

Wilcock, R.J., 1984. Methyl Chloride as a Gas-Tracer for Measuring Stream Reaeration Coefficients - I, Laboratory Studies. Water Research 18, 1, 53-57.

Wilson, G.T. and N. Macleod, 1974. A Critical Appraisal of Empirical Equations and Models for the Prediction of the Coefficient of Reaeration of Deoxygenated Water. Water Research 8, 6, 341-366.

Yu, S.L., T.L. Tuffy and D.S. Lee, 1977. Atmospheric Reaeration in a Lake. Office of Water Resources and Technology, U.S. Department of Interior.

Zison, S.W., W.B. Mills, D. Deimer and C.W. Chen, 1978. Rates, Constants and Kinetics Formulations in Surface Water Quality Modelling. Report Prepared for U.S. EPA, Tetre Tech, Inc., Lafayette, California.

Zogorski, J.S. and S.D. Faust, 1973. Atmospheric Reaeration Capacity of Streams, Part I, Critical Review of Methods Available to Measure and to Calculate the Atmospheric Reaeration Rate Constant. Environmental Letters 4, 1, 35-59.

CHAPTER 4

ADSORPTION AND ION EXCHANGE

by
Sven Erik Jørgensen

4.1 INTRODUCTION.

Adsorption is the transfer of components from the liquid phase onto the surface of a solid phase. The process may often be explained by an electrical attraction to the solid surface of components with a (minor) electrical charge. The adsorption results in the formation of a molecular layer of the adsorbate on the surface. Often an equilibrium concentration is rapidly formed on the surface and is sometimes followed by a slow diffusion into the particles of the adsorbent.

Ion exchange is an exchange of ions between a liquid and a solid phase. The exchange process may be explained in the same way as any other chemical process: the chemical energy at equilibrium after the process has been realized is lower than before the process was initiated. If a pure ion exchange process takes place the number of ions released is equivalent to the number of ions taken up by the process.

Pure adsorption or ion exchange is hardly observed in nature. A mixture of the two processes is most often observed.

Adsorption and ion exchange are significant processes in the environmental context and a description of the processes is often included in water quality modelling. Where water is in contact with suspended matter (organic matter or clay particles), sediment and biota, a significant transfer of matter by adsorption and ion exchange may take place. Chapter 17 deals with the exchange processes between water and sediment, where adsorption and, to a certain extent, ion exchange play an important role in this context.

A significant portion of impurities of water is found in the suspended matter, where the concentration of many impurities are magnitudes higher than in the water. Transport of pollutants in rivers and streams often takes place on suspended matter, either clay particles or organic matter. This implies that many pollutants, otherwise adsorbed or fixed by ion exchange on the sediment and therefore not be transported by clear water, are transported far by water with high turbidity.

Many organic compounds, including, many pesticides, are adsorbed on suspended matter or sediment. Heavy metals ions are adsorbed and/ or taken up by ion exchange by clay particles, which are often found to be the major component of suspended matter in river and streams .

Clay is however often a significant part of river and lake sediment and its iron content may often play an important role in the ability of sediment to bind phosphate. The difference between adsorption and ion exchange ability under aerobic and anaerobic conditions may often be explained by the tranformation of iron from oxidation stage 3 to 2. This is further discussed in chapter 17.

Overall there is little doubt that adsorption and ion exchange processes are important in the mass transfer of material in aquatic ecosystems and their formulation must therefore often be included in many water quality models.

4.2 MODELLING ADSORPTION AND ION EXCHANGE

Adsorption and ion exchange are fast processes. In water quality modelling the selected time interval is often either weeks, days or hours. It implies that these processes can be described by means of equilibrium equations. The conditions for the equilibrium may change very rapidly, but a new equilibirum is attained very rapidly, too - in minutes or hours at the most.

In some cases a water quality model may use a short time interval whereas the adsorption or ion exchange process is relatively slow. So there is a need for a description of the process rate. This chapter will therefore present not only **the equilibrium expressions** but also **the rate expressions**.

The most simple equilibrium expression uses a **linear adsorption isotherm**:

$$a = k_1 * C + k_2 \qquad (4.1)$$

where a denotes the concentration on the adsorbent, while C is the concentration in the liquid phase, and k_1 and k_2 are constants. At some adsorptions k_2 may be very small and equal to, or almost equal to, zero. In such cases k_1 expresses the number of times the concentration in the adsorbent is greater than the concentration in the liquid (water) phase and is then called the partition coefficient.

The **Freundlich adsorption isotherm** is expressed in the following equation:

$$a = k * C^n \qquad (4.2)$$

where

k and n = constants

a = the amount of solute adsorbed per unit weight

C = the equilibrium concentration of the solute in the liquid phase.

The values of k and n are given for several organic compounds in Table 4.1 (Rizzo et al., 1977).

Table 4.1
Freundlich's constant for adsorption of some organic compounds on activated carbon

Compound	k (mg/g)	n
Aniline	25	0.322
Benzene sulphonic acid	7	0.169
Benzoic acid	7	0.237
Butanol	4.4	0.445
Butyraldehyde	3.3	0.570
Butyric acid	3.1	0.533
Chlorobenzene	40	0.406
Ethylacetate	0.6	0.833
Methyl ethyl ketone	24	0.183
Nitrobenzene	82	0.237
Phenol	24	0.271
TNT	270	0.111
Toluene	30	0.729
Vinyl chloride	0.37	1.088

If n is equal to one in Freundlich's adsorption isotherm, k becomes the partition coefficient, as before.

Langmuir's adsorption isotherm is based on the following expression:

$$a = \frac{A_0 * C}{1 + b * C} \tag{4.3}$$

where

a and C = as defined above

b and A_o = constants

As can be seen, $a = A_o/b$ when $C - \infty$.

The Langmuir constant for several organic compounds which can be adsorbed on activated carbon has been found by Weber et al., 1964. Most types of waste water contain several substances which will be adsorbed in which case a direct application of Langmuir's adsorption isotherm is not possible.

Weber et al. (1965) have developed an equation, (4.4) and (4.5), for *competitive adsorption of two substances* (A and B). In other words competitive adsorption can be described in the same way as a competitive enzymatic reaction:

$$a_A = \frac{A_{Ao} \quad C_A}{1 + b_A * C_A + b_B * C_B} \tag{4.4}$$

$$a_B = \frac{A_{Bo} \quad C_B}{1 + b_A * C_A + b_B * C_B} \tag{4.5}$$

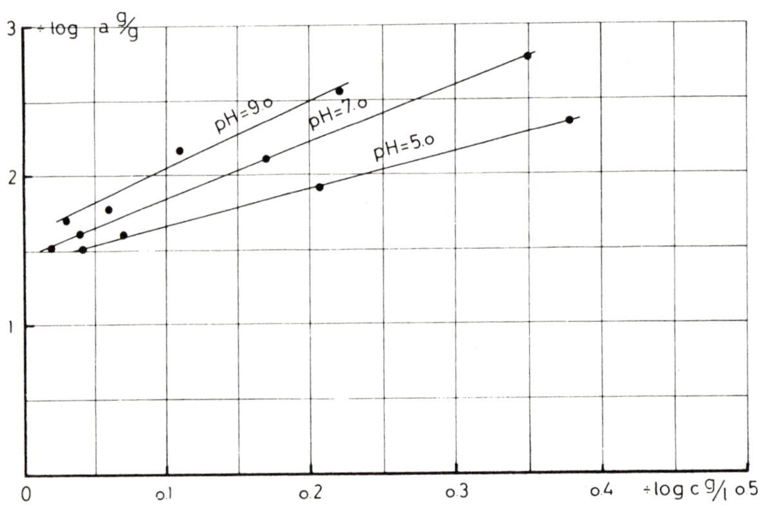

Fig. 4.1: Adsorption of lysine on a cellulose cation exchanger at three different pH-values (Jørgensen, 1976a)

When it is necessary to find whether Freundlich's adsorption isotherm or Langmuir's adsorption isotherm gives the best fit to a set of data, the two plots shown in Figs. 4.1 and 4.2 should be used. When a linear relation between log a and log C is obtained, Freundlich's adsorption isotherm is a good formulation of the set of data as:

$$\log a = \log k + n \log C \tag{4.6}$$

On the other hand, when the reciprocal values of a and C give a linear equation (see Fig. 4.2) Langmuir's adsorption isotherm gives a good formulation of the set of data, since

$$\frac{1}{a} = \frac{1 + b * C}{A_0 * C} \quad \text{or} \quad \frac{1}{a} = \frac{1}{A_0} * \frac{1}{C} + \frac{b}{A_0} \tag{4.7}$$

Ion exchange is known to occur with a number of natural solids, such as soil, humus, metallic minerals and clay. Clay, and in some instances other natural materials, can even be used for the demineralization of drinking water.

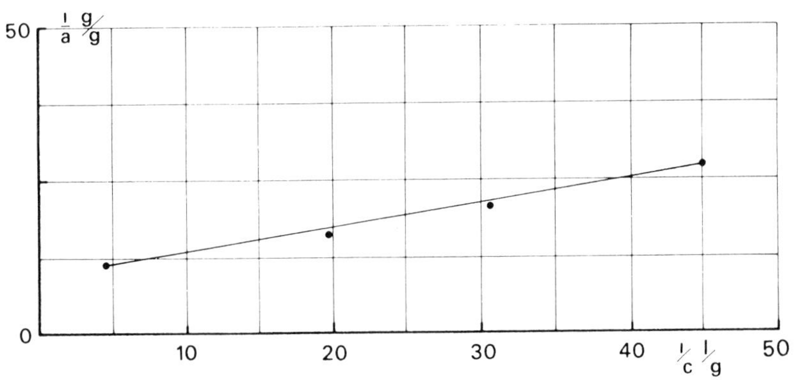

Fig. 4.2: Adsorption of uric acid on activated carbon (pH = 9.0) (Jørgensen, 1976b)

The cation exchangers contain functional groups such as sulphonic $R\text{-}SO_3H$ - carboxylic, R-COOH - phenolic, R-OH and phosphonic, $R\text{-}PO_3H_2$ (where R represents the matrix). It is possible to distinguish between strongly acidic cation exchangers derived from a strong acid such as H_2SO_4 and weakly acidic ones derived from a weak acid such as H_2CO_3. It is also possible to determine a pK-value for the cation exchangers in the same way as it is for acids generally.

This means:

$$R - SO_3H \;=\; SO_3^- + H^+ \tag{4.8}$$

$$\frac{[H^+] \; * \; [R - SO_3^-]}{[R - SO_3H]} = K \qquad pK = -\log K \tag{4.9}$$

Anion exchangers contain such functional groups as primary amine, $R\text{-}NH_2$, secondary amine, $R\text{-}R_1NH$, and tertiary amine $R\text{-}R_1R_2N$ groups and the quaternary ammonium group $R\text{-}R_1R_2R_3N^+OH^-$.

It can be seen that the anion exchanger can be divided into weakly basic and strongly basic ion exchangers derived from quaternary ammonium compounds.

The exchange reaction between ions in solution and ions attached to the matrix is generally reversible. The exchange can be treated as a simple stoichiometric reaction. For cation exchange the equation is:

$$A^{n+} + n(R^-)B^+ \;=\; nB^+ + (R^-)_nA^{n+} \tag{4.10}$$

The ion exchange reaction is selective, so that the ions attached to the matrix will have preference for one counter ion over another. Therefore the concentrations of different counter ions in the ion exchanger will be different from the corresponding concentration ratio in the solution.

According to the law of mass action, the equilibrium relationship for reaction (4.10) will give:

$$K_{AB} = \frac{a_B{}^n \; * \; a_{RA}}{a_A \; * \; a_{RB}{}^n} \tag{4.11}$$

where a_B and a_A are the activity of the ions B^+ and A^{n+} in the solution and correspondingly a_{RB} and a_{RA} are the activities of the ion exchanger in B- and A-form, respectively.

The Fig. 4.3 plot is often used as an illustration of the preference of an ion exchanger for a particular ion. As seen, the percentage in the resin is plotted against the percentage in solution.

The selectivity coefficient, K_{AB}, is not actually constant, but is dependent upon experimental conditions. A selectivity coefficient of 50% in solution is often used, $^{a}{}_{50\%}$.

If we use concentration and not activity, it will involve, for n = 1:

$$C_B = C_A \tag{4.12}$$

$$a_{50\%} = K_{AB,50\%} = \frac{C_{RA}}{C_{RB}} \tag{4.13}$$

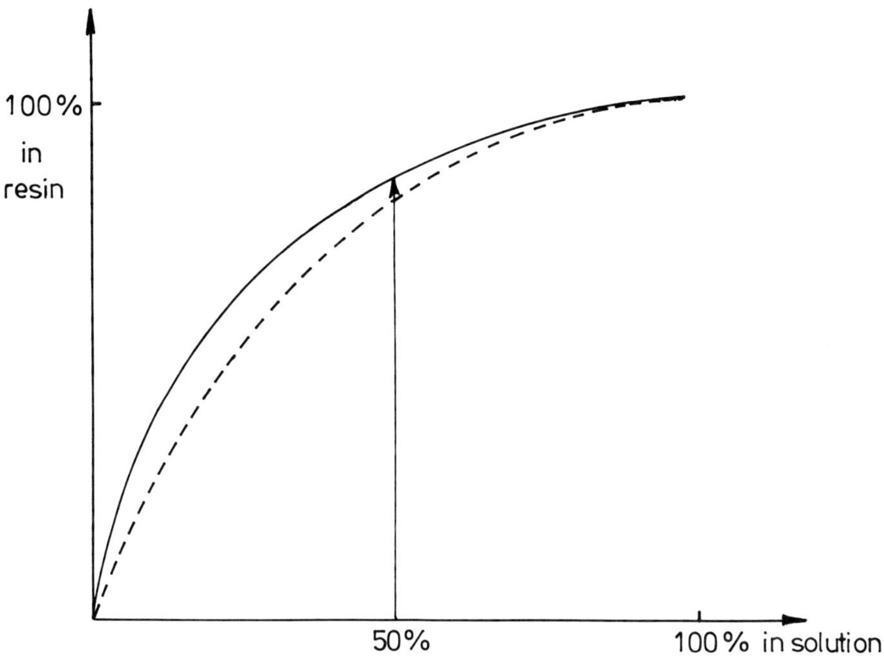

Fig. 4.3: Equilibrium plot % in solution versus % in ion exchanger. The dotted line indicates the case when the activity is = 1.0.

The plot Fig. 4.3 can be used to read $a_{50\%}$.

The selectivity of the ion exchange material for the exchange of ions is dependent upon the ionic charge and the ionic size. An ion exchanger generally prefers counter ions of high valence. Thus, for a series of typical anions of interest one would expect the following order of selectivity:

$PO_4^{3-} > SO_4^{2-} > Cl^-$.

Similarly for a series of cations:

$Al^{3+} > Ca^{2+} > Na^+$.

Complexation might cause difficulties by removal of iron, manganese, copper or zinc by an ion exchange process. Natural waters often contain chelating agents and chelation drastically changes the behaviour of a metal ion in water. Manahan et al. (1973) have found a copper equivalent chelating capacity of 0-3.5 mg/l in several water samples.

The adsorption rate can be limited by external diffusion, internal diffusion or by a combination of external and internal diffusion. The external diffusion controls the transfer of solute from the water to the boundary layer of fluid immediately adjacent to the external surface of the adsorbent. The external diffusion is governed by molecular diffusion and, in turbulent flow, by eddy diffusion.

The process can be described by the following equation:

$$V_a = k_e * a (Y - Y^+) \qquad (4.14)$$

where V_a = the rate of adsorption; Y = the concentration of the adsorbate in the fluid and Y^+ = the concentration of the adsorbate in the fluid in equilibrium with the existing concentration in the adsorbent. k_e = the external mass transfer coefficient.

Internal diffusion processes control the transfer of solid from the exterior of the adsorbent to the internal surface (Weber and Rumer, 1965). This condition is represented by the following equation:

$$V_a = k_i * a * \Sigma * (X^x - X) \qquad (4.15)$$

where Σ = the interparticle void ratio; X^x = the concentration of adsorbate in the solid phase that is assumed to be in equilibrium with the coexisting liquid phase at concentration, Y; X = the actual concentration of adsorbate in the solid phase.

If the internal and the external diffusion occur in comparable rates the respective mass transfer coefficient, measured individually, may be added (King, 1965):

$$\frac{1}{K} = \frac{1}{k_e} + \frac{1}{k_i} \qquad (4.16)$$

The kinetic processes involved in the ion exchange reaction are similar to those involved in the adsorption of substances from a solution by an adsorbent. When the rate is controlled by *a film diffusion mechanism,* **the**

rate of exchange will be:

$$-\frac{dC}{dt} = k_f * \frac{a}{V} (C - C^+) \tag{4.17}$$

where

C = the actual concentration in solution

k_f = the film transfer coefficient

V = the volume of solution

a = the effective area

C^+ = the equilibrium concentration

If, on the other hand, *the pore diffusion is rate limiting,* a diffusion model based on Fick's second law can be utilized to describe the reaction. The general expression for this type of reaction can be set up as follows:

$$\frac{\partial c}{\partial t} = \frac{1}{r^2} * \frac{\partial}{\partial r} (r^2 * D * \frac{\partial c}{\partial r}) - \frac{\partial q^+}{\partial t} \tag{4.18}$$

where r = the ion exchanger particle radius

D = the diffusion coefficient

q^+ = the concentration of exchanged ions of the ion exchanger in equilibrium with the concentration C

4.3 THE APPLICATION OF ADSORPTION AND ION EXCHANGE SUB-MODELS IN WATER QUALITY MODELLING

Adsorption and ion exchange are essential processes in the sediment-water exchange models and will therefore be treated further in this context in chapter 17.

The processes may furthermore be of importance in water quality modelling, when the concentration of suspended matter in the water is sufficently high to imply that significant amounts of the modelled components are adsorbed or present on the exchange complex.

Models of pollution by surfactants, pesticides and heavy metals often include adsorption and ion exchange processes, as these materials represent easily sorbed chemicals.

In the model of copper pollution of the Upper Nile Lake System the input for the lake system of copper was mainly as copper adsorbed on clay particles. The equilibrium between adsorbed and dissolved copper should therefore be included in the copper model for the lakes, see Jørgensen et al. (1982). The model is rather simple, but the adsorption process is anyhow of sufficient importance to be included, as shown in the conceptual diagram Fig. 4.4.

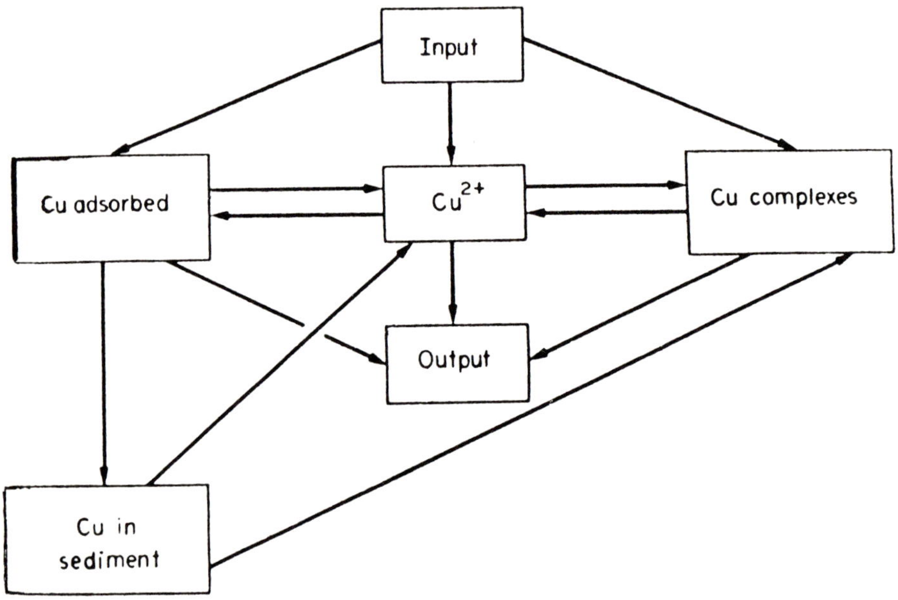

Fig. 4.4: Simple copper model, conceptual diagram.

For the same reasons these processes were included in the model of toxic metal speciation in acid aquatic systems developed by Fontaine (1984) and the heavy metal transport model in an arctic fjord developed by Nyholm et al (1984).

Examples of models of toxic organic components are to be found in the Mirex behaviour model in Lake Ontario, developed by Halfon (1984), the model of hydrophobic organic pollutants by Schwarzenbach and Imboden (1984), and the model of toxins in the Tamar estuary, England, developed by Harris et al. (1984).

The effects of toxic substances on aquatic biota are often determined by the concentration of the dissolved fractions. It is therefore of some significance to formulate the equilibrium between the sorbed and dissolved components, i.e. by use of adsorption isotherms. This was clearly demonstrated in the model developed by Schwarzenbach and Imboden (1984).

The removal of components from the water phase to the sediment is often essentially sedimentation of adsorbed (or ion exchanged) material. The transport from the water phase to the sediment can only be described properly by use of adsorption or ion exchange isotherms. This was shown to be essential for the eutrophication model of Lake Balaton. The sediment is calcium rich and has a high adsorption capacity for phosphate. The lake is shallow and strong wind is able to stir up the upper layer of the sediment. It implies that an essential part of the dissolved phosphate is adsorbed on the suspended matter and, when the wind slows down and the suspended matter settles, the phopsphate is tranported from the water phase to the sediment. Modelling exercises for Lake Balaton have shown that it is important for the development of a good eutrophication model to include these adsorption processes and the consequences for transport of phosphate from water to sediment.

4.4 PARAMETER ESTIMATION IN THE ADSORPTION - ION EXCHANGE SUBMODEL

The concentrations of pollutants in the environment are most often very low and it is therefore possible to apply a partition coefficient with acceptable approximations. Table 4.2. gives some adsorption values of organic compounds under various circumstances. Table 4.3. gives the equilibrium constants and thereby also the selectivity for several pairs of ion exchangers.

Table 4.2

Adsorption determinations Ref: Jørgensen et al. (1979)

Component	Adsorbent	Value	Condition
Alkylbenzene sulphonate	Algae (55 mg/l)	0.2-0.8 mg/l	20 mg/l in solution
Alkyl polyethoxylate	Algae (55 mg/l)	1-7 mg/l	20 mg/l in solution
m-aminophenol	Bentonite	023 mg/g ads.	pH = 7.7, 295°K
m-chlorophenol	Bentonite	0.81 mg/g ads.	pH = 7.7, 295°K
nonyl-phenol polyethoxylate	Algae (55 mg/l)	1-13 mg/l	20 mg/l in solution
2,4,5 - trichlorphenol	Bentonite	0.64 mg/g ads.	pH = 7.7, 295°K
O-phenyl-phenol	Bentonite	0.47 mg/g ads.	pH = 7.7, 295°K
2,4 - dichlorphenol	Bentonite	0.96 mg/g ads.	pH = 7.7, 295°K
2,6 - dimethylphenol	Bentonite	0.08 mg/g ads.	pH = 7.7, 295°K
3,4 - dimethylphenol	Bentonite	0.13 mg/g ads.	pH = 7.7, 295°K

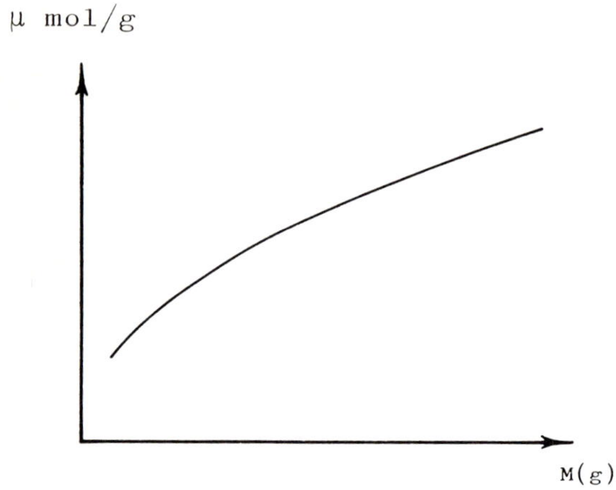

Fig. 4.5: Effect of molecular weight on capacity for adsorption for several sulphonated alkylbenzenes.

However, adsorption isotherms and equilibrium constants for ion exchange are only known for a very limited number of processes of interest for environmental modelling. Therefore estimation methods are needed.

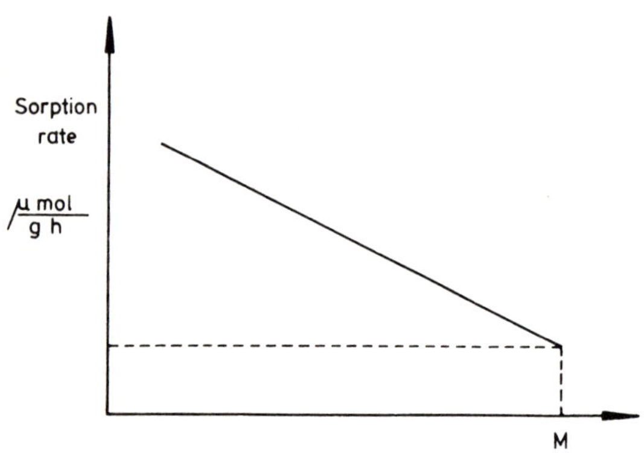

Fig. 4.6: Sorption rate plotted against molecular weight.

To a certain extent, it is possible to predict the adsorption ability of a particular component. The solubility of the dissolved substance is by far the most significant factor in determining the intensity of the driving forces. The greater the affinity of a substance for the solvent, the less likely it is to move towards an interphase to be adsorbed. For an aqueous solution this means that the more hydrophylic the substance is the less likely it is to be adsorbed. Conversely hydrophobic substances will be easily adsorbed from aqueous solutions. Also compounds with an amphoteric nature have maximum adsorption at the isoelectric point.

pH strongly influences the adsorption as hydrogen and hydroxide ions are adsorbed and the charge of the other ions are influenced by the pH of the water. For typical organic pollutants, the adsorption increases with increasing pH.

Normally the adsorption reactions are exothermic, which means that the adsorption will increase with decreasing temperature, although small variations in temperature do not tend to alter the adsorption process to a significant extent.

The extent of adsorption is proportional to the surface area. To be able to compare different adsorbents, a specific surface area, defined as that portion of the total surface area, that is available for adsorption per unit of adsorbent, is used. This means that the adsorption capacity of a nonporous adsorbent should vary inversely with the particle diameter, while for highly porous adsorbents the capacity should be almost independent of the particle diameter. However, for some porous material such as activated carbon, the breaking up of large particles to form smaller ones opens some tiny sealed channels in the column, which then become available for adsorption (Weber and Morris, 1964).

The nature of the adsorbate also influences the adsorption. In general an inverse relationship can be anticipated between the extent of adsorption of a solute and its solubility in the solvent (water) from which adsorption occurs. This is the so-called **Lundilius' rule**, which may be used for the semi-quantitative prediction of the effect of the chemical character of a solute on its uptake from solution (water) (Lundilius, 1920).

Ordinarily the solubility of any organic compound in water decreases with increasing chain length because the compound becomes more hydrophobic as the number of carbon atoms becomes greater. This is **Traube's rule**.

Adsorption from an aqueous solution increases as homologous series are ascended, largely because the expulsion of increasingly large hydrophobic molecules from water permits the formation of an increasing number of water-water bonds. Fig. 4.5 shows the effect of molecular weight on the capacity for adsorption for several sulphonated alkylbenzenes. As can be seen, the figure illustrates very well the Traube's rule mentioned above. The molecular weight also affects to the rate of uptake of solutes by activated carbon, if the rate is controlled by intraparticle transport. Data are plotted in Fig. 4.6 for the rates of adsorption of a series of sulphonated alkylbenzenes of different molecular size. It can be seen that the molar rate of uptake decreases with increasing molecular weight.

It is also in accordance with Lundilius' rule that increasing ionization means decreasing adsorption (when water is the solvent). For instance, the adsorption of several substituted benzoic acids varies inversely with the dissociation constant (Langloid, 1955).

Many organic compounds, e.g. sulphonated alicyclic benzenes, have a molecular structure consisting of both hydrophilic and hydrophobic groups. The hydrophobic parts will be adsorbed at the surface and the hydrophilic parts will tend to stay in the water phase.

Smith et al. (1977) have shown, in a limited number of case studies, that for a given sorbent the logarithm of the partition coefficient and the logarithm of the solubility linear are related (see Fig. 4.7). Although this relationship seems to be generally valid, compounds that interact via an ion exchange process probably would not fit this plot.

Recently, much effort has been directed towards elucidating the sorption of hydrophobic organic pollutants by sediments, soils, and aquifer materials. For neutral compounds (i.e. compounds exhibiting no ionizable functional groups) which include polycyclic aromatic hydrocarbons (Karichkoff et al. (1979), Means et al. (1980)), halogenated hydrocarbons (Chiou et al. (1979), Schwarzenbach and Westall (1981)), and a series of pesticides (Briggs (1981)), approximately linear sorption isotherms have been found for the range of concentrations typically encountered in natural waters:

$$C_s^z = K_p^z(s)\, C_l^z \tag{4.19}$$

The partition coefficient $K_p^z(s)$ has been shown to be primarily dependent on the lipophilicity of the compound, as expressed, e.g. by its octanol/water partition coefficient K_{ow}^z, and on the organic carbon content $f_{oc}(s)$ of the sorbent. Mathematical relationships between $K_p^z(s)$, $f_{oc}(s)$ and K_{ow}^z have been derived for various sets of compounds and natural sorbents:

$$K_p^z(s) = f_{oc}(s)\, K_{oc}^z = f_{oc}(s) * b * (K_{ow}^z)^a \tag{4.20}$$

Values reported for a and b include a = 1.00, b = $0.48*10^{-6}$ for polycyclic aromatic hydrocarbons (Karickhoff et al.,1979; Means et al.,1980), a = 0.52, b = $4.4*10^{-6}$ for a variety of pesticides (Briggs, 1981), and a =

0.72, b = $3.2*10^{-6}$ for alkylated and chlorinated benzenes (Schwarzenbach and Westall, 1981). From the presently available data it can be concluded that the values of a and b are primarily determined by the type of compounds (i.e., compound class(es)) on which the relationship is established, and only to a much smaller extent by the type of natural sorbents used. Thus, the reported relationships are very useful for predicting equilibrium partition coefficients of a great number of neutral hydrophobic organic compounds between water and natural sorbents of very different origins. Predictions of K_p^z (s) values are possible within a factor of 2. It should be noted that equation (4.20) is applicable only to sorbents exhibiting organic carbon contents greater than about 0.1% (f_{oc} (s) ≥ 0.001). For organic-poor sorbents, interactions of the chemical and inorganic matrix of the sorbent may become important (e.g., Schwarzenbach and Westall, 1981).

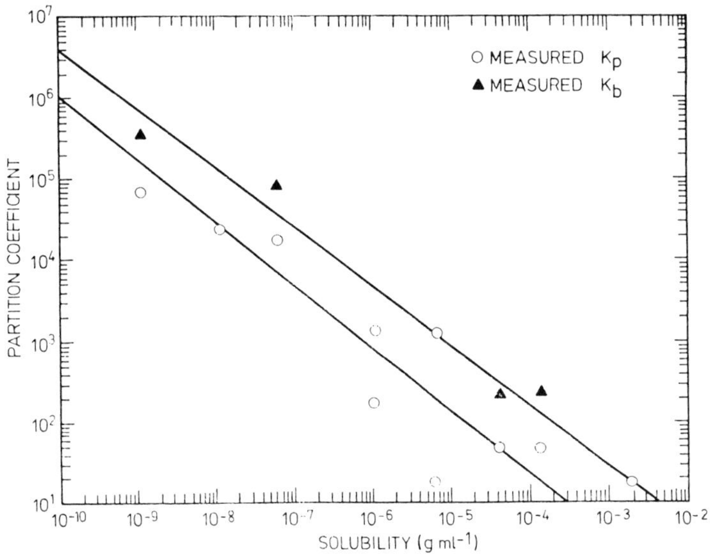

Fig. 4.7: Solubility versus partition coefficient on Coyte Creek sediments (K_p) and on a mixed population of bacteria (K_b).

4.5 CONCLUSIONS AND FURTHER RESEARCH NEEDS

Adsorption and ion exchange submodels are essential in the modelling of the exchange processes between water and sediment, as it will be demonstrated in chapter 17. But these submodels are also needed in many other aquatic models, particuarly for toxic substances where the amounts of toxic substances on the suspended matter are of great importance for the transport processes as well as for toxicity effects.

Consequently, good submodels for these processes are essential for the development of water quality models in general. Present submodels for these processes, including approximation by use of a partition coefficient, are reasonable good, but our knowledge of the parameters of the submodels is very limited and further research in this direction is absolutely needed. Determinations of more adsorption isotherms and ion exchange equilibrium constants in nature are also needed.

Estimation methods for these parameters are available but much more experience in the application of these methods is essential before it is possible to give a clear picture of the more general applicability of the methods. Probably and hopefully new and better estimation methods taking more factors into considerations will be developed during the coming years.

REFERENCES

Briggs, G.G., 1981. Theoretical and experimental relationships between soil adsorption, octanol/water partition coefficients, water solubilities, bioconcentration factors, and the parachor. J. Agric. Food Chem., 29: 1050.

Chiou, C.T., L.J. Peters and V.H. Fried, 1979. A physical concept of soil-water equilibria for nonionic compounds, Science, 206: 831.

Fontaine, T.D., 1983/84. Application of risk and uncertainty analysis techniques to a heavy metal speciation model. Ecological Modelling, 22: 101-108. Elsevier Science Publishers, B.V., Amsterdam.

Halfon, E., 1984. Error analysis and simulation of Mirex behaviour in Lake Ontario. Ecol. Modelling. 22: 213-253.

Harris, J.R.W., A.J. Bale, B.L. Bayne, R.C.F. Mantoura, A.W. Morris, L.A. Nelson, P.J. Radford, R.J. Uncles, S.A. Weston and J. Widdows. A preliminary model of the dispersal and biological effect of toxins in the Tamar estuary, England. Ecol. Modelling. 22: 253-285.

Jørgensen, S.E., 1976a. Reinigung häuslicher Abwässer durch Kombination eines chemischen Fällungs- und Ionenaustausch Verfahren. Thesis Karlsruhe University.

Jørgensen, S.E., 1976b. Recovery of phenols from industrial waste water. Prog. Wat. Tech., 8: Nos. 2/3, 65-79.

Jørgensen, S.E., (editor-in-chief; editorial board: M.B. Friis, J. Hendriksen, L.A. Jørgensen, S.E. Jørgensen and H.F. Mejer), 1979. Handbook of environmental data and ecological parameters. International Society of Ecological Modelling, Copenhagen.

Jørgensen, S.E., L. Kamp Nielsen, L.A. Jørgensen and H.F. Mejer, 1982. An environmental management model of the upper Nile Lake system. ISEM Journal, 4: 5-72.

Karickhoff, S.W., D.S. Brown and T.A. Scott, 1979. Sorption of hydrophobic pollutants on natural sediments. Water Res., 13: 241.
King, 1965. Mass transfer during short surface exposures in counter-current flow. Ind. Eng. Chem. Fundam, 4c: 125.

Langloid, G., 1955. Adsorption of fatty acids, which are partially dissociated in solution. Mem. Serb. Chim. Etat., Paris, 40: 83.
Lundelius, E.F., 1920. Adsorption and solubility. Kolloid Z., 26: 145.

Manahan, S.E. and M.J. Smith, 1973. The importance of chelating agents in natural waters and waste waters. Water and Sewage, 102-106.
Means, J.W., S.G. Wood, J.J. Hassett and W.L. Banwart, 1980. Sorption of polynuclear aromatic hydrocarbons by sediments and soils. Environ. Sci. Technol., 14: 1524.

Schwarzenbach, R.P. and J. Westall, 1981. Transport of nonpolar organic compounds from surface water to groundwater. Laboratory sorption studies. Environ. Sci. Technol., 15: 1360.
Schwarzenbach, R.P. and D.M. Imboden, 1984. Modelling concepts for hydrophobic pollutants in lakes. Ecol. Modelling. 22: 171-213.
Smith, J.H. et al., 1977. Environmental pathways of selected chemicals in freshwater systems. Part I, EPA 600/7-77-113.

Weber, W.J., Jr. and J.C. Morris, 1964. Adsorption of biochemically resistant materials from solution. Env. Health Series, AWTR-9.
Weber, W.J., Jr. and J.C. Morris, 1965. Intraparticle transport of sulfonated alkylbenzenes in a porous solid. Diffusion with nonlinear adsorption. Wat. Resources Research, 1: 365.

CHAPTER 5

HEAT EXCHANGE

by
G.T. Orlob and N. Marjanovic

5.1 INTRODUCTION

The rate of change in heat content of a water body is determined primarily by the rates of heat transport into and out of the water body by the physical processes of advection and effective diffusion and the heat exchange between the atmosphere and the water through the air-water interface. Additional exchanges of heat results from conduction between solid boundaries and the water and from biological and chemical processes, however these are usually so small relatively to the primary mechanisms that they are neglected in most models.

A general statement for conservation of thermal energy in a control volume V_j within a larger body of water, as illustrated in Figure 5.1, is

$$\frac{\partial H_j}{\partial t} = \underbrace{(h_i - h_o)_j}_{\text{local advection}} + \underbrace{h_{szj}}_{\text{heat exchange}} + \underbrace{(h_{aj} - h_{a,j+1})}_{\text{advection}} + \underbrace{(h_{dj} - h_{d,j+1})}_{\text{effective diffusion}} \tag{5.1}$$

where

H_j = cpV_jO_j is the content of the j-th element (J)

c = specific heat [J M^{-1} °C^{-1}]

p = density [M L^{-3}]

O_j = temperature (°C)

h_i = cp Q_i O_i, the heat advected by inflow [J T^{-1}]

h_o = cp Q_o O_i, the heat withdrawn by outflow [J T^{-1}]

h_{szj} = $(q_{sz}a_z - q_{s(z+\Delta z)}\, a_{z+\Delta z})$, the direct insolation to volume V_j; [J T^{-1}]

q_{sz} = $(1 - \beta)\, q_{sn}e^{-nz}$, the solar radiation flux at depth z, [J L^{-2} T^{-1}]

q_{sn} = net solar radiation flux penetrating the surface, [J L^{-2} T^{-1}]

β = ratio of radiation adsorbed at the surface to the net incoming radiation

n = bulk extinction coefficient [L^{-1}]

$h_{w,j}$, $h_{w,j+1}$ = $cpQ(z)O_z$ represent the heat advected along the vertical axis [J T^{-1}]

$h_{d,j}$, $h_{d,j+1}$ = $cpE_z a_z \partial O/\partial z$ represent the heat diffused along the vertical axis [J T^{-1}]

E_z = coefficient of vertical effective diffusion [L^2 T^{-1}]

a_z = area normal to the vertical axis at depth z, [L^2]

For the special case of the control volume at the surface, i.e., x = o, heat exchange is given by

$$h_{soj} = [q_n - (1 - \beta)\, q_{sn}]\, a_o$$

where q_n = net heat exchange at the air-water interface [J T^{-1}] (see equation 5.9).

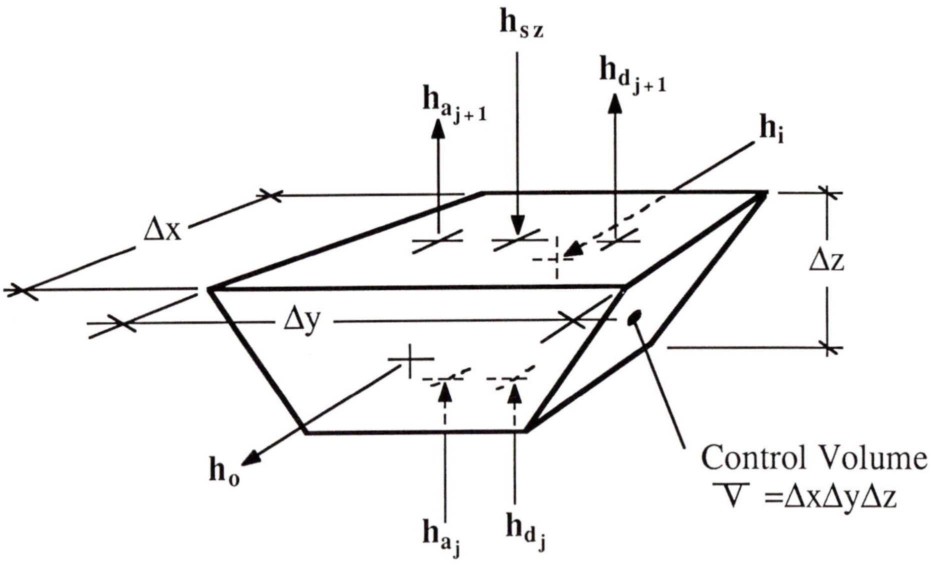

Fig. 5.1 Control volume for conservation of heat energy

The primary advective and effective diffusion transport are shown as occurring only along the vertical (z) axis in this representation, for the sake of clarity, but it is implicit that these mechanisms are operative along both horizontal axes as well.

5.2 EVALUATION OF HEAT BUDGET TERMS

5.2.1 Advection

Advective transport is determined by translation of the mean gradient of temperature with the mean cross sectional velocity. Hence, for advection along the z axis in a control volume of thickness Δz

$$cp \frac{\partial VO}{\partial t} = - cp\ w\ \Delta\ x\ \Delta ydz\ \left(\frac{\partial O}{\partial z}\right) \tag{5.2}$$

which for a discrete volume V_j becomes:

$$\frac{V_jO_j}{\partial t} = - (w\ a_zO)_j\ -\ (w\ a_zO)_{j+1} \tag{5.3}$$

or

$$\frac{V_jO_j}{\partial t} = - (QzO)_j\ -\ (Q_zO)_{j+1} \tag{5.4}$$

where w and Q_z are the mean cross section velocity and discharge along the z axis, respectively.

5.2.2 Effective Diffusion

Transport of heat energy by mixing processes analogous to molecular diffusion is represented by

$$cp \frac{\partial\ (VO)}{\partial t} = cp\ E_z\ \Delta\ x\ \Delta ydz\ \left(\frac{\partial^2 O}{\partial Z^2}\right) \tag{5.5}$$

which for a discrete volume V_j becomes:

$$\frac{V_jO_j}{\Delta t} = (E_z\ a_z\ (\frac{dO}{dz}))_j\ -\ (E_z\ a_z\ (\frac{dO}{dz}))_{j+1} \tag{5.6}$$

The effective diffusion coefficient E_z is considered to represent the sum of the processes of molecular diffusion, turbulent diffusivity and advective dispersion, i.e.

$$E_z = D_{mz} + e_z + E_{adz} \tag{5.7}$$

As a practical matter E_z, or its counterparts for the x and y axes, are usually determined empirically by field observation using tracers, or in the modelling process itself during calibration. Molecular diffusion is usually inconsequential in most cases of environmental modeling and is generally neglected. Turbulent diffusivity can be determined from detailed knowledge of velocity profiles, e.g. in stream systems, but for larger water bodies where motions are not governed entirely by boundary stresses it may be indeterminate. Advective dispersion is a consequence of the modelling approximation of the mean cross sectional velocity as representative of variations in velocity across the section. It is inherent in any model and can only be diminished in magnitude by more explicit and detailed representation of the veolcity field. All of these factors, together with certain numerical approximations associated with computational techniques, are combined to introduce "effective diffusion" into the model. More details concerning the nature of these factors are presented elsewhere in this book.

5.2.3 Direct Insolation

Insolation by short wave radiation from the overlying atmosphere occurs at any depth within the extinction limit. It may also occur as a result of reflection from solid surfaces within the water body, e.g. the bottom of a shallow stream. The rate of heat transfer is a function of the incident short-wave solar radiation, the extinction properties of the fluid and the depth below the surface, as given by

$$q_{sz} = q_{so}\ (1 - \beta)\ e^{-nz} \tag{5.8}$$

where q_{sz} and q_{so} are shortwave radiation fluxes in Joules per day per unit area at depth z(m), and passing through the surface, respectively, and n is an extinction coefficient (m^{-1}). The constant β is the proportion of q_{so} assumed to be absorbed in the surface layer, usually 1 metre in thickness.

Heat transfer to the surface layer by short wave insolation, βq_{so} is supplemented by long-wave atmospheric radiation and conduction from the atmosphere and depleted by back radiation, evaporation and sensible heat transfer. The net heat exchange resulting from the aggregation of these processes is described in detail in the following section.

5.3 HEAT EXCHANGE AT AIR-WATER INTERFACE

Heat exchanges through the air-water interface are dependent upon both the internal hydro-mechanical behaviour of the water body and the physics of its interaction with the overlying air mass. Meteorological factors, such as solar radiation, both short and long wave, wind, humidity, atmospheric pressure, air temperature, and cloudiness, figure prominently in many of the physical processes involved. The aggregate effect of the most important of these processes is represented in the general heat flux equation:

$$q_n = q_{sn} + q_{at} - q_{ws} - q_e \pm q_h \qquad (5.9)$$

where

q_n = net heat energy transfer across the air-water interface $[J\ L^{-2}\ T^{-1}]$

q_{sn} = net solar radiation heat flux

q_{at} = net atmospheric radiation heat flux

q_{ws} = water surface radiation heat flux

q_e = evaporation heat flux

q_h = sensible heat flux

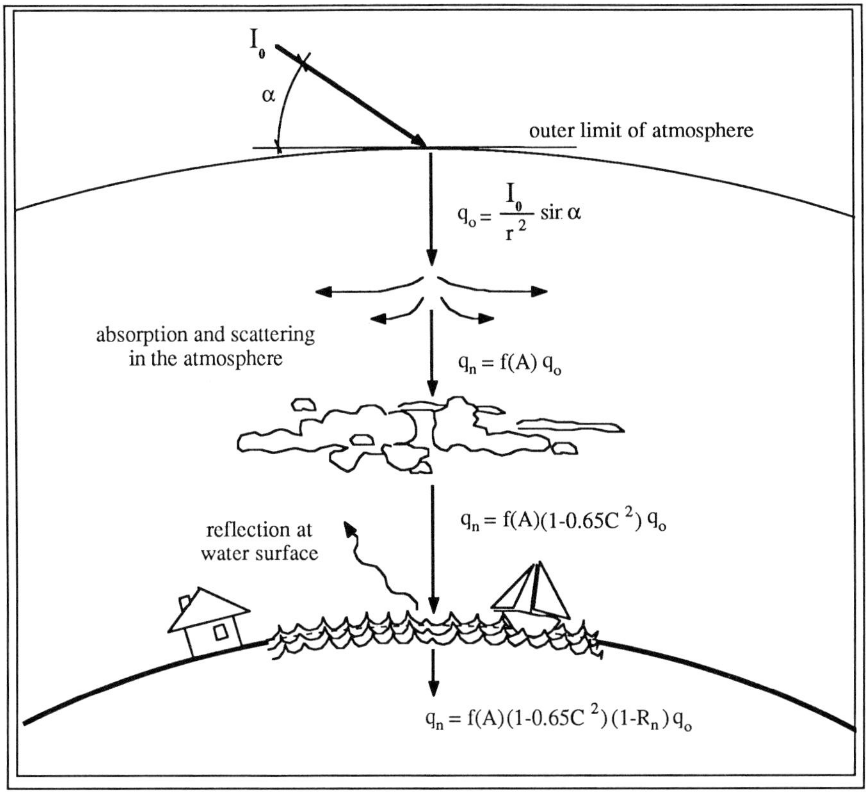

Fig. 5.2 illustrates the essential processes considered in this computation.

Extraterrestrial solar radiation.

The computation of solar radiation received at the top of the earth´s atmosphere, i.e., extraterrestrial solar radiation, is based on the solar constant, the rate of radiation emitted by the sun, trigonometric relationships that determine the intensity of the direct solar beam, and on the horizon plane of the observer.

Due to the rotation of the earth on its axis and about the sun, the angle of the sun´s rays to the horizontal plane varies continually. This angle is a function of the latitude of the observer and the time of year. Solar radiation intensity at the top of the atmosphere for a given location and time can be expressed as:

$$q_o = \frac{I_o}{r^2} \sin a \qquad (5.10)$$

where

q_o = solar radiation intensity $[J\ L^{-2}\ T^{-1}]$

I_o = solar constant, approximately 2.00 cal * cm^{-2} * min^{-1} (8.39 J $cm^{-2} min^{-1}$) equal to the solar radiation intensity at normal incidence at the top of the atmosphere for the mean distance between the earth and the sun

r = radius vector, i.e. the ratio of the actual distance between the earth and the sun to the mean distance; $0.98833 < r < 1.0167$ and $0.9669 < r^2 < 1.0337$

a = solar altitude in radians, a function of latitude, sun´s declination and local hour angle of the sun; $0 < a < \pi/2$.

The solar altitude is derived from spherical trigonometry as

$$\sin a = \sin ø \sin g + \cos ø \cos g \cos h \qquad (5.11)$$

where

ø = latitude of the observer in radians

g = declination of the sun, in radians

h = local hour angle (LHA) of the sun, in radians.

In practice, the declination of the sun g and the ratio r for a particular day may be assumed to remain constant. Approximate values, sufficiently accurate for heat transfer computations, can be calculated by the means of the following expressions:

$$r = 1 + 0.017 \cos \frac{2\pi}{365} (186 - d) \qquad (5.12)$$

$$\text{and } g = 23.45 \frac{\pi}{180} \cos \frac{2\pi}{365} (172 - d) \tag{5.13}$$

where
d = Julian day (1 to 365)

The local hour angle of the mean sun is the angle measured westward around the axis of the celestial sphere from the upper meridian of the observation point to the meridian of the sun, as can be seen in Figure 5.3.

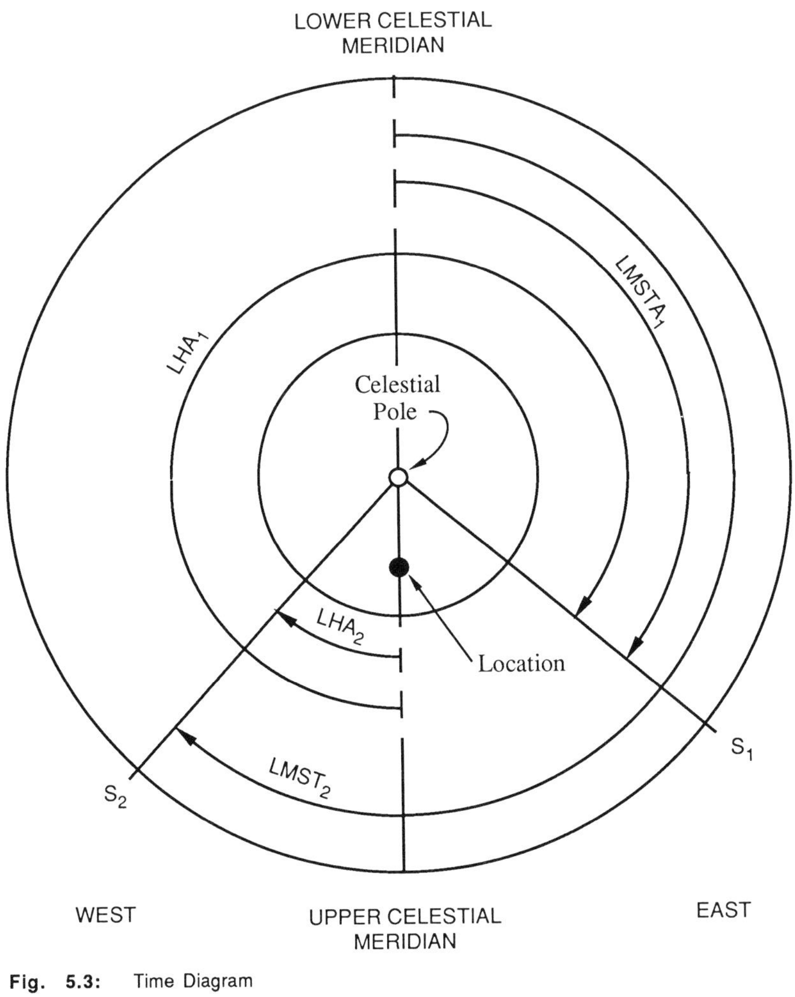

Fig. 5.3: Time Diagram

Thus the local hour is greater than 12 when the sun is east of the local meridian, that is before the sun passes the local meridian (local noon). When the sun is at the local meridian, it is local noon and LHA = 0. In the afternoon, the local hour angle of the sun is less than 12.

The angle measured westward around the axis of the celestial sphere from the lower meridian of the observation point to the precise meridian of the sun is called the Local True Solar Time (LTST). Since the earth´s angular motion around the sun is irregular, the apparent speed of the sun is not constant and, for the purpose of time counting, the "true sun" is replaced by an imaginary "mean sun". The corresponding time is called local mean solar time (LMST).

Mean solar time of 1200, or solar noon, corresponds to LHA = 0, so the relationships in Fig. 5.3 can be established between time and hour angle.

a) when the sun is east of the upper celestial meridian of the location:

$$LMST = LHA - 12 \tag{5.14}$$

b) when the sun is west of the upper celestial meridian of the location:

$$LMST = LHA + 12 \tag{5.15}$$

where
LMST = local mean solar time, hours
LHA = local hour angle of the mean sun, hours.

The relationship between standard time of the time zone of the observer and local mean solar tiem is given by the expression:

$$LMST = ST + DTSL$$

where
ST = standard time of the time zone of the observer, hours
DTSL = time difference between local and standard meridians, hours

The time difference between the local and standard meridians can then be calculated as:

$$DTSL = \frac{f}{15}(LSM - LLM) \tag{5.17}$$

where
LSM = longitude of the standard meridian, in degrees

LLM = longitude of the local meridian, in degrees
f = -1 for west longitudes
= 1 for east longitudes

Due to **the variable apparent movement of the sun**, the sun is either behind true solar noon, or ahead of it. The time difference between true solar noon and mean solar noon varies over the year between 0 and ± 16 minutes (or 0 and ± 4°). This time difference is called the Equation of Time. Therefore, local mean solar time has to be corrected by adding the Equation of Time:

$$LMST = ST + DTSL + ET \tag{5.18}$$

where
ET = equation of time, hours

The equation of time for each day is listed in solar ephemeris tables, such as provided by U.S. Nautical Almanac Office. While for certain purposes it may be appropriate to apply this correction, it is usually neglected in heat budget modelling in view of its small magnitude and the greater uncertainty in estimation of other variables influencing net short wave radiation, e.g. cloudiness. By substituting equations 5.16, 5.17 and 5.18 into 5.14 and 5.15 the local hour angle (LHA) can be computed by means of the following expressions:

a) *for the sun east of the local meridian:*

$$LHA = ST + 12 + DTSL + ET \tag{5.19}$$

b) *for the sun west of the local meridian:*

$$LHA = ST - 12 + DTSL + ET \tag{5.20}$$

The values for LHA, obtained from these equations are in hours; they can be converted into radians for substitution in equation 5.11 by:

$$h\,[rad] = \frac{\pi}{12}\ LHA\ [hours] \tag{5.21}$$

For a time period $\Delta t = t_2 - t_1$, the total extraterrestrial radiation can be computed by integrating equation 10 after appropriate substitutions

$$Q_{o\Delta t} = \int_{t_1}^{t_2} q_o * dt = \frac{I_o}{r^2} \int_{t_1}^{t_2} (\sin ø \ \sin g + \cos ø \ \cos g \ \cos h)\, dt \tag{5.22}$$

where

$Q_{o\Delta t}$ = total extraterrestrial solar radiation for the time period Δt [J L^{-2}]

t_1, t_2 = time limits in standard time of the observer, hours

Equation (5.22) can be most conveniently integrated if the first term of the equation is integrated over time and the second, over the hour angle (the integration limits and the differentials are transformed into hour angles). By substituting $dt = 12/\pi \; dh$, hourly totals of extraterrestrial solar radiation can be calculated:

$$Q_{o\Delta t} = \frac{I_o}{r^2} \left[(t_2 - t_1) \sin ø \; \sin g + \frac{12}{\pi} \cos ø \; \cos g \; (\sin h_2 - \sin h_1)\right] \quad (5.23)$$

The hour angles h_1 and h_2 should be calculated by using equations (5.19) and (5.20), with hour angles between 0 and 2π.

Daily totals or daily averages of solar radiation can be calculated by integration of equation (5.23), taking the time period Δt equal to the duration of possible sunshine. The time of sunrise is defined as the time when the centre of the sun is at the altitude of the horizon. For a flat, unobstructed horizon, the solar altitude at this moment a_{sr} is equal to zero. If the horizon is obstructed, or if refraction is important, a_{sr} can be greater or less than zero.

Since the sun is east of the local upper celestial meridian at sunrise, the hour angle $LHA_{sr} > 12$ (in hours); or $\pi < h_{sr} < 2\pi$, in radians, as shown on Figure 5.3. From equation (5.11), the hour angle of sunrise can be calculated as follows:

$$h_{sr} = \arccos \left\{ \frac{\sin a_{sr} - \sin ø \; \sin g}{\cos ø \; \cos g} \right\} = h_{sr}' \quad (5.24)$$

where

a_{sr} = solar altitude at sunrise, or altitude of the horizon

g = average daily declination, in radians

h_{sr} = hour angle of sunrise, measured westward from the upper celestial meridian, $\pi < h_{sr} < 2\pi$

h'_{sr} = complementary sunrise hour angle, $h_{sr}' < \pi/2$

If $\cos h_{sr}$ is negative, then $\pi < h_{sr} < 3\pi/2$. The total hour angle is in this case:

$$h_{sr} = \pi + | h_{sr} | \quad (5.25)$$

where

$$| h_{sr} | = \text{arc cos } | h_{sr} | \tag{5.26}$$

If cos h_{sr} is positive, then $3\pi/2 < h_{sr} < 2\pi$, and the total hour angle is

$$h_{sr} = 2\pi - h_{sr} \tag{5.27}$$

Also, by means of equation (5.24) the hour angle of sunset can be calculated. By assuming $a_{ss} = a_{sr}$, the hour angle of sunset can be expressed as:

$$h_{ss} = \text{arc cos } \frac{\sin a_{ss} - \sin ø \sin g}{\cos ø \cos g} \tag{5.28}$$

The sunset hour angle should be between 0 and π. If equation (5.28) yields a negative value, substract the absolute value from π to obtain the full hour angle in radians, which should be greater than $\pi/2$. From equations (5.19) and (5.20) the standard time of sunrise is calculated as

$$ST_r = \frac{12}{\pi} h_{sr} - 12 + DTSL - ET \tag{5.29}$$

and the standard time of sunset is

$$ST_s = \frac{12}{\pi} h_{ss} + 12 + DTSL - ET \tag{5.30}$$

The duration of possible sunshine, S_o, is therefore the difference between STS and STR:

$$S_o = STS - STR = 24 + \frac{12}{\pi} (h_{ss} - h_{sr}) \approx \frac{24}{\pi} h_{ss} \tag{5.31}$$

By calculating h_{sr} and h_{ss} and substituting them in equation (5.23) for h_1 and h_2 , the daily total of extraterrestrial solar radiation can be calculated as:

$$Q_{o1} = \frac{24}{\pi} \frac{I_o}{r^2} (h_{ss} \sin ø \sin g + \cos ø \cos g \sin h_{ss}) \tag{5.32}$$

where

Q_{o1} = daily total of extraterrestrial solar radiation, [J L^{-2}]

Ground solar radiation under a clear sky

The solar radiation intensity received at the top of the atmosphere is attenuated on its path through the atmosphere by scattering and absorption upon encountering gas molecules, water and dust particles. This attenuation can be described approximately by **the Lambert - Beer Law:**

$$q_z = q_o e^{-nz} \tag{5.33}$$

where

q_z = intensity of the solar radiation after passing through the atmosphere, $[J\ L^{-2}\ T^{-1}]$

q_o = intensity of solar radiation at the outer atmosphere, $[J\ L^{-2}\ T^{-1}]$

n = extinction coefficient, $[L^{-1}]$

z = "path length"; depth of the atmosphere, measured from the outer limit towards the earth's surface, [L]

Strictly, this relationship is valid *only for monochromatic light* and the intensity of solar radiation calculated by means of equation (5.33) can be considered only as an approximate value of the attenuated total spectral solar radiation flux.

The ratio q_z/q_o is called the transmission coefficient and depends on the type of air and the "path length", which can be expressed by the relative optical air mass m. Atmospheric transmissions of the solar beam originally published by Kimball (1927, 1928 and 1930) have been used in computations of solar radiation by many authors. Solar radiation reaching the ground is calculated according to the TVA (1972) recommendation as follows.

a. The intensity of the direct solar beam after attenuation by scattering in a moist, dust free atmosphere is given by:

$$q_{sc} = q_o * a' \tag{5.34}$$

b. Since scattering takes place in all directions, it is assumed that half of the scattered radiation reaches the ground, the other half is scattered back to space. Thus, the scattered radiation reaching the ground through a dust free atmosphere is:

$$q_i = 0.5\, q_o * (1 - a') \tag{5.35}$$

c. The direct solar beam intensity after scattering and absorption in a dust free, moist atmosphere is:

$$q_{sd} = q_o * a'' \tag{5.36}$$

d. The total radiation received at the ground under a moist, dust free atmosphere is calculated as a sum of direct and scattered (diffuse) radiation:

$$q_s = q_{sd} + q_{si} = q_o (a'' + 0.5 - 0.5 a') \quad (5.37)$$

where

q_s = extraterrestrial solar radiation intensity in $[J\ L^{-2}\ T^{-1}]$

a' = mean atmospheric transmission coefficient for cloudless, dust free moist air, after scattering only

a'' = mean atmospheric transmission coefficient for cloudless, dust free moist air after scattering and absorption

Both a' and a'' are functions of the optical air mass m and the moisture content of the atmosphere, w. These parameters will be discussed later.

e. Additional attenuation of solar radiation occurs due to scattering and absorption by dust. The transmission coefficient of a moist, dusty atmosphere can be expressed by:

$$a = a'' - d \quad (5.38)$$

where

a = transmission coefficient of a natural moist and dusty atmosphere

d = total depletion coefficient of the direct solar beam by scattering and absorption due to dust.

The depletion coefficient due to dust is a function of air mass m and the time of the year. Some typical values are given in Table 5.1.

Table 5.1
Total dust depletion coefficient for two levels of optical air mass, m.
$d = d_s + d_a$

Season	Washington, D.C.		Madison, Wisc.		Lincoln, Nebr.	
	m = 1	m = 2	m = 1	m = 2	m = 1	m = 2
Winter	-	0.13	-	0.08	-	0.06
Spring	0.09	0.13	0.06	0.10	0.05	0.08
Summer	0.08	0.10	0.05	0.07	0.03	0.04
Fall	0.06	0.11	0.07	0.08	0.04	0.06

Source: TVA (1972)

The depletion coefficient can be described as the sum of two components:

$$d = d_s + d_a \quad (5.39)$$

where

d_s = depletion coefficient of the direct solar beam by dust scattering

d_a = depletion coefficient of the direct solar beam by dust absorption

Assuming that half of the radiation scattered by dust is directed toward the ground, the diffuse radiation should be increased by $0.5{*}d_s{*}q_o$ and the direct radiation would then be reduced by $d{*}q_o$ where d_s is that part of the total dust depletion coefficient d (Table 5.2) that is attributed to scattering. The balance, d_a , is attributed to absorption.

f. Radiation reflected from the ground back to the sky is scattered once again on its way through the atmosphere and it is assumed that half goes back to earth and half to space. The resulting increment of diffuse radiation can be described as:

$$q_{gr} \; * \; 0.5\,(1 - a + d_s) \qquad (5.40)$$

where q_{gr} is total reflection of direct solar radiation flux at the ground, given as:

$$q_{gr} = R_g \; * \; q_s \qquad (5.41)$$

where

q_s = total solar radiation flux received at the ground

R_g = total reflectivity of the ground in the vicinity of the site as given in Table 5.2.

Table 5.2 Source: TVA (1972)
Estimated values of total reflectivity of the ground, R_g

Ground Condition	R_g
Meadows and fields	0.14
Leaf and needle forest	0.17 - 0.09
Dark, extended mixed forest	0.045
Heath	0.10
Flat ground, grass covered	0.25 - 0.33
Flat ground, rock	0.12 - 0.15
Sand	0.18
Vegetation early summer, leaves with high water content	0.19
- late summer, leaves with low water content	0.29
Fresh snow	0.83
Old snow	0.42 - 0.70
Water surface	$R_g = R_t$

g. The total radiation received at the ground under a moist, dusty atmosphere is composed of direct and diffuse radiation. Direct radiation can be described as:

$$q_{sd} = q_o \, (a'' - d) \tag{5.42}$$

and diffuse radiation as:

$$q_{si} = 0.5 \, q_o \, (1 - a' + d_s) + 0.5 \, q_{gr} \, (1 - a' + d_s) \tag{5.43}$$

The total solar radiation flux received at the ground per unit area can be obtained by summation of equations 5.42 and 5.43,

$$q_s = q_o \, [a'' + 0.5 \, (1 - a' - d) - 0.5 \, d_a] + 0.5 \, q_{gr} \, (1 - a' + d_s) \tag{5.44}$$

Substitution of equation 5.41 into 5.44 yields:

$$q_s = q_o \left\{ \frac{a'' + 0.5 \, (1 - a' - d) - 0.5 \, d_a}{1 - 0.5 \, R_g \, (1 - a' + d_s)} \right\} \tag{5.45}$$

which is the expression for the total solar radiation reaching the ground after scattering, absorption and the effect of dust in a moist, cloudless atmosphere. The value in brackets {} is considered a function of atmospheric quality, the f(A) indicated in Figure 5.2.

Ground solar radiation under a cloudy sky

Many investigations have been performed to relate radiation from a cloudy sky to radiation from a clear sky. But, usually only the most obvious factor responsible for the reduction of solar radiation, cloud cover, is taken into account. Other factors, such as thickness, type and reflectivity of clouds, and time periods for which the value of cloud cover is representative, have been neglected. Among the great number of empirical formulae, suggested by different investigators, the TVA (1972) suggests the following:

$$\frac{q_c}{q_s} = 1 - 0.65 \, C^2 \tag{5.46}$$

where

q_c = ground solar radiation intensity under a cloudy sky, $[J \, L^{-2} \, T^{-1}]$

q_s = ground solar radiation intensity under a clear sky

C = cloud cover in tenths of sky covered, from 0.0 to 1.0

Multiplying by the effect of cloudy sky, as described in equation 5.46, the net solar heat flux reaching the water surface under cloudy sky conditions, q_{sc}, can be calculated by means of the following expression:

$$q_{sc} = q_o (1 - 0.65 C^2) \{ \frac{a'' + 0.5 (1 - a' - d) - 0.5 d_a}{1 - 0.5 R_g (1 - a' + d_s)} \} \quad (5.47)$$

Effect of reflection at water surface

The values obtained by means of either quation 5.45 or 5.46 must be corrected for the effect of reflection at the water surface. Net solar radiation flux penetrating the water surface under clear sky conditions, q_{so}, is given by:

$$q_{so} = q_s - q_{sr} \quad (5.48)$$

where

q_s = total solar radiation flux received at the ground under clear sky conditions $[J\ L^{-2}\ T^{-1}]$

q_{sr} = total reflection of all direct and diffuse radiation incident on the water surface.

As given by Anderson (1954), q_{sr} can be related to q_s:

$$R_t = \frac{q_{sr}}{q_s} \quad (5.49)$$

where

R_t = total reflectivity of the water surface, a function of the solar altitude, a, of the form

$$R_t = A\, a^b \quad (5.50)$$

where a is the solar altitude in degrees and A and b are empirical constants, estimated to have values of about 1.18 and -0.77, respectively. For example, for a solar altitude of 90° (sun directly overhead) the value of R_t for clear sky conditions from equation 5.50 would be about 0.04, and for an altitude of 30°, about 0.09. Slightly lower values are obtained for cloudy skies. In practice it is common to adopt values of R_t in the range of 0.03 to 0.06 as representative of average conditions, independent of the altitude of the sun or cloudiness of the sky. Thus, the net solar radiation passing through the water surface is obtained by combining the effects of atmospheric absorption and scattering, cloud cover interception and reflection,

$$q_{sn} = \{\frac{a'' + 0.5\,(1 - a' - d) - 0.5\,d_a}{1 - 0.5\,R_g\,(1 - a' + d_s)}\}\,(1 - 0.65\,C^2)\,(1 - R_t) \qquad (5.51)$$

The mean atmospheric transmission coefficients a' and a" have been shown by Kimball (1930) to be functions of the optical air mass, m, and mean precipitable water content, w. To utilize Kimball's data these quantities must first be estimated.

The optical air mass, m, can be estimated with the empirical formula of Kasten (1964) adjusted for pressure differences from sea level,

$$m = \frac{P_z}{P_o}\,\{\frac{1}{\sin a + 0.15\,(a + 3.885)^{-1.253}}\} \qquad (5.52)$$

where P_z and P_o are barometic pressures at the specified location and sea level, respectively, and a is the solar altitude in degrees.

The mean clear sky precipitable water content, w, is determined from the TVA (1972) equation:

$$w = 0.85\,\exp(0.11 - 0.0614\,O_d) \qquad (5.53)$$

where O_d is the monthly mean dew point.

Using Kimball's empirical data for a' and a" as functions of m and w, WRE (1968) derived the following estimating equations

$$a' = m\,\exp -(0.465 + 0.134W)(0.129 + 0.171e^{-0.88m}) \qquad (5.54)$$

and

$$a'' = m\,\exp -(0.465 + 0.134W)(0.179 + 0.421e^{-0.72m}) \qquad (5.55)$$

5.3.2 Atmospheric Radiation

Some short-wave radiation from the sun plus radiation emitted by the ground or water surfaces enters the earth's atmosphere and is partly absorbed by water vapor, carbon dioxide, ozone and other atmospheric gases. These constituents, in turn, emit long wave radiation back to the ground and water surfaces, and outward to space. Such radiation is called atmospheric radiation. It is a function of absolute air temperature, cloudiness and water surface reflectivity. The factors influencing atmospheric radiation are illustrated in Figure 5.4.

A great number of investigations have been concerned with developing practical expressions for atmospheric radiation. The comparison conducted and reported by TVA (1972) showed that there is a remarkably close agreement between the results obtained by the formulae of several different investigators, so it is difficult to determine the best. The following expression is suggested as the most appropriate for calculation of atmospheric radiation under the cloudy sky conditions:

$$q_{at} = 0.937 \times 10^{-5} * o * T_2^6 (1 + 0.17 C^2) (1 - R_a) \tag{5.56}$$

where

q_{at} = long wave radiation from a cloudy sky, $[J\ L^{-2}\ T^{-1}]$

o = Stefan-Boltzman constant; $o = 4.875 \times 10^{-8}$ kcal*$m^{-2}hr^{-1}\ °K^{-4}$ (2.0411×10^{-7} kJ$m^{-1}hr^{-1}\ °K^{-4}$)

$T_2 = 273.2 + O_{a2}$, in °K; O_{a2} = dry bulb air temperature at 2 m above the ground, in °C

C = cloudiness, a decimal function of sky covered by clouds, between 0 and1

R_a = reflectivity of water surface for atmospheric radiation; $R_a \approx 0.03$.

The constant in equation 5.56 is Swinbank's coefficient (Swinbank, 1963) introduced as a conversion factor to account for using the sixth power of the absolute temperature rather than the fourth, as for an ideal "black body". Note also that the cloudiness factor recognizes the fact that a cloudy sky emits greater longwave radiation by a maximum of 17 percent. The factor 0.17 represents an average condition corresponding to alto cumulus cloud formations. For cirrus clouds the factor may be reduced to 0.04 and for stratus clouds increased to 0.24.

5.3.3 Water Surface Radiation

The long-wave radiation flux emitted by a water body with a surface temperature T_o is given by:

$$q_{ws} = e_W o\, T_o^4 \tag{5.57}$$

where

q_{ws} = water surface (back) radiation $[J\ L^{-2}\ T^{-1}]$

e_W = emissivity of the water surface, i.e. ratio of actual radiation to that of an ideal black body, ≈ 0.96

T_o = absolute temperature of the water surface, °K

o = Stefan-Boltzman Constant, 4.875×10^{-8} kcal*$m^{-2}hr^{-1}\ °K^{-4}$ (2.0411×10^{-7} kJ$m^{-1}hr^{-1}\ °K^{-4}$)

Factors influencing water surface radiation are illustrated in Figure 5.4.

5.3.4 Evaporative Heat Flux

The evaporation heat flux is determined by the rate of mass transfer of water from liquid to gaseous state times the latent heat of vaporization

$$q_e = p_w L E \tag{5.58}$$

where

q_e = evaporative heat flux, cal*cm^{-2}day^{-1}

p_w = density of water, g*cm^{-3}

L = latent heat of vaporization, cal*g^{-1} = 597.1 - 0.57 O_o;

O_o is water surface temperature, °C

E = evaporation rate, g*cm^{-2}*day^{-1} ≈ cm*day^{-1}

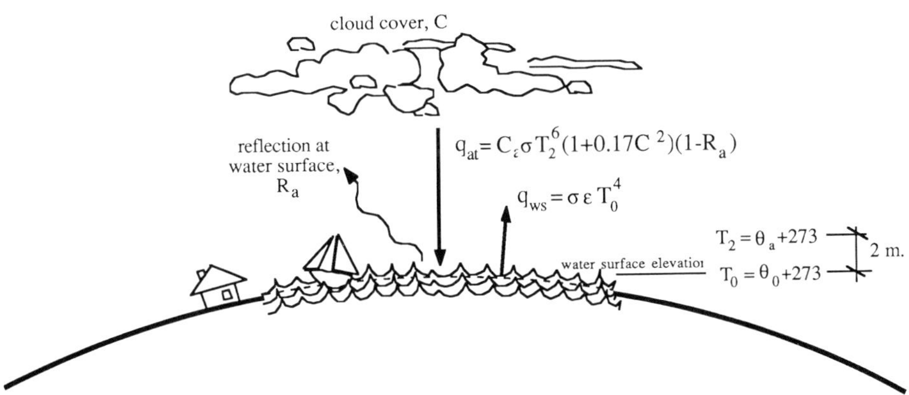

Fig. 5.4: Atmospheric and water surface radiation.

Estimation of evaporation rate for use in equation 5.58 is the most uncertain aspect of the heat budget calculation, despite the efforts of many investigators to refine and quantify the many factors that influence mass transfer through the air-water interface. Among the factors that must be considered are:

* wind velocity and turbulence
* air temperature
* water temperature
* atmospheric pressure
* vapor pressure of water
* relative humidity
* atmospheric stability
* water velocity and turbulence

The most direct approach to determining E is a water budget, assuming all other quantities in the budget equation can be directly measured. Unfortunately this is seldom the case, so alternatively a "model" of the real water body, usually an evaporation pan, is used, wherein

$$E = k_p E_p \tag{5.59}$$

where k_p is a pan coefficient, the ratio of model to prototype evaporation, and E_p is the measured pan evaporation. The equation has practical utility since pan evaporation rates are often included in standard meteorological observations. The coefficient k_p is determined empirically by comparison with evaporation rates from standardized pans e.g. "Class A Pan", to rates determined in control experiments by the water or energy budget methods. Typically, k_p varies from 0.65 to 0.80, depending on climatic conditions, prototype geometry, exposure and position of the pan relative to the prototype.

The literature on evaporation is replete with empirical representations for evaporation rate as a function of the most commonly measured meteorological quantities. For the most part they are of the general form developed from the Lake Hefner studies (Marciano and Harbeck, 1954)

$$E = N\, u(z)\, (e_s - e_z) \tag{5.60}$$

where

N = a constant, usually determined empirically

$u(z)$ = wind velocity, a function of elevation above water surface

e_s = saturation vapour pressure (usually at water surface temperature)

e_z = vapour pressure above water surface, a function of elevation

After a detailed review of available equations and the basis for their derivation, TVA (1972) recommended a relationship based on the Marciano-Harbeck equation, which after conversion to appropriate SI units is

$$E = 2.65 \times 10^{-4} * u_4\, (e_o - e_2) \tag{5.61}$$

where

E = evaporation rate cm*day^{-1}

u_4 = wind velocity at 4 metres above water surface, km* day^{-1}

e_o = saturation vapour pressure at water surface temperature, mm Hg

e_2 = vapour pressure at height of 2 metres above water surface, mm Hg

5.3.5 Sensible Heat Flux

Sensible heat is transferred between air and water by conduction and transferred away from the air-water interface by the same mechanisms as for evaporation. It is convenient to relate sensible and evaporative heat fluxes by Bowen's ratio

$$B = \frac{q_h}{q_e} \tag{5.62}$$

which can be computed from

$$B = 0.464 \frac{P}{P_o} \left(\frac{O_o - O_a}{e_o - e_a}\right) \tag{5.63}$$

where

P/P_o = ratio of actual to standard barometric pressure

O_o = water surface temperature, °C

O_a = air temperature, °C

e_o = saturation vapour pressure at water surface temperature, mm Hg

e_a = vapour pressure of the air, mm Hg

By substitution of equations 5.63 and 5.60 into 5.58 and 5.62 and solving for q_h the following is obtained:

$$q_h = 0.464\, N \frac{P}{P_o}\, u_z\, (O_o - O_a)\, p_w L \tag{5.64}$$

which can be used to predict sensible heat transfer. The value of the constant N can be adjusted according to the Lake Hefner method (Marciano and Harbeck, 1954), depending on the levels at which u_z and O_a are observed. For the conditions represented by equation 5.61 the sensible heat flux is

$$q_h \;=\; [1.23 \times 10^{-4}]\; u_4\; (O_o - O_2)\; p_w L \qquad (5.65)$$

where

u_4 = wind velocity at 4 metres, km*day^{-1}

O_o = water surface temperature, °C

O_z = air temperature at 2 metres above the water surface, °C

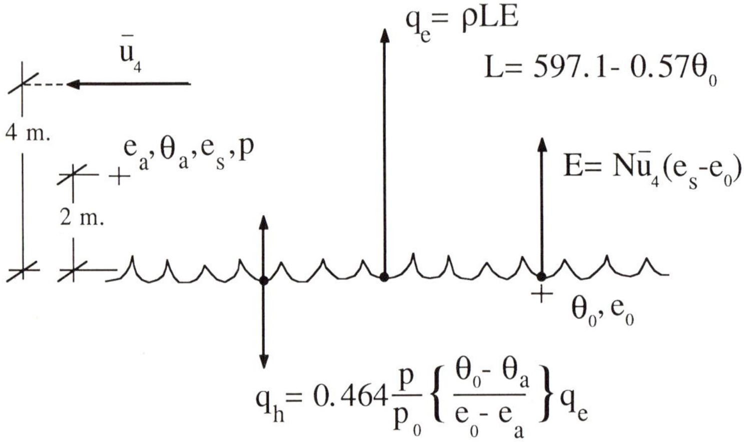

Fig. 5.5: Evaporative and sensible heat exchange.

5.3.6 Heat Flux Through Water Surface

Summing the five basic heat flux terms, the net heat flux through the air-water interface becomes

$$q_n \;=\; [\, f(A)\, (1 - 0.65\, C^2)\, (1 - R)\,]\; \frac{I_o}{r^2}\; \sin a$$

$$+\; 0.937 \times 10^{-5}\; o\, T_2^{\,6}\; (1 + 0.17\, C^2)\, (\, 1 - R_a)$$

$$-\; e_W\; o\; T_o^{\,4}\; -\; \{\, [\, 2.65 \times 10^{-4}]\; u_4\; (e_o - e_2)$$

$$\pm\; [1.23 \times 10^{-4}]\; u_4\; (O_o - O_2)\, \}\; p_w\; L \qquad (5.66)$$

5.3.7 Other Heat Losses

In addition to the five principal heat fluxes represented in equation 5.66, heat may be transfered into the water column by condensation from an overlying super-saturated air mass or into or out of the system by advective transfers associated with precipitation, evaporation and/or condensation. The condensation flux can be described by an expression of the same general form as equation 5.61. Heat gain by condensation is usually relatively small, therefore, it can be neglected in energy budget calculations, or considered as part of net evaporative heat loss.

Heat exchanges associated with advective transport of water through the interface, e.g. precipitation, are usually neglected, except in the most detailed studies.

5.3.8 Penetration of Heat Into the Water Column

The radiation flux passing through the water surface diminishes with depth by both scattering and absorption mechanisms, phenomena collectively referred to as light attenuation. Absorption is defined as diminution of light energy with depth as it is transformed into heat energy of the water mass. Each wavelength of the solar spectrum is attenuated differently so that the spectrum at a particular depth, z, is different from that of radiation entering the water surface.

The absorption spectrum for pure water shows an almost complete absorption of all radiation with wavelengths greater than about 1.2 μ within the first few centimetres below the water surface. Nearly all radiation outside the visible range (from 0.36 μ to 0.76 μ) is absorbed within the first metre or so, so that only visible light penetrates to greater depths in impoundments. Since the visible spectrum corresponds closely with that of short-wave insolation it is assumed that only short-wave radiation penetrates beyond the surface layer.

The percentage of total incoming solar radiation which penetrates through the first metre below the surface is low, between 30 and 40 percent of that entering the surface. The ratio between solar radiation absorbed in the first metre below the water surface and total incoming radiation, for different types of sea and lake waters are given in Table 5.3 [TVA (1972)].

After initial absorption in the 1-metre surface layer, leaving *only monochromatic* visible light, attenuation follows closely the relationship described by the Lambert-Beer Law:

$$q_{sz} = q_{sn} \, e^{-nz} \qquad (5.67)$$

where

q_{sz} = solar radiation intensity at depth z, in (cal/m^2*hour)

q_{sn} = net solar radiation penetrating the surface, in (cal/m^2*hour)

n = bulk extinction coefficient, in m^{-1}

z = depth below surface, in m

Table 5.3
Radiation absorbed in the first metre below the water surface (After TVA, 1972)

Water type	Fraction of incoming radiation absorbed in first metre, ß
Pure water	0.61
Clear oceanic water	0.62
Average oceanic water	0.65
Average coastal sea water	0.73
Turbid coastal sea water	0.77
Lake Mendota, Wisconsin	0.79
Trout Lake	0.89
Big Ridge Lake	0.75
Fontana Lake, TVA	0.70

Application of the Lambert-Beer Law usually assumes homogenity of optical properties of the water column, i.e., n = constant.

The exact analysis of solar radiation absorption in water is an extremely complex problem. Penetration of direct and diffuse radiation would have to be treated separately. Also, analytical approaches depend largely upon the local water properties, characterized by variations in time and space of the concentration of particles suspended in water and, particularly, of dissolved compounds or color. Therefore, empirical methods provide the only practical means to describe the complex radiation absorption phenomena.

However, *experimental evidence* indicates that equation 5.67 is not generally satisfactory for the description of radiation attenuation in the upper stratum, because the uppermost layer absorbs proportionately larger amounts of solar radiation. For practical purposes, in model development, it may be sufficient simply to substract a fixed fraction of the total incoming radiation as if it were totally absorbed at the surface and to attenuate the residual according to **the Lambert-Beer Law**.

The resulting expression is:

$$q_{sz} = (1 - \beta)\, q_{sn}\, e^{-nz} \tag{5.68}$$

where

ß = ratio of radiation absorbed in surface layer (1 metre) to net incoming radiation.

Field data indicate a wide range in both ß and n, depending on the presence of suspended particulates and dissolved compounds. In pure water, or exceptionally clear lake water such as that of *Lake Tahoe (California),* for example, ß may be as low as 0.30, while for exceptionally turbid waters virtually all of the shortwave radiation may be absorbed in the surface layer. In natural impoundments in a state of advanced eutrophication, where algal blooms are frequent, extinction coefficients of the order of 0.5 per metre might be expected. At the other extreme, in extremely clear waters of oligotrophic lakes (like Lake Tahoe, Crater Lake in Oregon and Lake Baikal in the USSR) values as low as 0.02 per metre are possible. A more accurate way of describing solar radiation absorption phenomena in such lakes would be to divide the incoming radiation into components by wavelengths, each with its own characteristic extinction coefficient and ß. However, experience in modelling such lakes has shown that such precision is not essential for simulation of an annual temperature cycle. Equation 5.68 with appropriate values of ß and n is satisfactory for most practical situations.

5.4 APPLICATIONS

The processes of heat exchange at the air-water interface are inherently complex, requiring even for crude estimation the measurement of a variety of governing parametres. To apply equation 5.66, for example, one must have available temporal measurements of at least four meteorological quantities: wind velocity, cloudiness, dry bulb air temperature, and relative humidity (or wet bulb air temperature). Additionally, one must have reasonable estimates of empirical coeffients like the Lake Hefner N, emissivity and reflectance. Alternatively, direct measurement of insolation and evaporation may reduce the uncertainties associated with empirical estimators. Temporal averaging can further simplify calculations. In certain cases an equilibrium temperature approach, i.e. $q^n = 0$, $O_o = O_e$, may be convenient for a rough estimation of heat energy transfer rates.

The general schema exemplified by equation 5.1, using the heat exchange estimator of equation 5.66, has been used successfully in a large number of applications. The reader is referred to Orlob (1984) for examples.

5.5 ACKNOWLEDGEMENTS

The general approach presented here for quantification of heat exchange at the air-water interface derives primarily from the pioneering work on modelling of the thermal structure of deep reservoirs by the Tennessee Valley Authority Engineering Laboratory (TVA, 1972), then under the direction of Rex A. Elder, the consulting firm Water Resources Engineers, Inc. (WRE, 1968) with which the senior author was associated, and the Ralph Parson Hydraulic Laboratory of MIT under the direction of Dr. D.R.F. Harleman.

SELECTED REFERENCES

Anderson, E.R., 1954. Energy Budget Studies, in Waterloss Investigations - Lake Hefner Studies, Tech. Rep. USGS Prof. Paper 269.

Bolsenga, S.R., 1964. Daily Sums of Global Radiation for Cloudless Skies, U.S. Army Material Command, Cold Regions Research and Engineering Laboratory, Hanover, N.H. Techn. Rept 160, Nov.

Kasten, F., 1964. A New Table and Approximation Formula for Relative Optical Air Mass, U.S. Army Material Command, Cold Regions Research and Engineering Laboratory, Hanover, N.H. Techn. Rept 136, Nov.

Kimball, H.H. "Measurements of Solar Radiation Intensity and Determination of its Depletion by the Atmosphere", Monthly Weather Review, V25 (1927), V56 (1928) and V58 (1930).

Marciano, T.T. and G.E. Harbeck, Jr. "Mass Transfer Studies", in Waterloss Investigations, Lake Hefner Studies, USGS. Prof. Paper 269, 1954.

Orlob, G.T. Water Quality Modeling: Streams, Lakes and Reservoirs, IIASA State of the Art Series, Wiley Interscience, London, 1984.

Raphael, J.M. "Prediction of Temperature in Rivers and Reservoirs", J. Power Div. ASCE V88 PO2: 151-181, July 1962.

Swinbank, W.C. "Long-wave Radiation from Clear Skies", Q. Journal of the Royal Met. Soc. of London, V89, July 1963.

TVA, "Heat and Mass Transfer Between a Water Surface and the Atmosphere, Engineering Lab. Report No. 14, Norris, Tenn., April 1972.

WRE Inc., "Prediction of Thermal Energy Distribution in Streams and Reservoirs", Final Report, California Dept. of Fish and Game, 88 pp, June 1967 (Revised Aug. 1968).

CHAPTER 6

SEDIMENTATION

by
Sven Erik Jørgensen

6.1 THE ROLE OF SEDIMENTATION IN MODELLING AQUATIC ECOSYSTEMS

Suspended matter is removed from water by sedimentation or settling. Precipitates, coagulates and phytoplankton might settle. Settling rates depend on the difference in density between the suspended matter and the water, the size and shape of the matter, the viscosity of water, the turbulence and velocity of the flow field. In addition, the physiological state of the phytoplankton cells also plays an important role.

Several physiological factors controlling the sinking rate have been suggested:

1. physiological state: cells in the stationary phase sink two to four times faster than cells in the exponential phase of growth (Smayda and Boleyn, 1965, 1966a + b)
2. nutrient depletion (Scavia et al., 1976)
3. light (Steele, 1974)
4. viscosity of water: is a function of temperature in fresh water (Scavia, 1980 and Terra Tech, 1980)
5. turbulence.

The suspended matter is removed to the sediment, where it may undergo several transformations and processes, which are covered in chapter 17. By turbulence and flow field the suspended matter might be resuspended and thereby be transported from the sediment back to the water column.

These processes are of great importance in ecological modelling, as a significant mass transfer from water to sediment (in many cases) can be explained by these processes. Consequently, many models have to include a proper description of sedimentation.

In many cases it is not too difficult to describe the sedimentation itself, but it is far more difficult to account for the influence of the hydrological flow pattern. Therefore theoretical approaches based upon physical considerations should almost always be accompanied by measurements of sedimentation rates by use of sediment traps and by the determination of net sedimentation rates by direct or indirect measurements of the increase of sediment per unit of time. This latter determination is often carried out by use of tracers, for instance by use of isotopes.

6.2 MODELS OF SEDIMENTATION

Removal by settling is most often described as a first order reaction:

$$\frac{dm}{dr} = -s * m \tag{6.1}$$

where m is the concentration of suspended matter and s is the rate of removal by sedimentation, s is the ratio between the settling rate, V_s, and the depth D:

$$s = \frac{V_s}{D} \tag{6.2}$$

Discrete settling

The settling of a discrete non-flocculating particle in a dilute suspension can be described by means of classical mechanics. Such a particle is not affected by the presence of other particles, and settling is therefore a function only of the properties of the fluid and the characteristics of the particles. As shown in Fig. 6.1 the particle is affected by three forces:

(1) Gravity, F_g; **(2)** the buoyant force, F_b and **(3)** the frictional force, F_f.

In accordance with **Newton's second law of motion**, the following equation can be set up:

$$m \frac{dv_s}{dt} = F_g - F_b - F_f \tag{6.3}$$

where v_s = the linear settling velocity of the particles, m = the mass of the particles and t = time.

Gravity is given by:

$$F_g = p * V * g \tag{6.4}$$

where p = the particle density, V = the particle volume and g = the acceleration due to gravity.

The buoyant force is:

$$F_b = p_e * V * g \qquad (6.5)$$

where p_e = the fluid density.

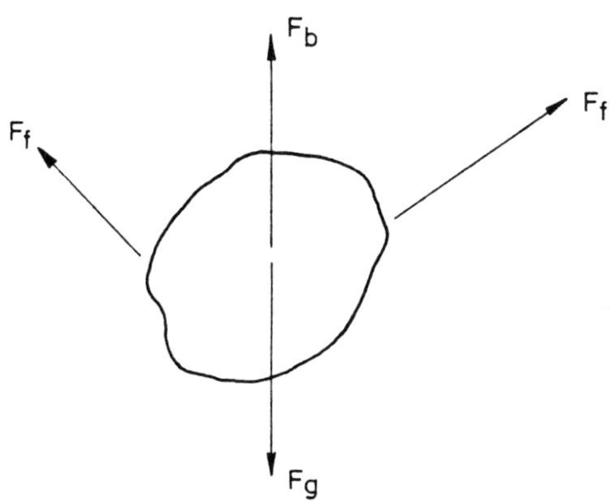

Fig. 6.1: The settling particle is affected by three forces: The gravity, F_g, the buoyant force, F_b and the frictional force, F_f.

The frictional force is a function of different particle parameters, such as roughness, size, shape and velocity of the particle, and of the density and viscosity of the fluid. It can be described by the following relationship:

$$F_f = \frac{C_d * A * p_e * v_s^2}{2} \qquad (6.6)$$

where C_d = Newton's dimensionless drag coefficient and A = the projected particle area in the direction of the flow. C_d varies with Reynolds number.

By substituting the equations (6.4), (6.5) and (6.6) in equation (6.3), an expression for the dynamic behaviour of the particles is obtained:

$$m * \frac{dv_s}{dt} = g(p - p_e)\,V - \frac{C_d * A * p_e * v_s^2}{2} \qquad (6.7)$$

After an initial transient period the acceleration becomes zero and the velocity is constant. This velocity can be obtained from equation (6.7):

$$v_s = \left(\frac{2g\,(p - p_e) * V}{C_d * p_e * A}\right)^{1/2} \qquad (6.8)$$

If the particles are spherical and the diameter is d, the V/A is equal to 2/3*d and equation (6.8) becomes:

$$v_s = \left(\frac{4g\,(p - p_e) * d}{3C_d * p_e}\right)^{1/2} \qquad (6.9)$$

Newton's drag coefficient C_d is, as mentioned, a function of **Reynolds number** and of the shape of the particle. The relationship between C_d and the Reynolds number for spheres and cylinders is given in Fig. 6.2.

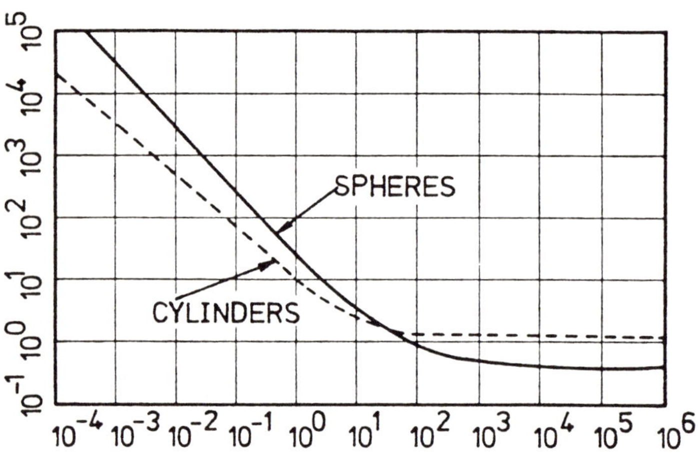

Fig. 6.2: Experimental variation of the drag coefficient with Reynolds number. After Fair et al. (1968).

When the Reynolds number is below 1, the relationship between C_d and Re can be approximated by $C_d = 24/Re$, where Re = Reynolds number defined as:

$$\frac{d * p_e * v_s}{\mu}, \quad \text{where } \mu = \text{the viscosity}$$

In this case (6.9) conforms with Stokes law:

$$v_s = \frac{g}{18\mu} (p - p_e) * d^2 \qquad (6.10)$$

From Fig. 6.2 it can be seen that C_d is approximately constant for turbulent flow in the range for Reynolds number between 1000 and 250,000. For this region the velocity v_s is given by:

$$v_s = 1.82 \left(\frac{(p - p_e) * d * g}{p_e} \right)^{1/2} \quad \text{(for spheres only)} \qquad (6.11)$$

Stokes law can be modified to account for non-spherical phytoplankton cells by use of an "equivalent radius" and shape factor in the formulation (Scavia, 1980):

$$V_s = \frac{2}{9} \frac{g R^2 (p_p - p_w)}{\P F_s} \qquad (6.12)$$

where V_s = settling velocity, length/time
g = acceleration of gravity, length/time^2
R = equivalent radius (based on a sphere of equivalent volume), length
p_p = density of the cell, mass/length^3
p_w = water density, mass/length^3
$\P$ = kinematic viscosity
F_s = shape factor

The shape factor has a value 1.0 and accounts for all factors, reducing the settling velocity. In a model of Lake Ontario Svavia (1980) used a shape correction factor of 1.3 for small diatoms, 2.0 for large diatoms and 1.0 for all other algal groups.

In practice, very few models use Stoke´s law as a model formulation (Scavia et al., 1976; Scavia, 1980; Park et al., 1980). Most models *lump many species into a few algal groups,* so representative values of the cell radius, shape factor, and cell density are difficult to define, making this level of detail unnecessary. Since the shape factor is really a calibration parameter, it is more direct or simpler to use the settling velocity as a calibration parameter. Also, Stoke´s law does *not* take into account turbulence and flow velocities which tend to keep algae in suspension or resuspend settled algae.

An additional factor which further complicates settling is the physiological state of phytoplankton cells mentioned above.

Settling rates are also partly dependent on the structure of the model. For example, one-dimensional layered lake models typically use settling velocities which are an order of magnitude lower than measured values or values used in two - or three-dimensional models which simulate hydrodynamic processes (Scavia and Bennett, 1980). This is probably because one-dimensional models do not adequately represent vertical transport process such as upwelling or entrainment of phytoplankton in large-scale circulations which effectively reduce the net settling rates (Scavia and Bennett, 1980).

Alternative and additional expressions for (6.1) and (6.2) may be used for instance:

1)

$$s * m = V_s * \frac{\partial Phyt}{\partial z} \tag{6.13}$$

where Phyt represents phytoplankton concentration, and z the depth.

2) Nyholm (1978) modelled several lakes by use of the same basic model:

$$V_S = \min \begin{cases} 0.2 \text{ m day}^{-1} \\ 0.05 * z \text{ m day}^{-1} \end{cases}$$

3) The settling rate is dependent on the temperature. Various expressions have been suggested to account for this relationship:

$$V_{S,T} = V_{S,T_r} * \sqrt{(T/T_r)} \tag{6.14}$$

where $V_{S,T}$ is the settling rate at the absolute temperature T, and V_{S,T_r} is the settling rate at the absoulte reference temperture T_r.

4) Tetra Tech (1980) uses:

$$f_s(T) = \frac{157.5}{0.069T^2 - 5.3T + 177.6} \tag{6.15}$$

where T = temperature in °C and $f_s(T)$ is a temperature adjustment function.

5) Scavia and Park (1976):

Scavia et al. (1976) have also expanded the settling rate formulation to account for variation in settling velocities due to the physiological conditions of the phytoplankton cells. The basic assumption is that the cells are healthiest and the settling rates smallest when neither light nor nutrients are limiting growth. The settling rates are therefore expressed as a function of the growth limitation factor f(L,P,N,C,Si). Potential formulations include (Scavia et al., 1976; Scavia, 1980):

$$s = \frac{V_{smax}(T_{ref})}{D} f_s(T) \left(\frac{K_{set1}}{f(L,P,N,C,Si) + K_{set2}} \right) \tag{6.16}$$

or

$$s = \frac{V_{smax}(T_{ref})}{D} f_s(T) \left[1 - K_{set2} f(L,P,N,C,Si) \right] \tag{6.17}$$

where $V_{smax}(T_{ref})$ = maximum settling velocity at reference temperature T_{ref} under poor physiological conditions, length/time

K_{set1}, K_{set2} = constants of the settling formulations

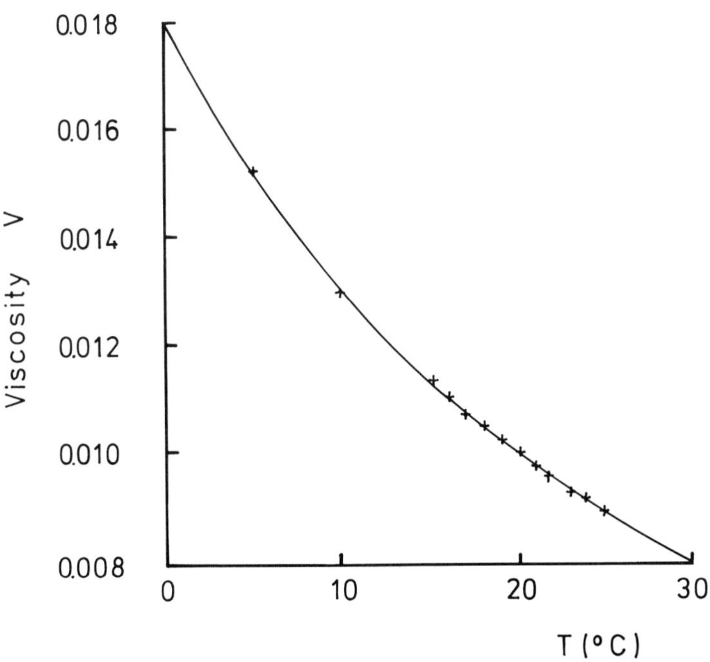

Fig. 6.3: Viscosity, v, plotted versus temperature. A regression analysis will show the following relationship: $v = 0.178 / (1 + 0.0337 * T + 0.00022 T^2)$

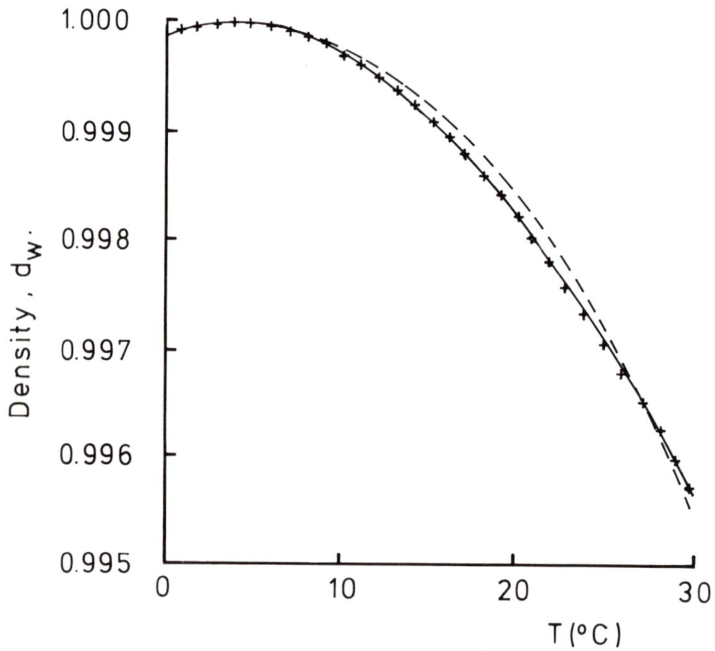

Fig. 6.4: Density of water, d_w, plotted versus temperature. A regression analysis will show the following relationship: $d_w = 0.999879 + 6.02602 * 10^{-5} * T^3$
$= 7.99470 * 10^{-6} * T^2 + 4.36926 * 10^{-8} * T^3$

6) Straskraba and Grauck (1985) suggest another method to consider the influence of temperature on the settling rate. They use the known relations between viscosity and density of water on the one side and the temperature on the other side; see Figs. 6.3 and 6.4. These relations have been used in the construction of Fig. 6.5, knowing that the sedimentation rate, U, is proportional to $d_a - d_w / v_w$. The larger d_a, the more strongly pronounced is the temperature dependence. d_a is the density of the particles, d_w the density of water and v_w is the viscosity of water.

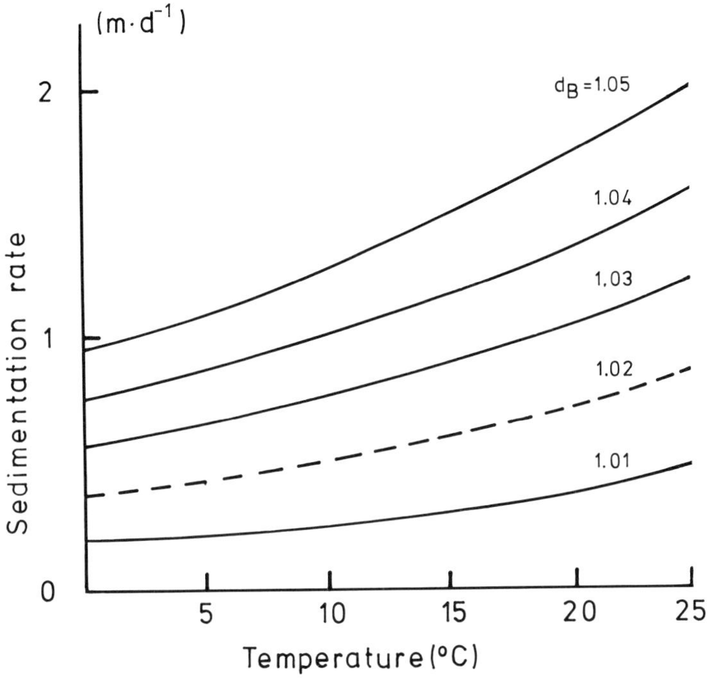

Fig. 6.5: Temperature dependence of the sedimentation rate of phytoplankton: $\mu = 0.333^*(d_B-d_W)/v_W$, where d_B is the density of phytoplankton, d_W is the density of water and v_W the viscosity of water. The dependence is shown for d_B = 1.01, 1.02, 1.03, 1.04 and 1.05.

In the case of higher concentrations the cells affect one another. The sedimentation rate for phytoplankton concentrations of about 300 μg/l chlorophyll was determined to be 60% of that for free sedimentation. Yet an increase in sedimentation is caused by coagulation of particles (Snodgrass and O'Melia, 1975).

7) A simple method to include the influence of physiological conditions on the settling rate would be to let the settling velocity at time t be related to the phytoplankton concentration at time $t-t_1$, as it is observed that the most pronounced settling takes place a few days (3-10 days) after the maximum phytoplankton concentration was recorded. This was attempted by Jørgensen (unpublished) and gave some minor improvements in the model results.

6.3 PARAMETER ESTIMATION

The estimation of the settling velocity as a parameter in ecological models may be crucial, as it determines the removal rate for a considered component, whether the component is suspended matter or phytoplankton. The sensitivity of this parameter to the phytoplankton concentration in a eutrophication model has been determined to be about -1.0, which means that if the parameter is increased 1% the phytoplankton concentration will decrease 1% (Jørgensen et al., 1978).

Settling velocity may be determined in three ways:

1. Values from previous models of the literature can be used to give a first estimation of the parameter. Tables 6.1 and 6.2 summarize values found in the literature. As can be seen, these values are indicated as ranges, and it is therefore necessary to calibrate the parameters by use of measured values for the stated variables.

2. Values from calculations based upon knowledge of the size can be used as first estimations. Due to the influence of the many factors, mentioned in section 6.1 and 6.2, a calibration is also required in this case. This method is hardly applicable for phytoplankton, because of their ability to change the specific gravity, but may be useful for other particles.

3. Measurements in situ by use of sedimentation traps. It is possible to determine the distribution of the material in inorganic and organic matter and partly also in phytoplankton and detritus, by analysis of chlorophyll (fresh material) phosphorus, nitrogen and ash. Measurements of phytoplankton settling velocities in the laboratory will hardly give a reliable value, as they do not take into account the various in situ factors already mentioned in section 6.1 and 6.2.

Table 6.1
Phytoplankton settling velocities

Algal type	Settling velocity (m/day)	References
Total phytoplankton	0.05 - 0.5	Chen & Orlob (1975); Tetra Tech (1976); Chen (1970); Chen & Wells (1975, 1976)
	0.05 - 0.2	O'Connor et al. (1975, 1981); Thomann et al. (1974, 1975); Di Toro & Matystik (1980); Di Toro & Connolly (1980); Thomann & Fitzpatrick (1982)
	0.02 - 0.05	Canale et al. (1976)
	0.4	Lombardo (1972)
	0.03 - 0.05	Scavia (1980)
	0.05	Bierman et al. (1980)
	0.2 - 0.25	Youngberg (1977)
	0.04 - 0.6 *	Jørgensen (1976); Jørgensen et al. (1978, 1979)
	0.01 - 4.0 *	Baca & Arnett (1976)
	0.1 - 2.0 *	Chen & Orlob (1975); Smith (1978)
	0.15 - 2.0 *	Duke & Masch (1973); Roesner et al. (1981)
	0.1 - 0.2 *	Brandes (1976)
Diatoms	0.05 - 0.4	Bierman (1976); Bierman et al. (1980)
	0.1 - 0.2	Thomann et al. (1979); Di Toro & Connolly (1980)
	0.1 - 0.25	Tetra Tech (1980); Porcella et al. (1983)
	0.03 - 0.05	Canale et al. (1976)
	0.3 - 0.5	Smayda & Boleyn (1965)
	2.5	Lehman et al. (1975)
	0.02 - 14.7 *	Collins & Wlosinski (1983)
Green Algae	0.05 - 0.19	Jørgensen et al. (1978)
	0.05 - 0.4	Bierman (1976); Bierman et al. (1980)
	0.02	Canale et al. (1976)
	0.8	Lehman et al. (1975)
	0.1 - 0.25	Tetra Tech (1980); Porcella et al. (1983)
	0.3	DePinto et al. (!976)
	0.08 - 0.18 *	Collins & Wlosinski (1983)
	0.27 - 0.89 *	Jørgensen et al. (1979)
Blue-green Algae	0.05 - 0.15	Bierman (1976); Bierman et al. (1980)
	0.08	Canale et al. (1976)
	0.2	Lehman et al. (1975)
	0.1	DePinto et al. (!976)
	0.08 - 0.2	Tetra Tech (1980); Porcella et al. (1983)
	0.10 - 0.11 **	Collins & Wlosinski (1983)
Flagellates	0.5	Lehman et al. (1975)
	0.05	Bierman et al. (1980)
	0.09 - 0.2	Tetra Tech (1980); Porcella et al. (1983)
	0.07 - 0.39 **	Collins & Wlosinski (1983)
Dinoflagellates	8.0	O'Connor et al. (1981)
	2.8 - 6.0 **	Collins & Wlosinski (1983)
Asterionella formosa	0.25 - 0.76 **	Jørgensen et al. (1979)
Chaetoceros lauderi	0.46 - 1.56 **	Jørgensen et al. (1979)

Table 6.1 (continued)

Algal type	Settling velocity (m/day)	References
Chrysophytes	0.5	Lehman et al. (1975)
Coccolithophores	0.25 - 13.6	Collins & Wlosinski (1983)
	0.3 - 1.5 **	Jørgensen et al. (1979)
Coscinodiscus lineatus	1.9 - 6.8 **	Jørgensen et al. (1979)
Cyclotella meneghimiana	0.08 - 0.31 **	Jørgensen et al. (1979)
Ditylum brightwellii	0.5 - 3.1 **	Jørgensen et al. (1979)
Melosira agassizii	0.67 - 1.87 **	Jørgensen et al. (1979)
Nitzschia seriata	0.26 - 0.50 **	Jørgensen et al. (1979)
Rhizosolenia robusta	1.1 - 4.7 **	Jørgensen et al. (1979)
Rhizosolenia setigera	0.22 - 1.94 **	Jørgensen et al. (1979)
Scenedesmus quadracauda	0.04 - 0.89 **	Jørgensen et al. (1979)
Skeletonema costatum	0.31 - 1.35 **	Jørgensen et al. (1979)
Tabellaria flocculosa	0.22 - 1.11 **	Jørgensen et al. (1979)
Thalassiosira Nana	0.10 -0.28 **	Jørgensen et al. (1979)
- Pseudonana	0.15 - 0.85 **	Jørgensen et al. (1979)
- Rotula	0.39 - 17.1	Jørgensen et al. (1979)

* Model documentation values
** Literature values
Other values: used in models.

Table 6.2
Detritus, settling rate

Item	Settling velocity (m/day)	References
Detritus	0.1 - 2.0	Jørgensen et al. (1979)
Nitrogen Detritus	0.05 - 0.1	Jørgensen et al. (1979)
Fecal Pellets (fish)	23 - 666	Jørgensen et al. (1979)

6.4 APPLICATION IN ECOLOGICAL MODELLING

Sedimentation is included in almost all environmental models of aquatic systems. It plays a role in BOD/DO models, as a substantial part of the BOD is removed by settling of suspended matter. Sedimentation is one of the most important processes in a eutrophication model as a significant amount of the phytoplankton is removed from the water column by settling. Models of toxic substances in aquatic ecosystems also consider this settling, as well as the significant amount of the toxic substances which may be suspended or adsorbed on the suspended matter.

6.5 CONCLUSIONS AND FURTHER RESEARCH NEEDS

Most experience in ecological modelling seems to indicate that the inclusion of hydrological and physiological processes and their influence on the settling velocity is very difficult and hardly necessary in most cases. It is crucial only to include these processes where they have particular importance for the problem (for instance phytoplankton concentration in a eutrophication model). A simple first-order reaction scheme for the sedimentation perhaps with addition of the temperature influence is a sufficient description in most cases. It will imply, that the influence of the above mentioned processes is subject to the calibration phase. As long as changes in the ecological structure are not included in the model, this relatively simple approach seems to give a complexity of the sedimentation submodel, which is in balance with the complexity of most other submodels.

This does not mean that we should not search for better descriptions. In cases where changes in ecological structures are significant, a better description of sedimentation might improve the model results considerably. Furthermore, our model descriptions are steadily improving and improvements of this particular submodel is also needed. Further research into the influence of physiological processes on sedimentation of phytoplankton seems needed, as it is known that the settling velocity of phytoplankton is highly time dependent. As mentioned in section 6.2 a timelag between settling and concentration of phytoplankton in eutrophication models produces some differences in the results.

REFERENCES

Baca, R.G. and R.C. Arnett, 1976. A Limnological Model for Eutrophic Lakes and Impoundments. Battelle, Inc., Pacific Northwest Laboratories, Richland, Washington.

Bierman, V.J., Jr., 1976. Mathematical Model of the Selective Enhancement of Blue-Green Algae by Nutrient Enrichment. In: Modelling Biochemical Processes in Aquatic Ecosystems. R.P. Canale (ed.). Ann Arbor Science Publishers, Ann Arbor, Michigan. pp. 1-31.

Bierman, V.J., Jr., D.M. Dolan, E.F. Stoermer, J.E. Gannon and V.E. Smith, 1980. The Development and Calibration of a Multi-Class Phytoplankton Model for Saginaw Bay, Lake Huron. Great Lakes Environmental Planning Study. Contribution No. 33. Great Lakes Basin Commission, Ann Arbor, Michigan.

Brandes, R.J., 1976. An Aquatic Ecologic Model for Texas Bays and Estuaries. Water Resources Engineers, Inc., Austin, Texas. For the Texas Water Development Board, Austin, Texas.

Canale, R.P., L.M. Depalma and A.H. Vogel, 1976. A Plankton-Based Food Web Model for Lake Michigan. In: Modelling Biochemical Processes in Aquatic Ecosystems. R.P. Canale (ed.). Ann Arbor Science Publishers, Ann Arbor, Michigan. pp. 33-74.

Chen, C.W., 1970. Concepts and Utilities of Ecological Model. ASCE Journal of the Sanitary Engineering Division, Vol. 96, No. SA5.

Chen, C.W. and G.T. Orlob, 1975. Ecologic Simulation for Aquatic Environments. Systems Analysis and Simulation in Ecology, Vol. 3. B.C. Patten (ed.). Academic Press, New York. pp. 476-588.

Chen, C.W. and J.T. Wells, Jr., 1975. Boise River Water Quality-Ecological Model for Urban Planning Study. Tetra Tech, Inc., Lafayette, California. For U.S. Army Engineering District, Walla Walla, Wash., Idaho Water Resources Board, and Idaho Dept. of Environmental and Community Services.

Chen, C.W. and J.T. Wells, Jr., 1976. Boise River Modelling. In: Modelling Biochemical Processes in Aquatic Ecosystems. R.P. Canale (ed.). Ann Arbor Science Publishers, Ann Arbor, Michigan. pp. 171-204.

DePinto, J.V., V.J. Bierman, Jr. and F.H. Verhoff, 1976. Seasonal Phytoplankton Succession as a Function of Phosphorus and Nitrogen Levels. In: Modelling Biochemical Processes in Aquatic Ecosystems. R.P. Canale (ed.). Ann Arbor Science Publishers, Ann Arbor, Michigan. pp. 141-169.

Di Toro, D.M. and J.F. Connolly, 1980. Mathematical Models of Water Quality in Large Lakes. Part II: Lake Erie. U.S. Environmental Protection Agency, Ecological Research Series. EPA-600/3-80-065.

Di Toro, D.M. and W.F. Matystik, Jr., 1980. Mathematical Models of Water Quality in Large Lakes. Part I: Lake Huron and Saginaw Bay. U.S. Environmental Protection Agency, Ecological Research Series. EPA-600/3-80-056.

Duke, J.H., Jr. and F.D. Masch, 1973. Computer Program Documentation for the Stream Quality Model DOSAG3, Vol. I. Water Resources Engineers, Inc., Austin Texas. For U.S. Environmental Protection Agency, Systems Develoopment Branch, Washington, D.C.

Fair, G.M., J.C. Geyer and D.A. Okun, 1968. Water Purification and Waste Water Treatment and Disposal. Water and Waste Water Engineering, vol. 2. John Wiley & Sons, Inc., New York.

Jørgensen, S.E., 1976. A Eutrophication Model for a Lake. Ecol. Modelling 2: 147-165.
Jørgensen, S.E., 1979. Lake Management. Pergamon Press, Oxford.
Jørgensen, S.E., H. Mejer and M. Friis, 1978. Examination of a Lake Model. Ecol. Modelling, 4: 253-278.
Jørgensen, S.E., H.F. Mejer, M. Friis, L.A. Jørgensen and J. Hendriksen (eds.), 1979. Handbook of Environmental Data and Ecological Parameters. ISEM, Copenhagen.

Lehman, J.T., D.B. Botkin and G.E. Likens, 1975. The Assumptions and Rationales of a Computer Model of Phytoplankton Population Dynamics. Limnol. and Oceanogr., 20(3): 343-364.
Lombardo, P.S., 1972. Mathematical Model of Water Quality in Rivers and Impoundments. Hydrocomp., Inc., Palo Alto, California.

Nyholm, N., 1978. A Simulation Model for Phytoplankton Growth and Nutrient Cycling in Eutrophic, Shallow Lakes. Ecol. Modelling, 4: 279-310.

O'Connor, D.J., D.M. Di Toro and R.V. Thomann, 1975. Phytoplankton Models and Eutrophication Problems. In: Ecological Modelling in a Resource Management Framework. C.S. Russell (ed.). Resources for the Future, Inc., Washington, D.C.
O'Connor, D.J., J.L. Mancini and J.R. Guerriero, 1981. Evaluation of Factors Influencing the Temporal Variation of Dissolved Oxygen in the New York Bight, PHASE II. Manhattan College, Bronx, New York.

Porcella, D.B., T.M. Grieb, G.L. Bowie, T.C. Ginn and M.W. Lorenzen, 1983. Assessment Methodology for New Cooling Lakes, Vol. 1: Methodology to Assess Multiple Uses for New Cooling Lakes. Tetra Tech, Inc., Lafayette, California. For Electric Power Research Institute. Report EPRI EA-2059.

Roesner, L.A., P.R. Giguere and D.E. Evenson, 1981. User's Manual for the Stream Water Quality Model QUAL-II. U.S. Environmental Protection Agency, Athens, Georgia. EPA-600/9-81-015.

Scavia, D., 1980. An Ecological Model of Lake Ontario. Ecol. Modelling, 8: 49-78.
Scavia, D. and R.A. Park, 1976. Documentation of Selected Constructs and Parameter Values in the Aquatic Model CLEANER. Ecol. Modelling, 2: 33-58.
Scavia, D. and J.R. Bennett, 1980. Spring Transition Period in Lake Ontario - A Numerical Study of the Causes of the Large Biological and Chemical Gradients. Can. J. Fish. Aquat. Sci., 37: 823-833.
Smayda, T.J. and B.J. Boleyn, 1965. Experimental Observations on the Flotation of Marine Diatoms. I. *Thalassiosira* cf. *nana, Thalassiosira rotula* and *Nitzschia seriata.* Limnol. Oceanogr., 10: 499-509.
Smayda, T.J. and B.J. Boleyn, 1966a. Experimental Observations on the Flotation of Marine Diatoms. II. *Skeletonema costatum* and *Rhizosolenia setigera.* Limnol. Oceanogr., 11: 18-34.
Smayda, T.J. and B.J. Boleyn, 1966b. Experimental Observations on the Flotation of Marine Diatoms. III. *Bacteriastrum hyalinum* and *Chaetoceros lauderi.* Limnol. Oceanogr., 11: 34-43.
Straskraba, M. and A.H. Gnauck, 1985. Freshwater Ecosystems, Modelling and Simulation. In: Developments in Environmental Modelling. Elsevier - Amsterdam, Oxford, New York, Tokyo.

Tetra Tech, Inc., 1976. Documentation of Water Quality Models for the Helms Pumped Storage Project. Tetra Tech, Inc., Lafayette, California. For

Pacific Gas and Electric Company.

Tetra Tech, Inc., 1980. Methodology for Evaluation of Multiple Power Plant Cooling System Effects, Volume V. Methodology Application to Prototype - Cayuga Lake. Tetra Tech, Inc., Lafayette, California. For Electric Power Research Institute. Report EPRI EA-1111.

Thomann, R.V., D.M. Di Toro and D.J. O'Connor, 1974. Preliminary Model Potomac Estuary Phytoplankton. ASCE, J. Environ. Engr. Div., Vol. 100, EE1: 699-715.

Thomann, R.V., D.M. Di Toro, R.P. Winfield and D.J. O'Connor, 1975. Mathematical Modelling of Phytoplankton in Lake Ontario, Part 1. Model Environment and Verification. Manhattan College, Bronx, New York. For U.S. Environmental Protection Agency, Corvallis, Oregon. EPA-600/3-75-005.

Thomann, R.V., J. Segna and R. Winfield, 1979. Verification Analysis Lake Ontario and Rochester Embayment Three-Dimensional Eutrophication Models. Manhattan College, Bronx, New York. For U.S. Environmental Protection Agency, Office of Research and Development, Duluth, Minnesota.

Thomann, R.V. and J.J. Fitzpatrick, 1982. Calibration and Verification of a Mathematical Model of the Eutrophication of the Potomac Estuary. For Department of Environmental Services, Government of the District of columbia, Washington, D.C.

Yongberg, B.A., 1977. Application of the Aquatic Model CLEANER to Stratified Reservoir System. Report #1. Center for Ecological Modelling, Rensselaer Polytechnic Institute, Troy, New York.

CHAPTER 7

COAGULATION

by
H.H. Hahn and S.E. Jørgensen

7.1. INTRODUCTION

Natural waters often contain colloids, which are too small for gravitational settling. The aggregation of these particles into large more readily settleable aggregates is termed coagulation.

Four mechanisms can be distinguished (Hahn and Stumm, 1968), leading to destabilization of colloids or fine suspension in saline water:

1. Neutralization of the negative charge by specific adsorption of positively charged species,
2. compression of double layer as a result of the ionic strength of the solution,
3. inter-particle bridging by adsorbed material,
4. enmeshment of clay and hydroxide particles.

It is suggested that in natural waters trace metals are present as soluble organic complexes in generally greater quantities than those present as inorganic ions. These complexes often interact with humic acids and coagulation occurs.

The increased coagulation which occurs in estuaries may lead to higher oxygen consumption in the bottom sediment, which could cause resolution of phosphates, iron and manganese oxides or even formation of sulphides. In such cases it is of great importance to include the process of coagulation in a eutrophication or oxygen model.

Two classes of collodial systems are recognized: hydrophobic and hydrophilic colloids. In colloids of the second kind there is a strong affinity between the particles and water, while in colloids of the first kind this affinity is absent. There exists a continuous transition between hydrophobic and hydrophilic colloids.

The aggregation of colloids is of great importance in the transport and distribution of matter in natural waters. Although dissolved substances tend to be distributed by convective mass transfer, the distribution of suspended matter is also influenced by the forces of gravity. Whether a particle will settle, depends on its density, its size and the water movement.

Obviously, colloidal reactions can profoundly influence the distribution in space and time of dissolved and suspended constituents. The various transportation processes of colloids and their effects on the ecology of natural water systems therefore need a description for answering the following typical questions:

1. What are the sizes of agglomerates found within the time of interest under the given conditions?
2. What is the rate of coagulation of various naturally occurring colloids at the given conditions (temperature, pH, ionic strength, other solutes, etc.)?
3. How far are discharged colloidal pollutants transported before they are removed by settling? Is it possible to find a variation of composition in sediments as it depends on the distance from the point of discharge?
4. What is the rate of decrease in the turbidity in a given water at given velocity or detention times?

7.2. MODELS OF THE COAGULATION PROCESS

7.2.1. Chemistry of Aggregation

The stability of suspended particles.

Many particles encountered in natural waters are so small that the effect of surface forces and inter-particle forces becomes apparent. If such particles carry a surface charge, for instance, then the forces resulting from the repulsion of charges of equal sign (on the particle surfaces) prevent a closer approach of these particles to each other. Such surface charges on particles encountered in natural systems might stem from:

- lattice imperfections
- adsorbed ions
- dissolution phenomena
- surface chemical reactions in general

Note: Most particulate matter in natural waters appears to have a negative charge at neutral pH-values (Hahn, 1967).

Small particles (of magnitude described above) will also exhibit forces of attraction upon a closer approach (*London - van - der - Waals*). Through non-compensated charge or potential effects, i.e. steric inhomogeneities, matter presents itself as disordered (*dipoles*). Ordering of these dipoles can lead to overall organization and thus to dipole effects manifested to the outside. Dipole elements in small particles might become ordered by the approach of such particles to each other. This means that decreasing distances between two particles will lead to increased order and consequently increased attraction.

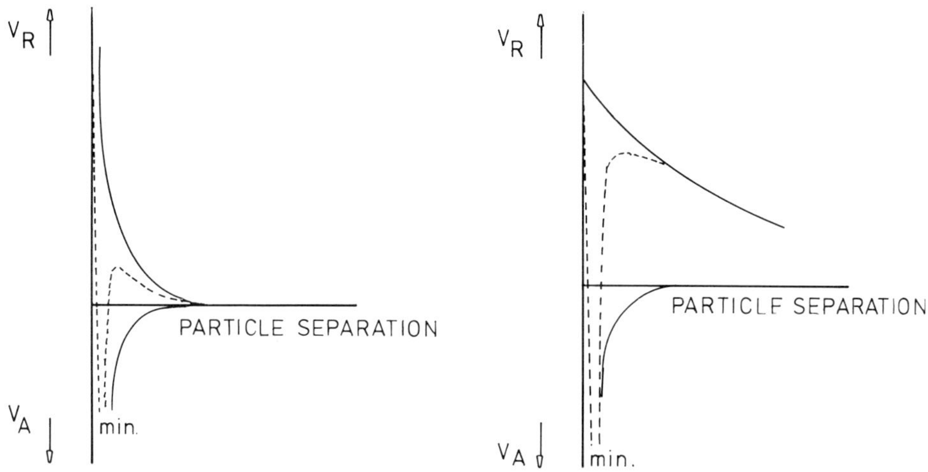

Fig. 7.1: Repulsion and attraction forces of two approaches particles (left: low salt concentration - right: higher salt concentration).

Fig. 7.1 (van Olpen, 1963) shows schematically the change in repulsive and attraction forces between two approaching particles (spheres). It is seen, that both energy components increase with decreasing particle distance. Since both energy terms change in a different way with particle distance the resultant or net energy also changes with particle distance.

Upon particle approach repulsive energy increases faster than the attractive energy. The result is an increasing energy barrier which disappears at very close particle distances.

The Destabilization of Suspended Particles

Double Layer Compaction

The "double layer" surrounding suspended particles results from the preferred adsorption of ions from the surrounding solute with a charge opposite to the one of the surface. At the same time the ions carrying a charge of the same sign as the particle surface will be found in considerably lower concentration the closer the approach to the solid surface. In this concept the ions with a charge opposite to the one of the solid surface are referred to as "*counter ions* ". The increase in concentration in *so-called counter-ions* leads to a much more intensive compensation of the surface charge, i.e. the (repulsive) effect of the surface charge is felt only at much shorter distances.

The effect of differently charged counter-ions can be expressed by changes in the approximate "*thickness* " of the so-called double layer (i.e. the two layers are the surface charge and the compensating counter-ion charge). At a counter-ion concentration of 0.01 mole/litre the thickness for a monovalent ion is 1000 A and for a divalent ion 500 A only. At a counter-ion concentration of 100 mole/litre the thickness for a monovalent ion is 10 A and for a divalent ion 5 A (van Olphen). This is quantified in the Schultze Hardy rule.

With increasing salt concentrations (i.e. counter-ion concentration) the diffuse part of the double layer is compacted. It is also observed that there is a finite thickness which cannot be reduced further. Even large concentrations of coagulating salts will not lead to a reversal of charge or potential. (Such charge or potential reversals are described later).

For practical purposes one can conclude that dosing problems do not exist if a certain minimum concentration is attained (the concentration is independent of the particle concentration), that overdosing does not lead to reduced efficiency, and that the coagula formed are of a relatively dense nature.

Adsorption coagulation

As described above there are surface chemical reactions causing dissolved species to be adsorbed to a surface or removed from it. Similarly charged species may be attracted to a surface for physical reasons. Such chemisorption (case 1) or physical adsorption (case 2) may cause a direct change in the surface charge respectively surface potential. These effects are to be distinguished from charge **compensation** as discussed in the previous paragraph. Examples of such changes in surface charge, as opposed to compensation of surface charge, can be represented as follows.

It is possible to describe the relationship between ions in solution and those attached to a surface in a quantitative way by an **adsorption isotherm**. (An adsorption isotherm describes the surface coverage as a function of the equilibrium concentration in the solution phase for a given and constant temperature.) Vice versa, it is also possible to clarify whether adsorption occurs by testing whether such adsorption isotherms do explain the observed behavior.

As the mechanism of adsorption states, the amount of material adsorbed increases with increasing amounts added (in an exponential way as illustrated for instance by the Langmuir isotherm). Thus, it is possible to adsorb such amounts of oppositely charged ions that the original surface charge of the particle is completely balanced, i.e. neutralized, and the particle appears to have a zero charge. Furthermore one must note that the surface concentration available, i.e. the surface of the particles in suspension, directly affects the amount of material adsorbed. At constant salt additions the amount of surface coverage (or change in charge) will decrease with increasing particle concentration (more precisely with increasing surface concentration).

At increased coagulant dosages, i.e. high equilibrium concentration of the oppositely charge ion, the extent of adsorption may be such that the original surface charge is not only eliminated but reversed. An originally negatively charged particle can thus become positively charged. This causes high stability, again. Such increase in stability with very high coagulant dosages of adsorbing ions resulting from charge reversal is referred to as **restabilization**.

From a practical point of view the following conclusions are significant

- **the necessary dosage depends on the original particle charge and the particle concentration**
- **the pH value is of great significance to the overall system**
- **constant dosage leads necessarily to fluctuating efficiency when applied under real world conditions**
- **high dosages may lead to restabilization**
- **the resulting coagula are dense (similar to the products of the process "double layer compaction").**

Flocculation

Flocculation is understood to be a process where aggregates of lower density are formed in a 3-dimensional way. Particles are not destabilized on an individual basis and do not collide individually as described in the concept of coagulation. In this case, the particles become incorporated into a three-dimensional network (the picture of a sponge may illustrate this).

Flocculation is accomplished be long-chain molecules or ions. Frequently organic polymers are used for flocculation. Along with the molecular weight, the charge of such molecules is of great significance in their use as flocculants. There are (see DVGW) **a.** cationic polymers **b.** anionic polymers and **c.** non-ionic polymers.

To explain the observations made with polymer induced aggregation a so-called **bridging-model** has been invoked. Long chain molecules adsorb (for reasons described above) with one end to one particle and with the other end to another one. Thus, there must always be free adsorption sites on the particle surface in order to allow bridge formation. At higher flocculant concentration surface coverage becomes so high that free adsorption sites for molecules extending from other particles no longer are available. This is shown schematically below.

There is also another mechanism that can be invoked for the explanation of aggregation phenomena observed with shorter chain molecules. This so-called **patch model** stipulates that the (in this case, oppositely) charged molecules will absorb onto the particle surface and change the surface charge in a similar way to that described above.

From a practical point of view the following may be said about the use of flocculants:

- **the amount of chemicals added is crucial to the success of the process**
- **the pH-value, controlling the surface charge and the adsorption process must be observed**
- **overdosing may cause restabilization**
- **the intensity of mixing in the phase of flocculant addition is to be controlled carefully**
- **the resulting products, i.e. the floccules, are usually less dense than the coagula.**

Particle Removal by Enclosure in Precipitation Products

Frequently metal ions (i.e. metal salts) are used in water and wastewater treatment. These metal ions will hydrolize. They will also react with the acid-base-system "water" and form hydroxo-complexes. These reactions are shown schematically below.

A solubility diagram of Fe(III) or Al that can be constructed from the respective equilibrium constant shows that at average dosages and the most frequently observed pH conditions Me-hydroxide formation will occur.

The way this equilibrium point is reached is significant. (It must be pointed out that in a strictly thermodynamic sense there will be no equilibrium between Me-hydroxide and the dissolved species in such short time as is provided in those technical systems that are discussed here.) If the domain of insolubility is reached for given total Me-concentration by increasing pH from originally acidic values to neutral values, then there will be kinetic intermediaries that resemble those positively charged *Me-hydroxo-complexes.* Analogous observations can be made when approaching the domaine of insolubility from very high pH values; which may lead to negatively charged intermediaries.

In all real systems such as waste water one will rarely be able to distinguish between coagulation/flocculation and precipitation. Both processes will occur simultaneously with a rate and to an extent that is determined by the composition of the dissolved phase.

The significance of the precipitation phenomena for waste water treatment processes lies in the fact that Me-hydroxide precipitates may and will incorporate suspended particles into the newly formed solid phase.

From practical points of view the following may be said about precipitation:

- **under most conditions hydroxide formation is favoured**
- **hydroxide formation means additional consumption of chemicals**
- **hydroxide formation may support suspensa removal, in particular such suspensa that might not respond to classical aggregation processes**
- **the resulting products do not show great strength**
- **the resulting products have a higher water content**

7.2.2. Physics of Aggregation

Particle Collision due to Brownian Motion and Velocity Gradients (Von Smoluchowski, 1917).

Suspended particles come into closer contact with each other, i.e. collide through Brownian motion (of significance for small particles only and therefore not considered to any extent in practical applications) and through laminar and turbulent velocity gradients (of importance for larger suspensa). Frequently the assumption holds that colloid destabilization proceeds at a faster rate than particle transport. Particle transport is therefore to be considered as the rate limiting step. The overall agglomeration of particles can be described kinetically by the frequency of particle collision. The original particle size distribution and the number of collisions per time unit leading to agglomeration must be known to describe the rate of the process (von Smoluchowski, 1917).

Laminar and turbulent flow of liquids causes velocity gradients in a liquid if it is a viscous medium. Particles suspended in this medium do not all move with the same velocity but rather move relative to each other and collide. One must distinguish here between laminar flow with linear velocity gradients and turbulent flow with energy dissipation in turbulent eddies (see paragraph 7.3.2). In the latter case the size of the suspended particle relative to the turbulent eddies (bearing in mind that there is a whole spectrum of eddy sizes) determines in which manner particles are transported to each other. Here only the relatively simple, but unrealistic situation, where linear velocity gradients prevail will be discussed. The results of the derived model, though, have so far also been useful in explaining numerous observations made in turbulent media.

The number of collisions per unit time, leading to agglomeration, is determined from the number of particles entering a sphere of the size of the coagulation diameter, i.e. the sum of the diameters of the two spheres under consideration, during a given period of time:

$$N = 4/3\ \pi\ (Rij)^3 * ai * aj * du/dz \qquad (7.1)$$

where
Rij = coagulation diameter
ai = radius of the central particle
aj = radius of the particle moving towards the central one
du/dz = velocity gradient

If one compares this expression for the collision frequency with that one obtained for collisions due to Brownian motion (not shown in this discussion), it can be seen that for large particles, even at lower velocity gradients, the latter ones are determining the extent of collisions. At room temperature, and other conditions that might be considered standard, the ratio of collision rates due to Brownian motion and due to linear shear flow are 1 if the particle diameter is 1 micrometer and the velocity gradient is 1 per second.

Non-laminar Flow Conditions

Camp and Stein (1943) defined a mean velocity gradient in the following form which makes the above formulation applicable to non-laminar flow conditions:

$$G \approx du/dz = \sqrt{P / (v^*\mu)} \qquad (7.2)$$

where
P = power input (torque times rotational speed)
v = reactor volume
μ = absolute viscosity

According to the authors this mean velocity gradient describes the situation in a real-world reactor with satisfactory precision. (A cursory dimensional analysis will show that the concept is dimensionally correct.)

Effects of Reaction Time

The rate law described above shows clearly a linear dependence between the growth of aggregate size consequent on the disappearance of particles (measured as reduction in particle number), and reaction time. This is seen in figure 7.2.

It has been suggested by Camp that the product:

ENGERGY INPUT * REACTION TIME (7.3)

is constant. Furthermore it has been recommended by Camp to set this (dimensionless) "Camp number" for practical systems at values of

$$G * T = 10{,}000 \text{ to } 100{,}000 \tag{7.4}$$

Numerous observations have confirmed that this order of magnitude is a reasonable one.

The following conclusions can be derived for the practical application of the process:

1. **Brownian motion is of relatively little significance** for real-world aggregation systems
2. **It is always necessary to supply energy (for instance through wind) in order to have a reasonable reaction rate**
3. **A rate law based on the concept of linear velocity gradients**, i.e. laminar flow conditions, gives insight into the rate controlling parameters, even though laminar flow conditions might never be encountered in real systems
4. **Rate law and observations have suggested that increased energy input up to a certain level increases the reaction rate**
5. **Energy input and reaction time are complimentary parameters**, i.e. there exists an optimal (dimensionless) number for the product of energy input and reaction time.

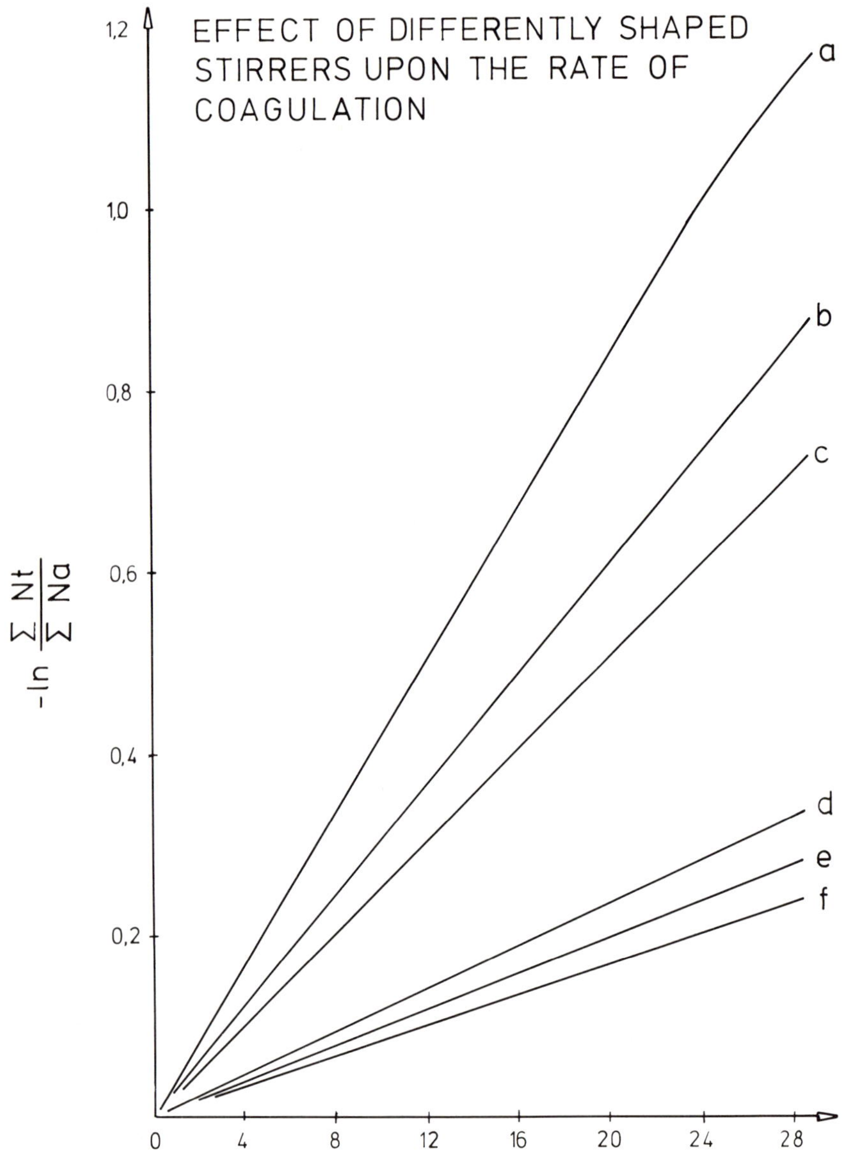

Fig. 7.2: Coagulation of kaolinite. a: anchor type stirrer, b: arid type stirrer, c: turbine type stirrer.

7.3. ASPECTS OF APPLICATION AND PARAMETER ESTIMATION

7.3.1. Simultaneous coagulation and precipitation

Coagulation is the formation of larger aggregates from solid substances, i.e. no change in phase. Precipitation, the formation of solid, non-dissolved species, implies a phase transition. In chemical terms coagulation and precipitation are distinctly different processes.

In waste water systems, however, particularly when metal ions (metal salts as coagulants) are used, both processes might occur simultaneously. To what extent these two different processes occur, and with what reaction rate they proceed depends upon the composition of the dissolved phase. This dissolved phase is very complex and changing in its nature in wastewater systems.

Figure 7.3 is a schematic representation of all processes that might occur in waste water when metal ions are added. It will be seen that in the presence of phosphate ions, for instance, precipitation will prevail. If, however, the hydroxide ions predominate, then metal hydroxides will be primarily formed. Under conditions of intermediate pH values, dissolved hydroxo-complexes will be formed leading to coagulation.

It is also important to note that all pathways indicated in Figure 7.3 will be followed simultaneously, but to a differing extent, and with possibly different reaction rates. And, furthermore, there are changes from one pathway to another, if the chemical situation allows this. Such switches are illustrated by the possibility, or even practical necessity, of completing precipitation by coagulation or flocculation.

7.3.2. The role of adsorption

Chemical addition so far discussed leads to a change of phase or to aggregation/floc formation. It does not allow per se the elimination or modification of substances that do not precipitate or coagulate. However, in practical applications it has been observed that constituents of the dissolved phase that do not belong to either of the above mentioned categories are also reduced in their concentration. The most plausible explanation is a **sorption process** favored by the very large specific surfaces formed, for instance in metal hydroxide precipitation and by use of clay minerals as precipitants (Overbeck, 1952, Jørgensen, 1971 and Culp et al., 1968).

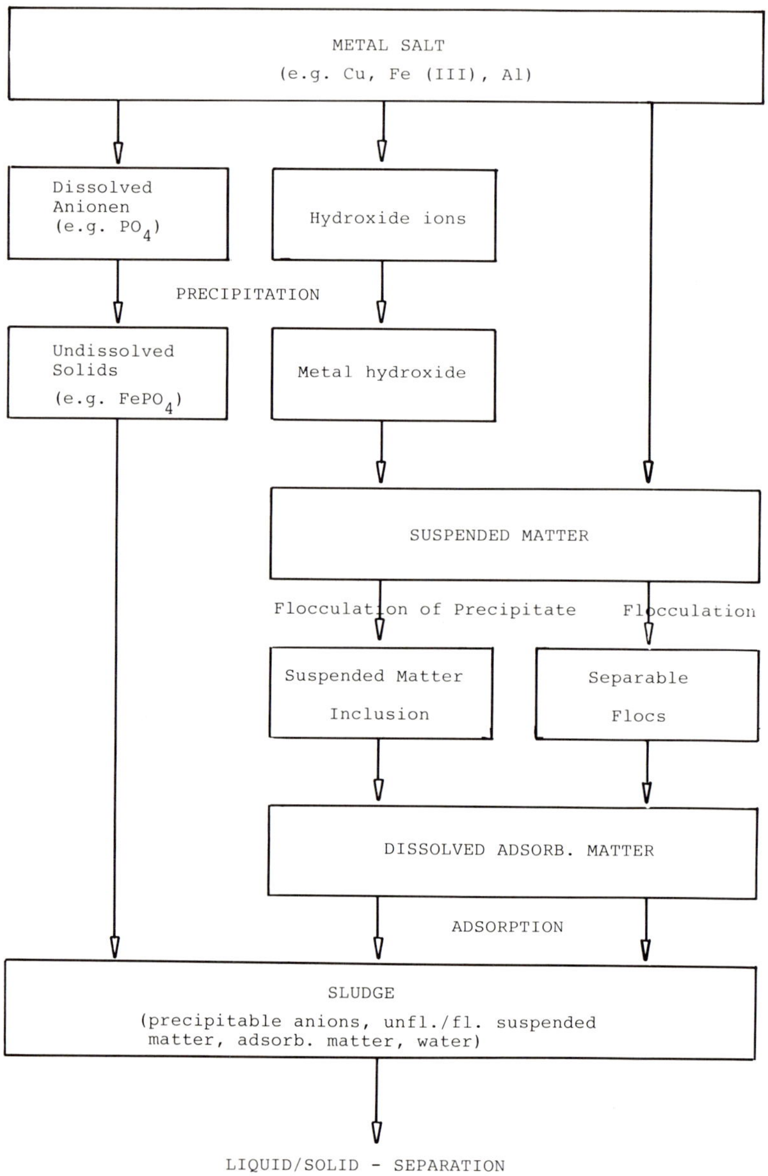

Fig. 7.3: Parallel reactions in chemical treatment by metal salts.

It is reasonable therefore to expect such adsorption processes to occur in situations where solid surfaces are formed or reformed and where surface active species are available (as is the case in waste water). An example of such elimination by adsorption on to coagulated solids is heavy metal removal in water systems.

7.3.3. Physical Aspects of Application

For illustration's sake a first order reaction rate process is assumed:

$$d(n) / dt = -k(n) \tag{7.5}$$

where n = particle or chemical agent concentration
k = reaction rate constant
d/dt = derivative with respect to reaction time

In plug flow (PF) this reaction will proceed to an extent that can be predicted from the integrated rate law.

$$N_t = N_0 e^{-kt} \tag{7.6}$$

where N_0 = particle or reagent concentration at time $t = 0$
N_t = concentration at time "t"
t = time

If this curve is drawn as function of the reaction time there is a logarithmic decrease in the final concentration with increased reaction time.

In completely mixed continuous flow (CMCF) the same reaction will proceed differently; in particular the reaction end-point, after specified reaction time has elapsed, will change in a different way from the above reactor. In other words: Increase of reaction time will lead to different results in terms of reaction end point in the CMCF case.

The effect can be seen from the following mass-balance equation:

INFLOW - CHANGE DUE TO REACTION - OUTFLOW = OVERALL CHANGE (7.7)

and for a steady state equilibrium with the assumption of no "overall change" the following relation holds:

$$Q * Ni - k\,Ne\,V - Q\,Ne = V\,dN/dt \tag{7.8}$$

where

Q = rate of flow through the reactor

Ni = influent concentration

Ne = effluent concentration (also corresponding to the concentration within the reactor)

Frequently for technical systems there is a steady state where $dNe/dt=0$ and consequently $Ne = Ni/(kT + 1)$. The effect of prolonged reaction or detention time upon the reaction end-point is significantly different from the logarithmic one for the PF seen above.

7.4. CONCLUSIONS AND FURTHER RESEARCH NEEDS

It is not possible today to build a workable model of coagulation processes in natural waters, and this has not been attempted in this chapter. It is, however, possible to quantify some of the factors that influence the coagulation process. It is necessary to use observations (see the questions in the Introduction) to state the role of the coagulation process in natural waters; to use these observations to give a description of the coagulation process and its interaction with other processes, (e.g. removal of pollutants by settling) and then describe in the total model, changes in the coagulation process by application of changes in the influencing factors.

It is, furthermore, **not possible** to give predictions on the particle diameters that result from the coagulation. Again, it is necessary to observe the particle diameters and/or the settling rate, and use the knowledge about those factors that might influence the particle diameters and thereby the settling rate, and to describe changes caused by other environmental factors.

This must not be interpreted as meaning the coagulation process is unimportant in environmental modelling. Settling of pollutants is often a very significant removal process for pollutants - not only for colloidal pollutants, but also the pollutants which are adsorbed on the coagulants.

Our experience in modelling the coagulation by this method, which might be termed "indirect modelling", is rather limited. More experience is required in the future, before this can lead to more profound conclusions.

We do have a certain knowledge about the chemical factors that determine the coagulation process in the laboratory. This knowledge should of course be used when modelling the coagulation process, but unfortunately experience of natural systems is very limited.

REFERENCES

Camp, T.R. and P.C. Stein, 1943. Velocity Gradients and Internal Work in Fluid Motion. J. Boston Soc. Civ. Eng. Vol. 30, pp. 219-238.

Hahn, H.H., 1967. Der Koagulationsprozess. In: Jahrbuch vom Wasser, vol. 33, pp. 172-191, Verlag Chemie Weinheim.

Hahn, H.H. and Stumm, W., 1968. The Role of Coagulation in Natural Waters. American Journal of Science, Vol. 268, pp. 354-368.

Jørgensen, S.E. & Johnsen I., 1989. "Principles of Environmental Science and Technology". Elsevier, Amsterdam, New York.

Van Olphen, H., 1963. Clay Colloid Chemistry. Interscience Publishers, New York/London.

Von Smoluchowski, M., 1917. Versuch einer mathematischen Theorie der Koagulationskinetik kolloidaler Lösungen. Z. Phys. Chem., Vol. 82, pp. 129-147.

CHAPTER 8

PRECIPITATION

by
S.E. Jørgensen and H.H. Hahn

8.1 INTRODUCTION

Precipitation in a strictly chemical sense is the transition of a substance from the dissolved state to the non-dissolved state upon the addition of other (dissolved) reagents that lead to the formation of precipitates.

In particular, if chemicals causing precipitation are added to water, other reactions may also take place such as for instance coagulation. Thus, in any application of the precipitation process it is very difficult to distinguish between these reactions. Only on the basis of a detailed knowledge of the composition of the (waste-) water matrix is it possible to describe the direction into which the process advances, i.e. which reaction is favoured or which reaction is suppressed.

Precipitation is accomplished by a reaction between a specific metal ion and an anion, for instance:

$$Cu^{2+} + CO_3^{2-} = CuCO_3 \qquad (8.1)$$

In surface water, and in the pore water of sediments, there is a predominance of the following anions: chloride, sulphate, carbonate, hydrogen carbonate, hydroxide, and under reducing conditions anionic species derived from hydrogen sulphide. The chlorides and sulphates of metals are readily soluble, whereas the carbonates, hydroxides and sulphides only dissolve with difficulty.

Hydroxides precipitate in several forms, which may behave quite differently in the aquatic environment with respect to the effects of co-precipitation or later redissolution. Precipitates may persist in metastable equilibrium with the solution and may slowly convert into the aged forms, thereby becoming more stable and inactive.

Heavy metal sulphides are practically insoluble at neutral pH. The solubility of carbonates in aqueous solution is highly dependent on the partial pressure of carbon dioxide, which is illustrated by the following reaction:

$$CaCO_3 + CO_2 + H_2O = Ca^{2+} + 2HCO_3^- \qquad (8.2)$$

All carbonates tend to be more soluble in the presence of carbon dioxide.

The solubility is highly dependent on pH, as the concentration of the precipitating anions hydroxide, carbonate or sulphide - decreases with decreasing pH due to reaction with the hydrogen ion:

$$OH^- + H^+ = H_2O \qquad (8.3)$$
$$CO_3^{2-} + H^+ = HCO_3^- \qquad (8.4)$$
$$S^{2-} + H^+ = HS^- \qquad (8.5)$$
$$HCO_3^- + H^+ = H_2CO_3 = H_2O + CO_2\,(g) \qquad (8.6)$$
$$HS^- + H^+ = H_2S\,(g) \qquad (8.7)$$

With increasing pH first carbonate and then the hydroxide become the stable phase for may metal ions. For negative values of the reduction potential, the sulphide remains the stable phase over a wide pH range for many metal ions. Some metal hydroxide might have ampholyte character and the following reactions might occur at higher pH, using zinc as an example (see also Fig. 8.1):

$$ZN(OH)_2\,(s) + OH^- = Zn(OH)_3^- \qquad (8.8)$$

The various interacting processes, which determine the solubility at different pH values can conveniently be illustrated in a graphical double logarithmic representation.

If one is dealing with waters containing very few components, then it is frequently possible to refer to precipitation processes in a stricter sense.

Precipitation can be expected to take place in estuarine zones, where an increase in ionic strength and a change in pH frequently occurs. pH of river water generally ranges between 6.5 and 7.5, and in seawater is approximately 8.1. Precipitation of iron takes place, for instance, during the hydrolysis of hydrates of iron(III)ions:

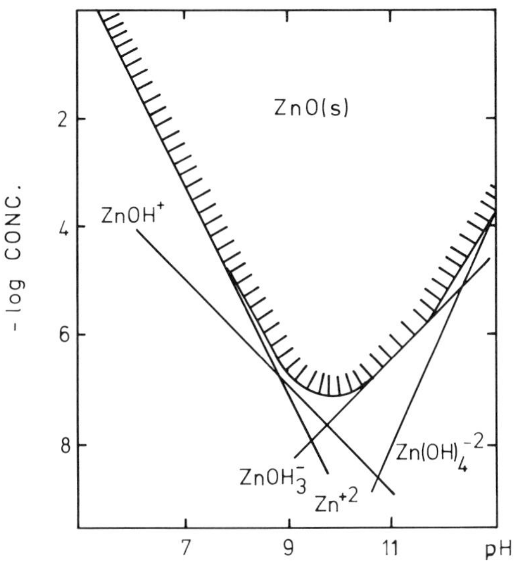

Fig. 8.1: Solubility of ZnO. At low pH Zn^{2+} determines the solubility. At higher pH $ZnOH^+$ takes over, then $Zn(OH)_3^-$ and finally $Zn(OH)_4^{2-}$

8.2. MATHEMATICAL MODELS OF THE PRECIPITATION PROCESS

It has been indicated that precipitation is a chemical reaction with a relatively fast rate. Thus, in most instances the rate itself is of no direct concern, and there exist no models to describe the rate aspects of the process.

The application of the process depends to a large degree upon the equilibrium situation that characterizes this process. Thus, there are mathematical models for the precipitation process that describe the equilibrium or the end-point of the reaction for specified boundary conditions. This type of model is to be presented below. However, it must be pointed out that for most practical applications such models derived from thermodynamic principles, have to be modified. In fact, there are situations where a quantitative application of these principles is no longer possible.

If precipitation is defined as the transformation of two or more dissolved components to a non-dissolved substance, the so-called precipitate, then dissolution processes and precipitation processes are similar reactions but of opposite direction. The solubility of a product, or vice versa the degree to which precipitation will control the dissolved species, is determined by the capacity of the solute to accomodate specific ions. This capacity is controlled by:

- the energy of bonding between the ions under consideration
- the dielectric characteristics of the solute
- the type and number of ions present in the system

The solubility of certain species or the relationship between two or more precipitating partners is furthermore controlled by third partners which lead to so-called side reactions. The solubility is also determined by temperature and ambient pressure.

There are a number of important concepts or definitions from the discipline of thermodynamics that describe directly, as well as indirectly, the quantitative models existing in this area.

The **MASS LAW** describes the solubility and the corresponding precipitation reaction in terms of a solubility product. As seen from the example below the solubility product describes the equilibrium concentrations of the precipitating ions, in particular the ion to be removed by precipitation.

$$[CA^{2+}] * [CO_3^{2-}] = K * [CaCO_3] = K_s \qquad (8.9)$$

STOCHIOMETRIC RELATIONSHIPS describe how many atoms, molecules or ions of one reaction partner react with corresponding forms of the other partner. The earlier example will illustrate this again:

$$Ca^{2+} + CO_3^{2-} = CaCO_3 \qquad (8.10)$$

The REACTION RATE with which the precipitation occurs, or with which a disturbed solubility/precipitation equilibrium is balanced again, is finite. However, in most instances of interest for the practical application of this process the reaction rate is so large that the available detention time, or reaction time, suffices to reach the equilibrium. It has been indicated above that either in the stage of mixing of chemicals with the waste water stream, or in the transport of the precipitating system from one reactor (mixing reactor) to the next one (in most instances a reactor for the liquid-solid separation) the flow or detention time is large enough. However, there are situations where a change in the stochiometric parameters in terms of an overdosing of the precipitation causing reagent leads to improved reaction rates and to increased efficiency. Efficiency in this instance is interpreted as reduced concentrations of the ion to be precipitated.

This stochiometric parameter in its practical implication can also be formulated as a quotient of concentration values. This is indicated in the example below.

$$Qu = Al^{3+} / PO_4^{3-} \tag{8.11}$$

On the basis of thermodynamic arguments this quotient should have a value of 1 in the instance shown above. However, practical observations have indicated that increased quotients may increase the reaction rate and the separation of the precipitate. (Usually each precipitation reaction is followed in its practical application by a liquid-solid separation step. Depending upon the specific gravity of the formed solids or depending upon the amount of solids formed, such separation steps can be sedimentation or flotation or filtration.)

It has been indicated that the equilibrium concentrations are a function of ambient pressure and temperature. Similarly the reaction rate is strongly affected by these parameters.

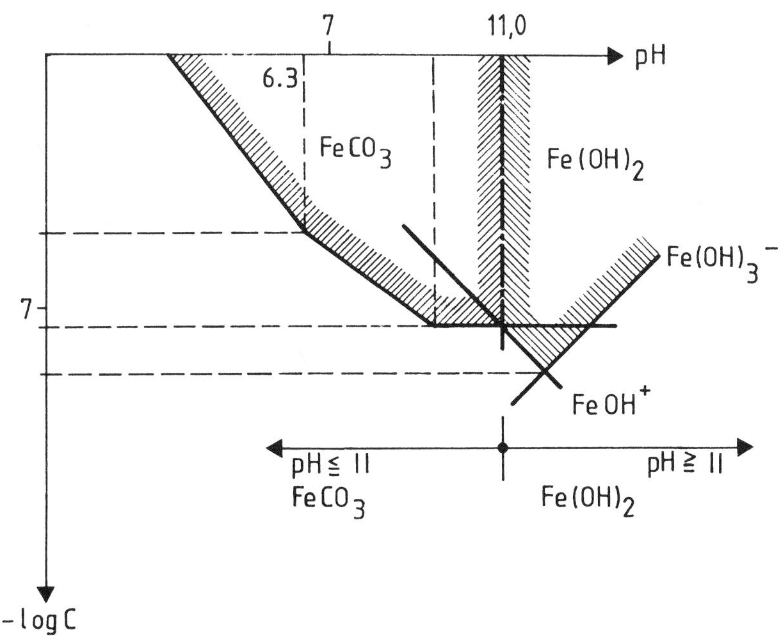

Fig. 8.2: Precipitation of Fe-carbonate and Fe-hydroxide occuring at different pH values. The lines in this diagram and Fig. 8.1 are deducted from the mass law.

One further variable is to be described or defined: **the pH value** of the precipitating system which is of utmost importance. In aqueous solutions all multivalent metal ions form hydroxides or carbonates. And the pH value describes the relative abundance of these species. It furthermore describes the distribution of all other species present in the system and with this the possibility of other side reactions (see Fig. 8.2).

As stated above these quantitative concepts can be used for the description and prediction of precipitating reactions. However, there are deviations in practice from these relatively simple models.

Some of these deviations are known and can be indicated in their relative effect upon the technical process. Before precipitates are formed there must be a certain supersaturation of the system. This necessary supersaturation can be substituted or aided by other particulate matter which serves as nucleation seeds.

In addition a precipitated system ages. This means that with increasing time elapsed there are further reactions predominantly in the direction of further precipitation and quasi-crystallisation.

Finally, one must mention the phenomenon of adsorption in connection with the precipitation process where a large amount of adsorptive surface is created. This adsorption phenomenon might even include the fixation of substances that are involved in the precipitation process itself.

8.3 PARAMETER ESTIMATION

The parameters contained in the above presented models are:

a. the solubility product
b. the stochiometric quotient

The estimation of these parameters, understood here as the recipe for selection, differs from parameter estimation in other model applica- tions. In essence these parameters controlling the solution/precipitation process can be found in standard literature. Or a more theoretical point of view consult thermodynamic textbooks, such as Latimer (1952) or Sillen and Martell (1971). For the translation of such basic data into information that may be in agreement with the limitations of practical systems it would be advisable to consult texts like Stumm and Morgan (1981) for natural systems orientation or texts like Weber (1972) or Benefield et al. (1982), or more technically oriented systems.

The possibility of determining such site and project specific parameters are not too good compared to other modelling situations. Thermodynamic data - and this is the major part of the information needed - are difficult to determine in situ and require systematic experimentation.

Table 8.1 gives the solubility products for a number of hydroxides, carbonates and sulphides of environmental interest.

Table 8.1
Negative Logarithms of Solubility Products of Heavy-Metal Hydroxides (Total Ionization Constants), Carbonates and sulphides (pH = 7 at 25°C)

Hydroxides		-log K_{sp}	sulphides	-log K_{sp}	Carbonates	-log K_{sp}
$Mn(OH)_2$	(aged)	10.9	MnS (pink)	9.6	$MnCO_3$	10.2
$Mn(OH)_2$	(cryst.)	12.7	MnS (green)	12.6		
$Cd(OH)_2$	(aged)	14.4	CdS	27.8	$CdCO_3$	11.3
$Fe(OH)_2$	(aged)	15.1	FeS	17.2	$FeCO_3$	10.5
$PbO + H_2O$	(red)	15.3	PbS (galena)	27.5	$PbCO_3$	13.1
$Co(OH)_2$	(blue)	14.2	CoS (alpha)	20.4	$CoCO_3$	12.8
$Co(OH)_2$	(pink, aged)	15.7	CoS (beta)	24.7		
$Zn(OH)_2$	(amorphous)	15.5	ZnS (wurtzite)	21.6	$ZnCO_3$	10.8
$ZnO + H_2O$	(aged)	16.8	ZnS (sphalerite)	23.6		
$Ni(OH)_2$	(active)	14.7	NiS (alpha)	18.5	$NiCO_3$	6.9
$Ni(OH)_2$	(aged)	17.2	NiS (gamma)	25.7		
$CuO + H_2O$	(active)	19.7	CuS	36.1	$CuCO_3$	9.6
$CuO + H_2O$	(tenorite)	20.5	Cu_2S	48.0	$Cu_2(OH)_2CO_3$	33.8
$HgO + H_2O$	(red)	25.4	HgS	52.4		
$Cr(OH)_3$	(aged)	27.4				
$Fe(OH)_3$	(aged)	39.1				

8.4. APPLICATION AND EXAMPLES

The precipitation of Cu^{2+} is used as example to illustrate the reactions of a metal ion in water.

According to Hartinger (1976) one has to distinguish between water systems with a high concentration of heavy metals and those where the concentration is lower. The following discussion centres on systems where the heavy metals concentration is lower.

The reaction proceeds schematically as follows:

$$Cu^{2+} + SO_4^{2-} = CuSO_4\,(s) \qquad (8.12)$$

With increasing concentration of hydroxide ions the sulphate is changed by and into a hydroxide.

$$CuSO_4 * 3Cu(OH)_2\,(s) + 2\,OH^- = 4\,Cu(OH)_2\,(s) + SO_4^{2-} \qquad (8.13)$$

It becomes apparent from these global equations that, depending upon the composition of the solution, different precipitates are formed. Furthermore it becomes apparent that the composition of the precipitates changes if the composition of the dissolved system changes - for instance, if the pH value changes.

Altogether the following precipitates are known in the precipitation reaction for the control of heavy metals (see Hartinger, 1976):

- hydroxide
- alkaline sulphate
- alkaline chloride
- alkaline nitrate

In addition to this it is known (Verband der chemischen Industrie) that heavy metals can be precipitated as sulphides.

Optimal pH-value

As indicated above the type of precipitate formed depends upon the composition of the solution, particularly with respect to the various anions. The global chemical equilibria presented above have shown furthermore that, depending upon the concentration of the hydroxide ions (pH), the concentration of the remaining dissolved species are larger or smaller. This means that pH plays a significant role. Furthermore it means that there must be a pH where the concentration of the remaining (precipitating) cation is lowest. This is termed the "optimal pH value".

In this example of copper precipitation (according to Hartinger, (1976)) the pH at which the reaction for the various precipitation agents begins is as follows:

Sodium hydroxide	pH from 5.8 to 5.9
Calcium hydroxide	pH around 5.9
Sodium carbonate	pH around 5.6

A significant improvement with respect to lowering the remaining copper concentration in solution is observed only at significantly higher pH-values. The concentration of copper in solution drops below 1 mg/l in the instance of precipitation with sodium or calcium hydroxide only above pH 7.5. Furthermore it must be kept in mind that this pH-value changes with time (due to the "ageing processes" mentioned above).

REFERENCES

Benefield, L.D., J.F. Judkins and B.L. Weand. "Process Chemistry for Water and Wastewater Treatment", Prentice Hall, Englewood Cliffs, 1982.

Hartinger, L. "Taschenbuch der Abwasserbehandlung", Band 1 Chemie, Carl Hanser Verlag, München, 1976.

Latimer, W.M. "The Oxidation States of the Elements and their Potentials in Aqueous Solutions", Prentice Hall, Englewood Cliffs, 1952.

Sillen, L.G. and A.E. Martell. "Stability Constants of Metal Ion Complexes", Supplement no. 1, Chemical Society, London, 1971.

Stumm, W. and J.J. Morgan. "Aquatic Chemistry", Second edition, A Wiley-Interscience Publication, John Wiley & Sons, New York, 1981.

Weber, W.J. Jr. "Physicochemical Processes", Wiley-Interscience, New York, 1972.

CHAPTER 9

COMPLEX FORMATION

by
Sven Erik Jørgensen and Arne Jensen

9.1 INTRODUCTION

Metal ions are able to form co-ordination complexes with a number of inorganic and organic ions and molecules. The metal ion is called the central atom and the substance that co-ordinates to the metal ion is termed a ligand.

The formation of complexes may often be explained by formation *of a covalent bond,* where the ligand is the electron donor and the metal ion the electron acceptor. This can be exemplified by the following process:

$$Cu^{2+} + 4NH_3 \rightleftarrows [H_3N{:}\overset{NH_3}{\underset{NH_3}{Cu}}{:}NH_3]^{2+} \qquad (9.1)$$

Some **ligands** have more than *one co-ordination group* in their structure and are then called bidentate, tridentate, tetradentate, and so on. Multidentate ligands may form what are termed chelate rings, and these often have a high stability, cf. Table 9.3. Some typical structural formulas are:

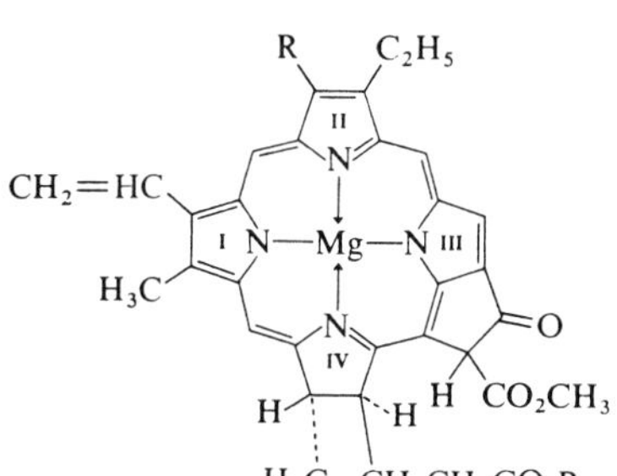

(a)

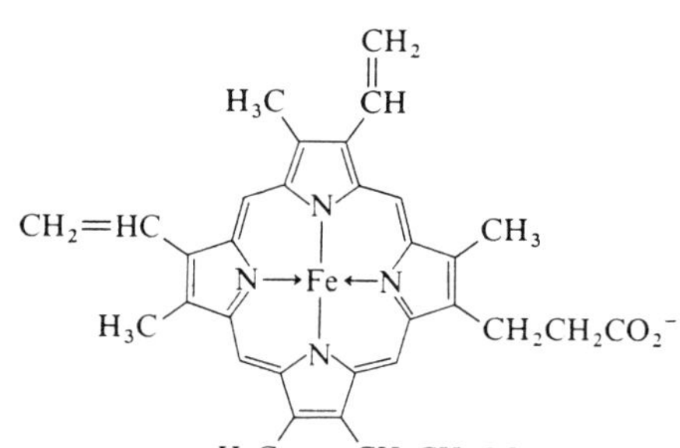

(b)

(9.2)

(a) One chlorophyll molecule. (b) One heme group

The mass equation is valid for these reactions as for other chemical processes. For the process:

$$Me^{n+} + L^{m-} \text{ — } MeL^{(n-m)+} \qquad (9.3)$$

where Me denotes a metal and L a ligand. In this case the mass equation assumes the following form:

$$\frac{[MeL^{(n-m)+}]}{[Me^{n+}]\,[L^{m-}]} = K \qquad (9.4)$$

K is termed stability constant, if the process - as here - is the formation of a complex.

The co-ordination number of a metal is the number of bonds formed from the metal ion to the donor atoms of the ligands. Even co-ordination numbers (2, 4, 6 and 8) are much more common than uneven ones, **co-ordination numbers 4 and 6 being the most frequently encountered.**

The co-ordination numbers of the metal ions having particular environmental interest are shown in Table 9.1.

Table 9.1
Co-ordination Numbers of Pertinent Metal Ions

Cu^{+}	2
Ag^{+}	2
Hg_2^{2+}	2, 4
Li^{+}	4
Be^{2+}	4, 6
Al^{3+}	4, 6
Fe^{3+}	4, 6
Cu^{2+}	4, 6
Ni^{2+}	4, 6
Hg^{2+}	2, 4
Fe^{2+}	6
Mn^{2+}	4, 6

The formation *of co-ordination complexes* are playing a major role in modelling the distribution as well as the effect of metal ions in aquatic ecosystems due to the following influences:

1) increase of metal solubility:

$$MeY(S) + L = MeL + Y \qquad (9.5)$$

where Me represents a metal, which is bound in the sediment at the compound MeY; L denotes a ligand,

2) alteration of the distribution between oxidized and reduced forms of metals. The equilibrium constant for the process

$$Me^{n+} + me^{-} — Me^{(n-m)+}$$

differs from the one for

$$MeL^{n+} + me^{-} — MeL^{(n-m)+}$$

if the complexes MeL^{n+} and $MeL^{(n-m)+}$ have different stability constants, which often is the case,

3) alleviation of toxicity due to alteration of the availability to aquatic life,

4) change in adsorption and ion exchange on sediment and suspended matter, and

5) change in stability of the metal containing colloids.

The presence of ligands in water will generally increase the transfer of metals from sediment, soil and suspended matter and thereby increase the solubility, but as complexes show lower toxicity than the metal ions, the harm to the aquatic life in most cases does not increase corresponding to the increased solubility.

The application of *co-ordination chemistry to environmental problems* is very complex; as a great number of ligands are present in aquatic eco-systems all the ligands competing simultaneously for information of complexes with the metal ions. The interactions between competing ligands are exemplified in Figs. 9.1 and 9.2, where concentrations of various mercury species are plotted against the chloride concentration at two different pH levels.

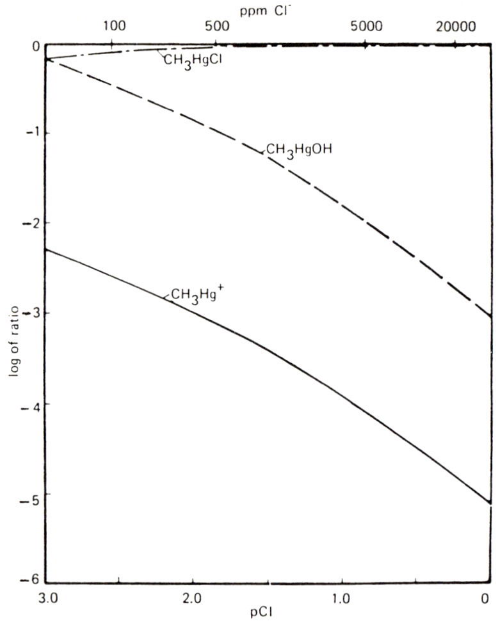

Fig. 9.1: Methylmercury species: total methylmercury versus Cl^- concentration at pH = 7.0.

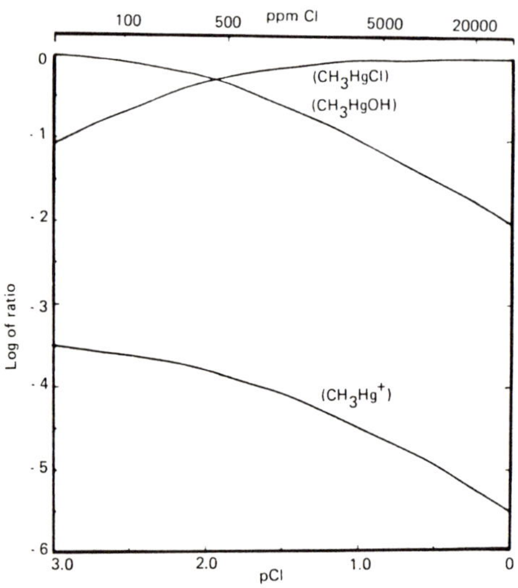

Fig. 9.2: Methylmercury species versus Cl^- concentration at pH = 8.

Table 9.2 shows the ligand concentrations in seawater which has a rather constant composition, while the ligand concentration in freshwater varies immensely and often is higher for the organic ligands. The most significant changes from a seawater to a freshwater ecosystem (at constant pH and redox potential) may be summarized in the following points (Sibley and Morgan, 1977):

1. considerably higher ionic strength in seawater,
2. lower contents of adsorbing and ion exchanging surfaces in seawater,
3. different concentrations of trace metals,
4. different concentrations of major cations and anions, and,
5. usually higher concentrations of organic ligands in freshwater systems.

The particulate components dominate for most metals in the freshwater models, while the complex bound components for most metals dominate in seawater, but the dissolved complexes vary considerably more in freshwater than in seawater. The general changes in chemical species of selected trace metals, from seawater to freshwater models, are shown in Fig. 9.3.

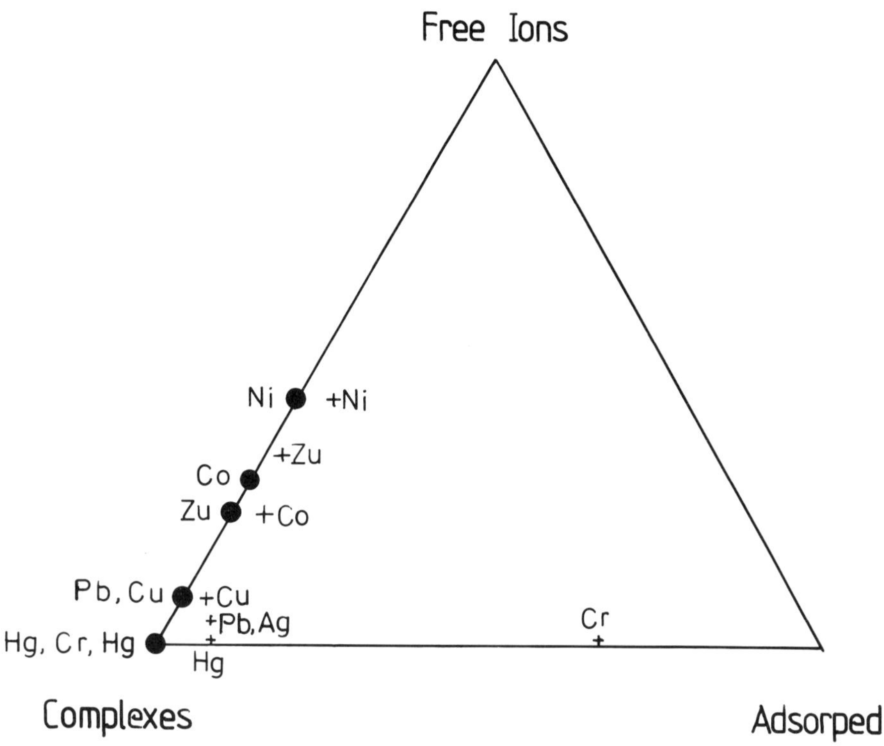

Fig. 9.3: Chemical species in typical seawater • and freshwater +. Reproduced from Sibley and Morgan (1977).

Humic acids and fulvic acids play an important role in the transportation of heavy metals in water, especially in freshwater. It has been estimated that at a total bonding capacity of 200-600 meq/100 g humic acid, approximately two thirds of the available positions exist as complexation sites (Rashid and King, 1971).

Synthetic complexing agents may in many cases play an important role and the presence of these ligands in aquatic ecosystems may have a serious impact on the heavy metal remobilization. The use of nitriloacetate (NTA) as an alternative to polyphosphates has been intensely discussed during the last few years (Thom 1971, Dunlap et al. 1971, Motolla 1974, Prakash 1976).

9.2 MODELS OF COMPLEX FORMATION

Fig. 9.4 shows a conceptual diagram of a metal speciation, which considers formation of labile and inert complexes. The formation of complexes is in most cases a rapid process compared with the time intervals used in a model describing the distribution and effect of metals. Consequently, the complex formation can be described by use of equilibrium expressions, and generally it is not necessary to take reaction kinetics into consideration.

For most complex equilibria, it will be possible to find the equilibrium constant, as will be demonstrated in section 9.3. Furthermore, its dependence on the temperature can easily be found by use of thermodynamic calculation, cf. chapter 10.

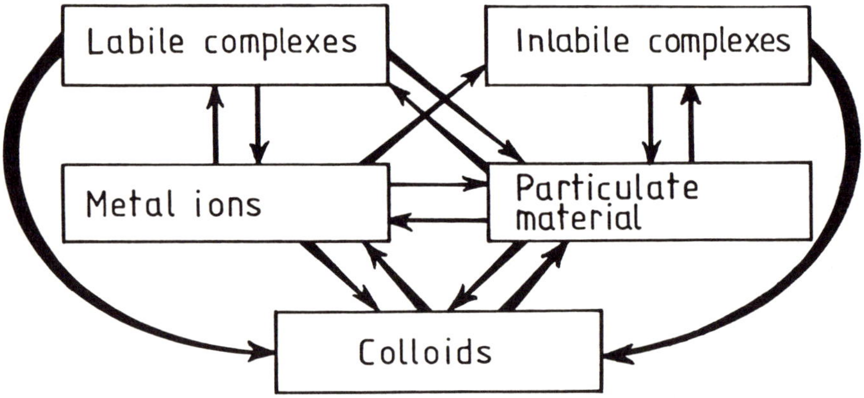

Fig. 9.4: Conceptual diagram of metal speciation.

Table 9.2 Concentration of ligand in interstitial seawater (M) (Reducing conditions)[a]

Ligand	Concentration	Ligand	Concentration
Total soluble carbonate	$8 \cdot 10^{-3}$	Glutamic acid	$1.09 \cdot 10^{-6}$
Total soluble borate	$6 \cdot 10^{-4}$	Glycerine	$4.0 \cdot 10^{-6}$
Total soluble silicate	$5 \cdot 10^{-4}$	Glycollic acid	$7.89 \cdot 10^{-6}$
Ammonia	$4 \cdot 10^{-4}$	Histidine	$2.58 \cdot 10^{-7}$
Nitrite	$7 \cdot 10^{-7}$	p-Hydroxybenxoic acid	$4.35 \cdot 10^{-7}$
Nitrate	$1.4 \cdot 10^{-6}$	Hydroxyproline	$3.05 \cdot 10^{-7}$
Orthophosphate	$2.5 \cdot 10^{-5}$	Lactic acid	$1.11 \cdot 10^{-7}$
Sulfide	$5 \cdot 10^{-4}$	Leucine	$7.63 \cdot 10^{-7}$
Sulfate	$2.8 \cdot 10^{-2}$	Lysine	$6.85 \cdot 10^{-7}$
Fluoride	$8 \cdot 10^{-4}$	Malic acid	$1.49 \cdot 10^{-5}$
Chloride	0.5	Methionine	$1.34 \cdot 10^{-7}$
Bromide	$8 \cdot 10^{-6}$	Ornithine	$8.47 \cdot 10^{-7}$
Iodide	$5 \cdot 10^{-7}$	Proline	$1.74 \cdot 10^{-7}$
Acetic acid	$2 \cdot 10^{-4}$	Serine	$1.90 \cdot 10^{-6}$
Alanine	$1.12 \cdot 10^{-6}$	Threonine	$8.40 \cdot 10^{-7}$
Arginine	$1.15 \cdot 10^{-7}$	Tryptophane	$9.80 \cdot 10^{-8}$
Aspartic acid	$1.2 \cdot 10^{-6}$	Tyrosine	$5.24 \cdot 10^{-7}$
Citric acid	$1.04 \cdot 10^{-6}$	Valine	$5.13 \cdot 10^{-7}$
Cysteine	$1.65 \cdot 10^{-7}$		

[a] A factor of 20 was applied, based on the fact that the N and P compounds in interstitial water are about 20 times more concentrated than those in average seawater.

The description of complex formation is further complicated by the occurrence of acid-base-, precipitation- and redox reactions. The two former reactions are taken into consideration by use of a coefficient ß, which is defined as the ratio between the reactive form of the ligand, or the metal ion, and the total concentration.

Let us consider the following reaction:

$$Me^{n+} + L^{m-} — MeL^{(n-m)+} \quad (9.6)$$

The following bireactions may occur:

$$p\,Me^{n+} + n\,Y^{p-} = Me_p Y_n \quad (9.7)$$

$$m^- + H^+ — HL^{(m-1)-}\ ;\quad HL^{(m-1)-} + H^+ — H_2L^{(m-2)-} \quad (9.8)$$

$ß_{Me}$ can be found from the solubility products, S, provided that $[Y^{p-}]$ is known, and molar concentrations are used in lieu of activities:

$$[Y^{p-}]^n\ [Me^{n+}]^p = S \quad (9.9)$$

$$\frac{[Me^{n+}]}{[Me_{total}]} = \beta_{Me} = p\sqrt{\frac{S}{[Y^{p-}]^n}} \cdot \frac{1}{[Me_{total}]} \qquad (9.10)$$

Correspondingly, β_L can be found, when pH is known; and concentrations are used as in equation (9.9):

$$\frac{[L_{total}]}{[L^{m-}]} = \frac{1}{\beta_L} = 1 + \frac{[H^+]^m}{K_1{}^* K_2 \ldots\ldots K_m} + \frac{[H^+]^{m-1}}{K_2{}^* K_3 \ldots\ldots K_m} +$$

$$+ \frac{[H^+]^{m-2}}{K_3{}^* K_4 \ldots\ldots K_m} + \ldots\ldots\ldots + \frac{[H^+]}{K_m} \qquad (9.11)$$

where K_m , K_{m-1} etc. are equilibrium constants for the stepwise ionization of H_mL.

β_L can also be determined from a logarithmic diagram, as shown in Fig. 9.5. Here the logarithm of the concentration of the various species is plotted against pH.

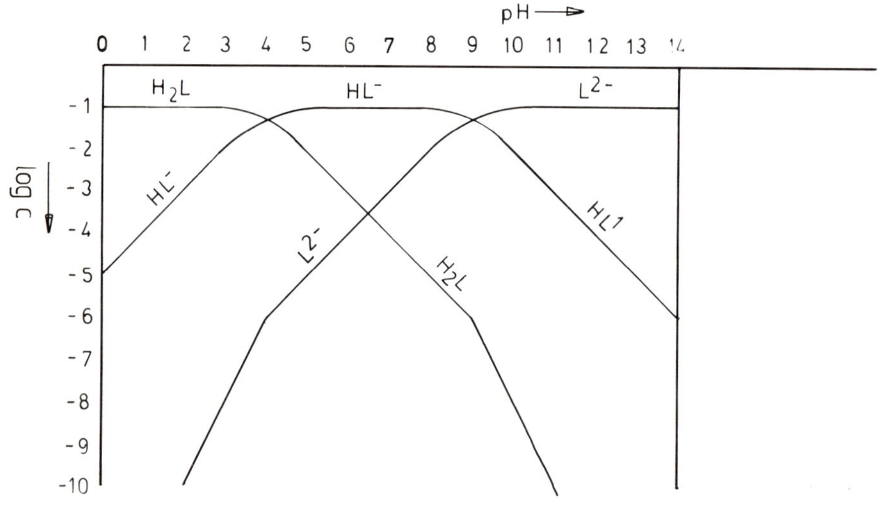

Fig. 9.5: Speciation of H_2L as function of pH. pK_a-values are 4.0 and 9.0. If β_L at pH = 5.0 must be determined for L^{2-}, we have $L^{2-} = 10^{-5}$ while C = total concentration of all species = 10^{-1} $\beta_L = 10^{-4}$

The shape of the curves can be seen from the following considerations:

process:

$$a \rightleftharpoons H^+ + b \tag{9.12}$$

The mass equation is:

$$K_a = \frac{[b]\,[H^+]}{[a]} \tag{9.13}$$

or

$$\log[a] = pK_a - pH + \log[b] \tag{9.14}$$

<u>At pH = pK:</u>

$$[a] = [b] = C/2 \tag{9.15}$$

where C is the total concentration, i.e. [a] + [b] = C.

<u>At low pH:</u>

$$[a] \sim C \quad \text{and} \quad \log[b] = \log C - pK_a + pH \tag{9.16}$$

<u>At high pH:</u>

$$[b] \sim C \quad \text{and} \quad \log[a] = \log C + pK_a - pH \tag{9.17}$$

Fig. 9.6 shows log ß = log $ß_Y$ as a function of pH, evaluated either from calculations based upon equation (9.11) or from a logarithmic diagram.

When $ß_{Me}$ and ß have been determined, a reformulation of equation (9.4) can take place, i.e.

$$\frac{[MeL^{(n-m)+}]}{[Me^{n+}]\,[L^{m-}]} = K = \frac{[MeL^{(n-m)+}]}{[Me_{total}] * ß_{Me} * [L_{total}] * L} \tag{9.18}$$

or

$$\frac{[MeL^{(n-m)+}]}{[Me_{total}]\,[L_{total}]} = K * ß_{Me} * ß_L = K_C \tag{9.19}$$

K_C is termed the conditional equilibrium constant. If K_C is known, it is easy to compute the concentrations of free metal ions and metal complexes.

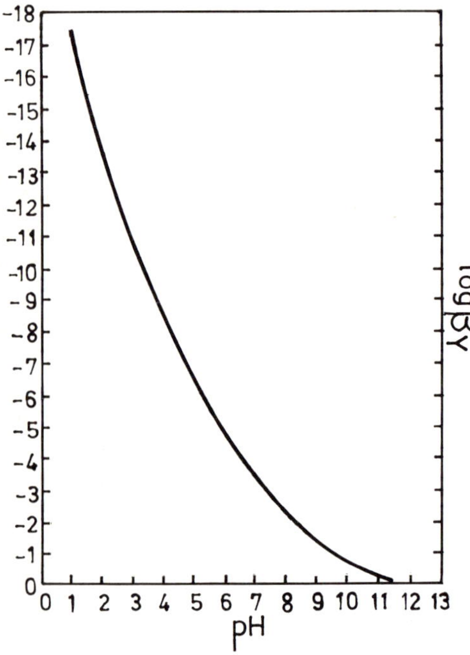

Fig. 9.6: Fraction β_Y of all forms of uncomplexed EDTA present as Y^{4-}, vs pH.

However, the condition must be known for determination of K_C, and the computations are of course only valid at the given conditions.

Interactions between a dissolution and a complex formation process are often of importance, when mobilization of metal ions from the sediment is modelled. It may be possible to make an acceptable description as follows:

$$Me_pYn + pL^{m-} \text{ — } nY^{p-} + p\,MeL^{n-m} \qquad (9.20)$$

The mass equation for process (9.20) is:

$$\frac{[MeL^{(n-m)+}]^p * [Y^{p-}]^n}{[L^{m-}]^p} = K' = S * K_c \qquad (9.21)$$

where S is the solubility product of Me_nY_p.

The interactions between redox- and complex processes may be illustrated by the following scheme:

$$
\begin{array}{ccccc}
Me^{n+} & + \; me^- & \overset{(K_1)}{\longrightarrow} & Me^{(n-m)+} & \\
+ & & & + & \\
L & & & L & \\
(K_2) & & & (K_4) & \\
MeL^{n+} & + \; me^- & \underset{(K_3)}{\longrightarrow} & MeL^{(n-m)+} &
\end{array}
\tag{9.22}
$$

As seen, the redox equilibrium is needed influenced by the formation of complexes, as mentioned in section 9.1.

It can be concluded from this presentation that it is possible to cope with several interacting processes simultaneously. A description is relatively easy to set up, provided that equilibrium is obtained rapidly, compared with other processes included in the model such as hydrodynamic and biological processes, which are often valid.

It is important here, as well as in other modelling tasks, to include only processes of importance for the overall picture of the problem. Many complexes have, however, high stability constants implying significant processes in the models of distribution and effect of metals. The relevant simplification is therefore quite often a question of which complex formations are most pronounced in a given case study.

$$
\frac{K_3}{K_1} = \frac{[MeL^{(n-m)+}]\,[Me^{n+}]}{[MeL^{n+}]\,[Me^{(n-m)+}]} * \frac{L}{L} = \frac{K_4}{K_2} \tag{9.23}
$$

$$
\frac{K_3}{K_1} \neq 1 \quad \text{if} \quad \frac{K_4}{K_2} \neq 1 \tag{9.24}
$$

Table 9.3
Gross stability constants for complex equilibria of environmental importance

Process			Log Ω
$Cd^{2+} + Cl^-$	=	$CdCl^+$	2.7
$Cd^{2+} + 2Cl^-$	=	$CdCl_2$	2.7
$Cd^{2+} + 3Cl^-$	=	$CdCl_3^-$	2.8
$Cd^{2+} + 4Cl^-$	=	$CdCl_4^{2-}$	2.9
$Cd^{2+} + 6Cl^-$	=	$CdCl_6^{4-}$	2.25
$Cd^{2+} + 2HS^-$	=	$Cd(HS)_2$	15.2
$Cd^{2+} + 3HS^-$	=	$Cd(HS)_3^-$	17.1
$Cd^{2+} + HS^-$	=	$CdHS^+$	7.9
$Cd^{2+} + OH^-$	=	$CdOH^+$	4.2
$Cd^{2+} + 2OH^-$	=	$Cd(OH)_2$	8.9

Table 9.3 (continued)

Process			Log Ω
$Cd^{2+} + 3OH^-$	=	$Cd(OH)_3^-$	9.6
$Cd^{2+} + 4OH^-$	=	$Cd(OH)_4^{2-}$	9.6
$Cd^{2+} + NH_3$	=	$Cd(NH_3)^{2+}$	2.7
$Cd^{2+} + 2NH_3$	=	$Cd(NH_3)_2^{2+}$	4.7
$Cd^{2+} + 3NH_3$	=	$Cd(NH_3)_3^{2+}$	6.2
$Cd^{2+} + 4NH_3$	=	$Cd(NH_3)_4^{2+}$	7.2
$Cd^{2+} + 5NH_3$	=	$Cd(NH_3)_5^{2+}$	6.9
$Cd^{2+} + 6NH_3$	=	$Cd(NH_3)_6^{2+}$	5.2
$Cd^{2+} + NTA^{3-}$	=	$Cd(NTA)^-$	9.6
$Co^{2+} + H_2O$	=	$Co(OH)^+ + H^+$	-9.6
$Cr^{3+} + OH^-$	=	$Cr(OH)^{2+}$	10.6
$Cr^{3+} + 2OH^-$	=	$Cr(OH)_2^+$	19.2
$Cr^{3+} + 4OH^-$	=	$Cr(OH)_4^-$	33.1
Cr^{3+} + citric acid H_3	=	Cr(citrate acid) + $3H^+$	-5.6
Cr^{3+} + serine$^-$	=	Cr(serine)$^{2+}$	8.8
$Cu^{2+} + OH^-$	=	$Cu(OH)^+$	6.0
$Cu^{2+} + 3OH^-$	=	$Cu(OH)_3^-$	15.2
$Cu^{2+} + B(OH)_4^-$	=	$Cu(B(OH)_4)^+$	7.1
$Cu^{2+} + 2B(OH)_4^-$	=	$Cu(B(OH)_4)_2$	12.6
$Cu^{2+} + Cl^-$	=	$CuCl^+$	1.6
$Cu^{2+} + CO_3^{2-}$	=	$CuCo_3$	6.8
$Cu^{2+} + 2CO_3^{2-}$	=	$Cu(Co_3)_2^{2-}$	10.0
$H_2S + CuS + HS^-$	=	$Cu(HS)_3^-$	-4.4
$CuS + 3HS^-$	=	$CuS(HS)_3^-$	-4.4
Cu^{2+} + histidine$^-$	=	Cu(histidine)$^+$	10.6 - 11.7
Cu^{2+} + 2histidine$^-$	=	Cu(histidine)$_2$	18.6 - 19.7
$Cu^{2+} + HCO_3^-$	=	$CuCO_3 + H^+$	-3.6
Cu^{2+} + leucine$^-$	=	Cu(leucine)$^+$	7.0
Cu^{2+} + 2leucine$^-$	=	Cu(leucine)$_2$	15.4
Cu^{2+} + valine$^-$	=	Cu(valine)$^+$	7.9
Cu^{2+} + 2valine$^-$	=	Cu(valine)$_2$	14.8
Cu^{2+} + glutamate$^-$	=	Cu(glutamate)$^+$	7.9
Cu^{2+} + 2 glutamate$^-$	=	Cu(glutamate)$_2$	14.8
Cu^{2+} + aspartate$^-$	=	Cu(aspartate)$^+$	8.6
Cu^{2+} + 2aspartate$^-$	=	Cu(aspartate)$_2$	15.3
Cu^{2+} + histidine$^-$	=	Cu(histidine)$^+$	10.6
Cu^{2+} + 2histidine$^-$	=	Cu(histidine)$_2$	18.6
Cu^{2+} + glycinate$^-$	=	Cu(glycinate)$^+$	8.6
Cu^{2+} + 2glycinate$^-$	=	Cu(glycinate)$_2$	7.0
Cu^{2+} + cysteinate$^-$	=	Cu(cysteinate)$^+$	19.5

Table 9.3 (continued)

Process			Log Ω	
Cu^{2+} + 2alaninate$^-$	=	Cu(alaninate)$_2$	15.5	
Cu^{2+} + 2serinate$^-$	=	Cu(serinate)$_2$	14.6	
Cu^{2+} + phthalate^{2-}	=	Cu(phthalate)	3.1	
Cu^{2+} + SO_4^{2-}	=	$CuSO_4$	2.3	
Fe^{3+} + $2OH^-$	=	$Fe(OH)_2^+$	20.4	
Fe^{3+} + $4OH^-$	=	$Fe(OH)_4^-$	35.3	
Fe^{3+} + citrate^{3-}	=	Fe(citrate)	28.1	
Fe^{3+} + $2B(OH)_4^-$	=	$Fe(B(OH)_4)_2^+$	15.6	
Fe^{2+} + Cl^-	=	$FeCl^+$	0.4	
Fe^{2+} + $2Cl^-$	=	$FeCl_2$	0.8	
Fe^{2+} + citrate^{3-}	=	Fe(citrate)$^-$	18.6	
Fe^{2+} + HPO_4^{2-}	=	$Fe(HPO_4)$	7.0	
Hg^{2+} + Cl^-	=	$HgCl^+$	6.7	
Hg^{2+} + $2Cl^-$	=	$HgCl_2$	13.8	
Hg^{2+} + $3Cl^-$	=	$HgCl_3^-$	14.8	
Hg^{2+} + $4Cl^-$	=	$HgCl_4^{2-}$	15.3	
Hg^{2+} + Cl^- + Br^-	=	HgClBr	2.0	
Hg^{2+} + cysteine	=	Hg(cysteine)$^{2+}$	43.6	
Hg^{2+} + glycinate$^-$	=	Hg(glycinate)$^+$	10.3	
HgS + S^{2-}	=	HgS_2^{2-}	0.57	
HgS + $2HS^-$	=	$HgS(HS)_2^{2-}$	-3.7	*
HgS + H_2S + HS^-	=	$Hg(HS)_3^-$	-3.5	
HgS + 2HS	=	$HgS(HS)_2$	-4.2	*
Ca^{2+} + CO_3^{2-}	=	$CaCO_3$	3.2	
Mg^{2+} + CO_3^{2-}	=	$MgCO_3$	3.4	
Mg^{2+} + OH^-	=	$MgOH^+$	2.6	
Ca^{2+} + OH^-	=	$CaOH^+$	1.3	
Na^+ + CO_3^{2-}	=	$NaCO_3^-$	1.2	
Na^+ + OH^-	=	NaOH	-0.7	
Na^+ + SO_4^{2-}	=	$NaSO_4^-$	0.7	
Mn^{2+} + Cl^-	=	$MnCl^+$	1.1	
Mn^{2+} + $2Cl^-$	=	$MnCl_2$	1.5	
Mn^{2+} + $3Cl^-$	=	$MnCl_3^-$	1.1	
Mn^{2+} + HCO_3^-	=	$MnHCO_3^+$	1.8	
Ni^{2+} + Cl^-	=	$NiCl^+$	0.7	
Ni^{2+} + $2SO_4^{2-}$	=	$Ni(SO_4)_2^{2-}$	2.2	
Ni^{2+} + OH^-	=	$Ni(OH)^+$	4.7	
Ni^{2+} + cysteinate$^-$	=	Ni(cysteine)$^+$	10.5	
Ni^{2+} + 2tyrosine	=	Ni(tyrosine)$_2^{2+}$	10.4	
Ni^{2+} + histidine$^-$	=	Ni(histidine)$^+$	9.8	

Table 9.3 (continued)

Process			Log Ω
Ni^{2+} + 2histidine$^-$	=	Ni(histidine)$_2$	17.8
Pb^{2+} + CO_3^{2-}	=	$PbCO_3$	7.4
Pb^{2+} + Cl^-	=	$PbCl^+$	1.6
Pb^{2+} + $2Cl^-$	=	$PbCl_2$	1.8
Pb^{2+} + $6Cl^-$	=	$PbCl_6^{4-}$	3.0
Pb^{2+} + $2CO_3^{2-}$	=	$Pb(CO_3)_2^{2-}$	9.9
PbS + H_2S	=	$PbS(H_2S)$	-6.8
$PbS(H_2S)$	=	$PbS(HS)^-$ + H^+	-6.9
Zn^{2+} + Cl^-	=	$ZnCl^+$	0.4
Zn^{2+} + $2Cl^-$	=	$ZnCl_2$	0.6
Zn^{2+} + OH^-	=	$Zn(OH)^+$	4.9
Zn^{2+} + $2OH^-$	=	$Zn(OH)_2$	12.9
Zn^{2+} + $3OH^-$	=	$Zn(OH)_3^-$	14.2
Zn^{2+} + $4OH^-$	=	$Zn(OH)_4^{2-}$	15.5
Zn^{2+} + phthalate^{2-}	=	Zn(phthalate)	2.9

9.3 PARAMETER ESTIMATION

Stability constants can be found in many chemical handbooks, e.g. the Handbook of Chemistry and Physics.

Table 9.3 gives this constant for a number of complex processes of environmental interest.

A first approximation can be found by the use of Pearson's classification of electron acceptors and donors. He distinguishes between hard and soft categories to explain the difference in the stability of complexes (Pearson 1968a, b). As a general rule it was found that hard acceptors prefer to bind to hard donors and soft acceptors to soft donors, when forming stable compounds. Table 9.4 gives an abbreviated list of metal acceptors and ligand donors. As a hard acceptor is characterized by low polarizability, low electronegativity and large positive charge density, similarly, a hard donor is characterized by low polarizability, but high electronegativity and a high negative charge density, whereas the opposite holds for a soft donor atom. Since electron acceptors are broadly classified as acids in accordance with Lewis' definition and electron donors are referred to as bases, the principle presented is known as the HSAB principle (hard and soft acids and bases principle).

The principle is conveniently used to select the complexes that are of importance to include in a model description. Some metals occur in the earthen crust as ores of oxides and carbonates, whereas other metals occur as sulphides. The explanation is that the hard acids such as Mg^{2+}, Ca^{2+}, Al^{3+} and Mn^{2+} form strong bonds with the hard bases, e.g. carbonate and oxide. Conversely, softer acids, Hg^{2+}, Ag^{+} and Pb^{2+}, prefer soft bases such as sulphides. Similarly, the formation of metal-humic complexes play no role for hard acids such as Mg^{2+}, Ca^{2+} and Mn^{2+} while these complexes are of importance for the transport in aquatic systems of Cd^{2+}, Sn^{2+} and Pb^{2+}.

Table 9.4
Classification in Accordance with the HSAB System

Hard acceptor	Intermediate	Soft acceptor
H^+, Na^+, K^+, Be^{2+}, Mg^{2+} Ca^{2+}, Mn^{2+}, Al^{3+}, Cr^{3+} Co^{3+}, Fe^{3+}, As^{3+}	Fe^{2+}, Co^{2+}, Ni^{2+} Cu^{2+}, Zn^{2+}, Pb^{2+}	Cu^+, Ag^+, Au^+, Tl^+ $Hg_2{}^{2+}$, Pd^{2+}, Cd^{2+}, Pt^{2+} Hg^{2+}, CH_3Hg^+
Hard donor	**Intermediate**	**Soft donor**
H_2O, OH^-, F^- Cl^- $PO_4{}^{3-}$, $SO_4{}^{2-}$, $CO_3{}^{2-}$, O^{2-}	Br^-, $NO_2{}^-$ $SO_3{}^{2-}$	SH^-, S^{2-}, RS^- CN^-, SCN^-, CO, R_2S, RSH, RS^-

9.4 APPLICATION OF COMPLEX FORMATION IN ENVIRONMENTAL MODELS

Only a few models consider formation of complexes, because it is presumed that this process is insignificant for the problem in hand. This assumption may hold for a number of case studies, for instance in the case that biomagnification or the hydrodynamic distribution is modelled. Examples are published by Thomann (1974 and 1984) and Fagerstrøm & Åsell (1973).

The uptake rate of metals by organisms is, however, often influenced highly by the presence of complexes as pointed out by Baccini (1976), a fact, which is often neglected by modellers. Fig. 9.7 shows a typical relationship between metal concentration in phytoplankton and water in absence and in presence of organic complexing ligands.

The relationship is linear, but in the latter case the increase in metal concentration in phytoplankton does not occur as long as the metal is bound to the organic ligand. Complex formation may also explain seasonal variation in metal contents of the biomass. Baccini & Robert (1976) have published an excellent illustration of seasonal variations of copper in biomass. During the period of high primary production, i.e. spring and summer, pH increases, which leads to a reduction of the biologically available copper. The transfer of copper to the biota is thereby reduced, while pH decreases during the winter months, leads to an increase in available copper due to a lower concentration in reactive species of copper complexing to organic ligands. Although the total copper concentration is constant (1-2) * 10^{-8} M, the pCu in biomass during the winter months is about 4.5, while during the spring and summer months pCu is 6.0 or above.

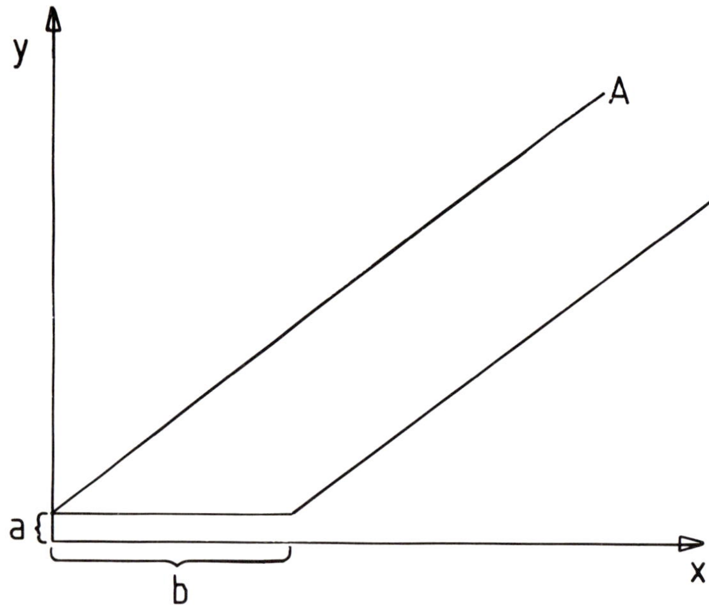

Fig. 9.7: Concentration of a trace metal in phytoplankton, Y, versus the concentration in water, X (mg/l water). A is the relationship for water in the absence of organic ligands, while B is for water in the presence of organic ligands equivalent to the metal concentration. a indicates the trace metal concentration found in phytoplankton from completely uncontaminated water.

Seip (1978) found a difference between uptake of zinc by algae in polluted and unpolluted water. In the latter the uptake was higher, which might be explained by a more pronounced complex formation in polluted water. Seip has developed a model for the uptake of zinc. Predictions were well in accordance with observations, but the uptake rate was found by calibration and no attempts were made to explain the difference in uptake rate by the differences in organic ligand concentrations.

Fig. 9.8 shows a simple effect model which considers the formation of complexes.

Figs. 9.9 and 9.10 give the result of this copper model. The role of formation of humic acid copper complexes is important and should not be neglected in models of metal mobilization in the sediment.

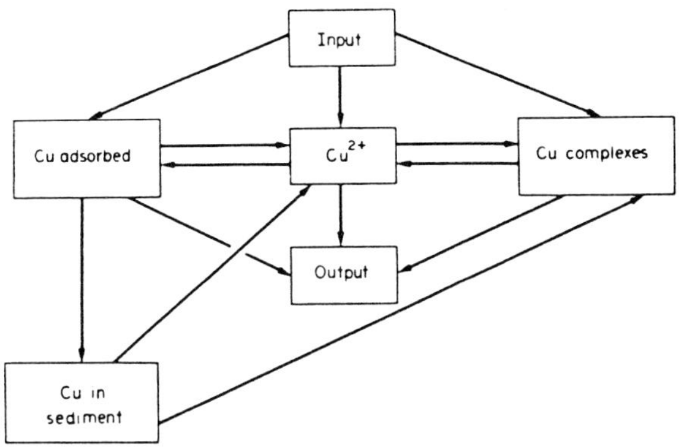

Fig. 9.8: Conceptual diagram of a simple copper model. Cu^{2+} is translated into an effect in the model.

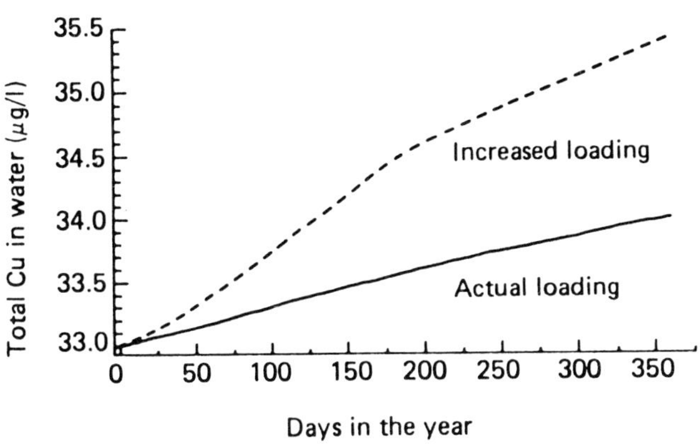

Fig. 9.9: A one-year simulation of total copper in the water phase in Lake Victoria (Entebbe) - CUT - at actual and at tenfold increased loading.

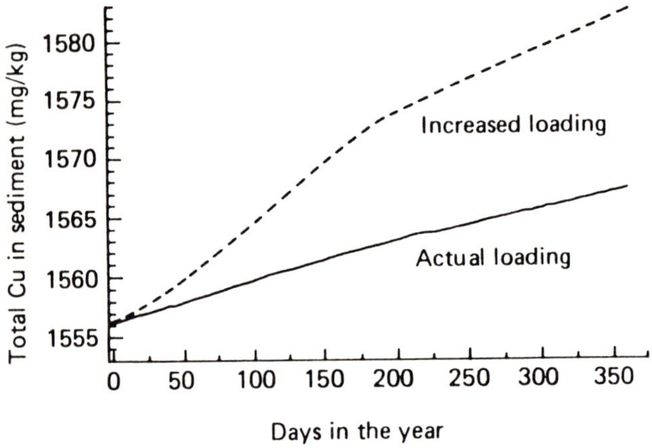

Fig. 9.10: A one-year simulation of total copper in the sediment in Lake Victoria (Entebbe) - SECU - at actual and at tenfold increased loading.

9.5 CONCLUSIONS

The literature shows only a few examples of models which consider complex formation as an important process to be included in the description. Many complexing agents are competing in nature. This renders it rather cumbersome to include complex processes in model constructions. However, these processes are crucial in models of metal mobilizations and of toxic effects of metals, and even in models of biotic uptake of metals, as exemplified in section 9.4.

A simplification is often possible in practical modelling, because in many cases it is feasible to exclude many complexing agents and only include a few very stable complexes in the model description. In such cases it is rather simple to take the complex formation into consideration, since the equilibrium constants in many actual cases can be found in the literature and an equilibrium description often will be sufficient (as mentioned in section 9.2).

It can, therefore, be recommended to use more complex processes only in the modelling of metal distributions and effects. More experience is needed to give a better estimation concerning the inclusion or omission of these processes. Until this experience has been gained, the process should be included wherever it seems appropriate and feasible.

REFERENCES

Baccini, P., 1976. Untersuchungen über den Schwermetallhaushalt der Seen. Schweiz. Z. Hydrol. 38: 121-158.

Baccini, P. and P.V. Roberts, 1976. Die Belastung der Gewässer durch Metalle. Beil. Forsch. Tech. Neue Zürcher Z. 18: 57-58.

Culp, G., S. Hansen and G. Richardson, 1968. High-rate sedimentation in water treatment works. J. Am. Water Works Ass., 60: 681.

Dunlap, W.J., R.L. Cosby, J.F. McNabb, B.E. Bledsoe, M.R. Scalf, 1971. Investigation concerning probable impact of nitrilotriacetic acid on ground waters. Environmental Protection Agency. Washington D.C, U.S. Gov. Printing Office.

Fagerstrøm, T. and B. Åsell, 1973. Methyl mercury accumulation in an aquatic food chain. A model and implications for research planning. Ambio 2: 164-171.

Jørgensen, S.E., 1971. Precipitation of proteins in waste water. Vatten, 27: 58-72.

Mottola, H.A., 1974. Nitrilotriacetate acid as a chelating agent. Toxicol. Environ. Chem. Rev. 2: 99-161.

Overbeek, J.Th.G., 1952. Kinetics of Flocculation. Chapter VII in colloid Science, vol. I, Irreversible Systems, H.R. Kruyt, Ed., Elsevier, Amsterdam.

Pearson, R., 1968a. Hard and soft acids and bases, HSAB, Part I. Fundamental principles. J. Chem. Educ. 45: 581-587.

Pearson, R., 1968ab Hard and soft acids and bases, HSAB, Part II. Underlying theories. J. Chem. Educ. 45: 643-648.

Prakash, A., 1976. NTA (nitrilotriacetic acid) - an ecological appraisal. Water Pollut. Control Direct. Can. Econ. Technol. Rev. Rep. EPS 3-WP-76-8.

Rashid, M.A. and L.H. King, 1971. Chemical characteristics of fractionated humic acids associated with marine sediments. Chem. Geol. 7: 37-43.

Seip, K.L., 1978. Mathematical model for uptake of heavy metals in benthic algae. Ecological Modelling 6: 183-198.

Thom, N.S., 1974. Nitrilotriacetic acid: Literature survey. Water Res. 5: 391-399.

Thomann, R.V., 1974. Physio-chemical and ecological modelling the fate of toxic substances in natural water systems. Ecol. Modelling. 22: 145-171.

Thomann, R.V. et al., 1974. A food chain model of cadmium in western Lake Erie. Water Research 8: 841-851.

CHAPTER 10

HYDROLYSIS AND CHEMICAL REDOX PROCESSES

by
Sven Erik Jørgensen and Arne Jensen

10.1 INTRODUCTION

The term hydrolysis is applied for processes which can proceed with water, hydrogen ions and hydroxyl ions. Redox processes cover those processes where transfer of electrons takes place.

Chemically these two types of processes are significantly different, but they are treated together in this chapter, because the environmental consequences of the two processes are somewhat similar. Furthermore, the description of the processes in models is almost the same, as the description is mainly dependent on the reaction rate and to a less extent on the process type.

In Table 10.1 are given a few illustrative examples of the two types of processes for organic as well as for inorganic reactants.

Table 10.1
Examples of hydrolysis and redox processes in the environment

$Fe^{3+} + 3H_2O$	=	$Fe(OH)_3 + 3H^+$
$Mn^{2+} + 2H_2O$	=	$Mn(OH)_2 + 2H^+$
$Fe^{3+} + e^-$	=	Fe^{2+}
$MnO_2 + 2e^- + 4H^+$	=	$Mn^{2+} + 2H_2O$
$CH_2Cl_2 + 2H_2O$	=	$CH_2(OH)_2 + 2H^+ + 2Cl^-$
$CH_3COC_4H_9 + H_2O$	=	$CH_3COOH + C_4H_9OH$
$1/4\ O_2 + Fe^{2+} + 3/2\ H_2O$	=	$FeO(OH)\ (s) + 2H^+$
$2(CH_3)_2S + O_2$	=	$2(CH_3)_2SO$
$CH_3CHO + 1/2\ O_2$	=	CH_3COOH

Hydrolysis may give rise to the solubility of metal ions, e.g.

$$Al(OH)_3 + OH^- \longrightarrow Al(OH)_4^- \quad (10.1)$$

$$Al(OH)_3 + 3H^+ + nH_2O \longrightarrow Al(H_2O)_{n+3}^{3+} \quad (10.2)$$

or the processes may imply formation of insoluble compounds, often hydroxides:

$$Fe^{3+} + 3H_2O \longrightarrow Fe(OH)_3\,(s) + 3H^+ \quad (10.3)$$

The increased solubility of heavy metals with decreasing pH due to formation of metal aqua ions has great environmental interest. As pH is decreased as a result of acidic precipitation in areas with low pH buffer capacities, the toxic effect of metal ions is increased significantly. As an example of these relations Fig. 10.1 illustrates the aluminium concentration in clearwater lakes in Sweden and Norway as functions of pH.

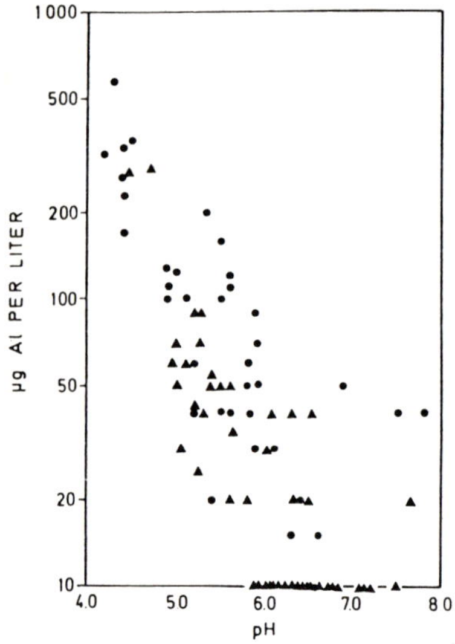

Fig. 10.1: Aluminium concentration in clearwater lakes in Sweden (circles) and Norway (triangles, unpublished data). Values plotted as 10 µg/litre are below the analytical detection limit (S.E. Jørgensen and I. Johnsen, 1988).

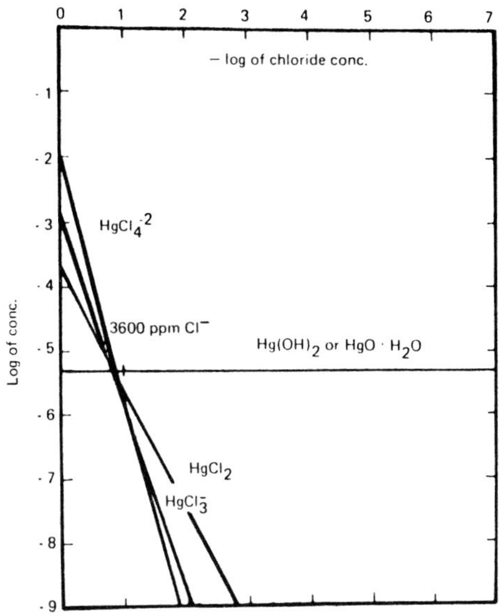

Fig. 10.2: Mercury complexes versus chloride concentration at 25°C and pH=7. (Jørgensen et al. 1979)

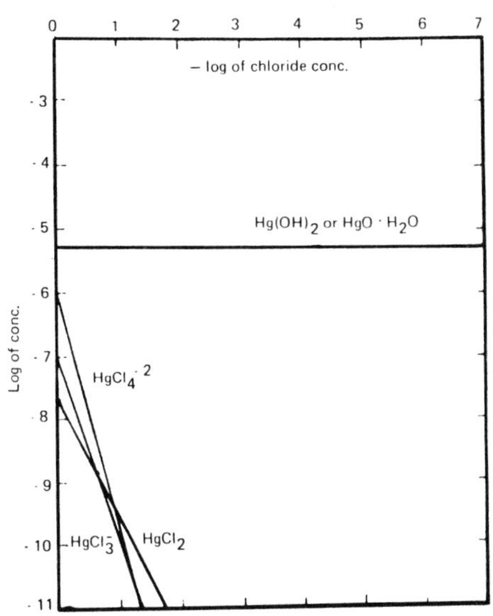

Fig. 10.3: Mercury complexes versus chloride concentration at 25°C and pH=9. (Jørgensen et al. 1979)

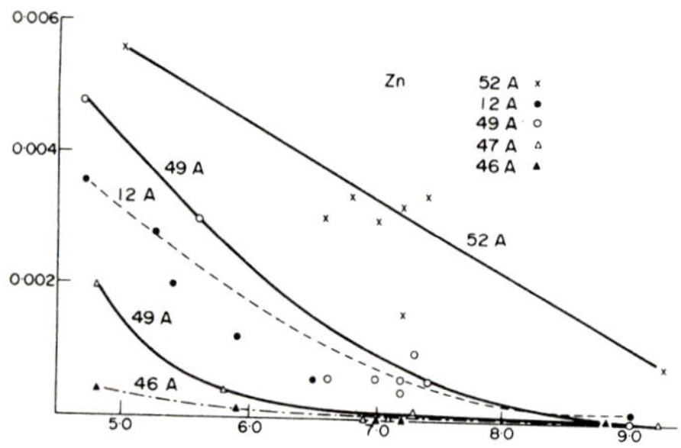

Fig. 10.4

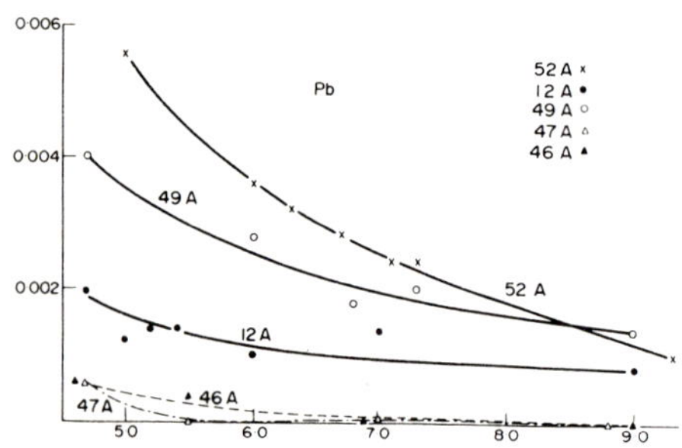

Fig. 10.5

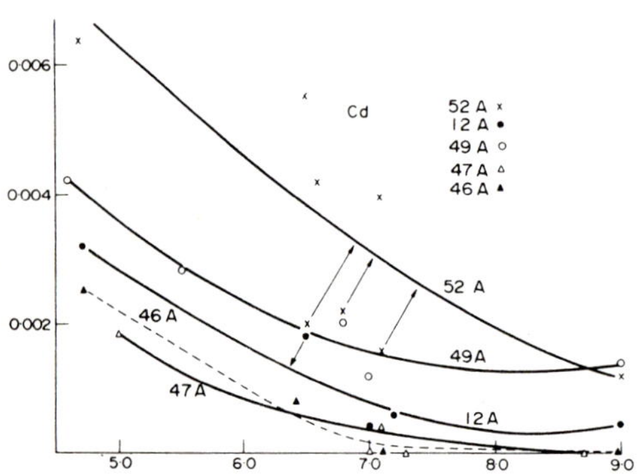

Fig. 10.6

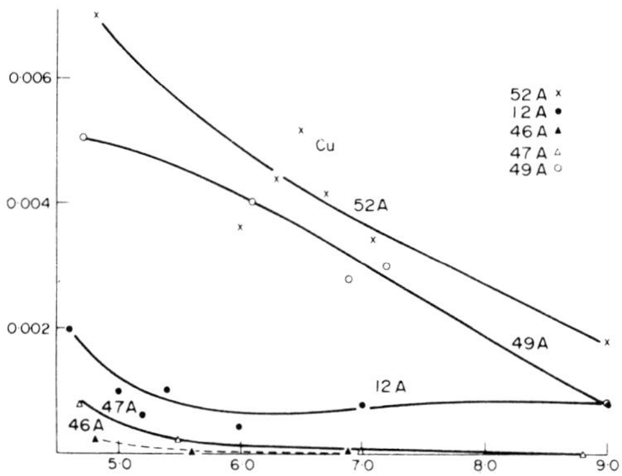

Fig. 10.7

Soil sample	Clay (%)	Humus (%)	CEC (meq/100 g)	Dry matter (g/litre)	Ignition residue (g/kg)
47A	33.2	3.6	27.5	800	740
46A	34.4	3.9	28.0	823	757
12A	22.8	11.8	24.5	702	631
49A	11.5	3.3	13.5	908	856
52A	2.4	2.1	5.1	989	959

Figs. 10.4-7: Distribution g/litre of soil water to g/kg bound in soil for different types of soilnamed D (see table) versus pH. The results Zn, Pb, Cd and Cu are shown (S.E. Jørgensen, 1975).

The hydrolysis interferes with the formation of complexes as illustrated in Figs. 10.2 and 10.3. As will be seen pH determines the concentration of chloride complexes. Figs. 10.4 - 10.7 illustrate other pertinent examples. A particular example is the heavy metals release from soil as function of pH. It is clear from these figures that lowering of the pH as a result of acidic precipitation will increase the heavy metal concentration in run-off.

Metal ions are able to form a number of species as a result of hydrolysis such as aqua, hydroxo, hydroxo-oxo and oxo complexes. This implies that multivalent metal ions are able to participate in a series of consecutive proton transfers:

$$Fe(H_2O)_6^{3+} = Fe(H_2O)_5OH^{2+} + H^+ = Fe(H_2O)_4(OH)_2^+ + 2H^+ =$$

$$Fe(OH)_3(H_2O)_3\ (s) + 3H^+ = Fe(OH)_4(H_2O)_2^- + 4H^+ \qquad (10.4)$$

The predominant species are determined by the dilution and pH, as is obvious from the following two examples:

$$Mg^{2+} + H_2O = MgOH^+ + H^+ \tag{10.5}$$

$$Mn^{2+} + H_2O = MnOH^+ + H^+ \tag{10.6}$$

Fig. 10.8 attempts to describe how the oxidation of the central atom determines the predominant species.

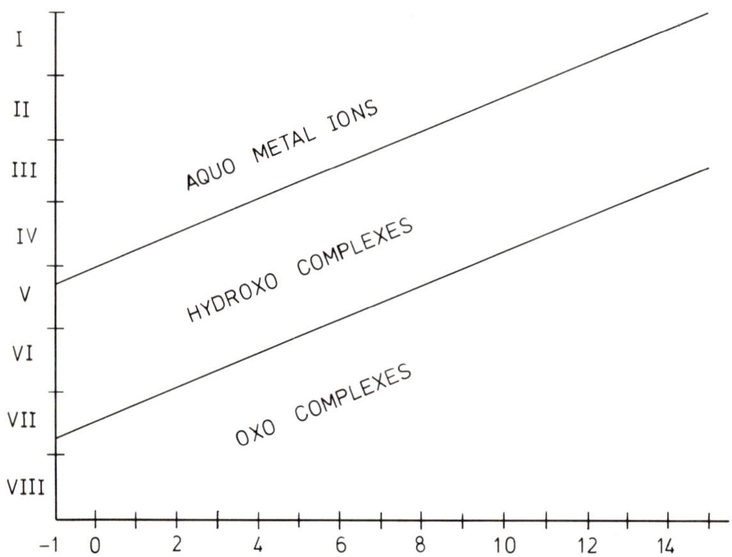

Fig. 10.8: pH range for occurrence of complexes for various oxidation states.

As seen from this discussion the hydrolysis of inorganic compounds is of major environmental interest. However, hydrolysis of organic compounds in the aquatic environment is of equal interest.

Organic pollutants can undergo reactions with water, resulting in introduction of a hydroxyl group into the chemical structure:

$$RX + H_2O = ROH + HX \quad (10.7)$$

$$RCOX + H_2O = RCOOH + HX \quad (10.8)$$

These reactions are catalyzed by oxonium and/or hydroxyl ions, which implies that the rate of hydrolysis, R, is given by the equation:

$$R = \sim \frac{dc}{dt} * K_H - [A] = K_A * [H^+] [A] + K_B [OH^-] [A] + K_N * [H_2O] * [A] \quad (10.9)$$

where K_H is a pseudo first order rate constant at a given pH, while K_A and K_B are second order rate constants. K_N is the second order rate constant for neutral reaction of a chemical compound with water, which may be expressed as a pseudo first-order rate constant. Equation (10.9) indicates that the rate of hydrolysis is strongly dependent on pH, unless K_A and K_B are equal to zero. Some chemicals including alkyl halides, were found (Mabey and Mill, 1978) to have hydrolysis rates which are independent of pH over the usual environmental pH-range (4-9), while other chemicals such as carboxylic acid esters have hydrolysis rates which are highly dependent on pH.

The mechanisms of hydrolysis including predictive test methods to estimate kinetic rates of hydrolysis of various compounds, have been studied by Mabey et al. (1978), Wolfe et al. (1978) and Tinsley (1979). Table 10.2 gives some examples of hydrolysis rates of some halogenated compounds.

Many inorganic ions are dominant participants in environmental **redox processes** and Table 10.3 shows the equilibrium constant of pertinent redox processes in aquatic conditions.

It is obviously of importance to describe the stability relationships for various inorganic components. This can be done in a simple graphic representation using pe-pH diagrams. Such diagrams show in a comprehensive way how protons and electrons simultaneously shift the equilibrium under various conditions of pe and pH.

pe is defined as the negative logarithm of the relative electron activity.

$$pe = -\log[e] \quad (10.10)$$

which is a parallel to the pH definition:

$$pH = -\log[H^+] \quad (10.11)$$

Table 10.2
Hydrolysis rates at 25°C and pH 7 of some halogenated compounds

Compound	Rate constants (1/sec) K_N	$K_B[OH^-]$	K_h	$t_{1/2}$
CH_3F [a]	$7.44 \cdot 10^{-10}$	$5.82 \cdot 10^{-14}$	$7.44 \cdot 10^{-10}$	30 yr
CH_3Cl [a]	$2.37 \cdot 10^{-8}$	$6.18 \cdot 10^{-13}$	$2.37 \cdot 10^{-8}$	339 days
CH_3Br [a]	$4.09 \cdot 10^{-7}$	$1.41 \cdot 10^{-11}$	$4.09 \cdot 10^{-7}$	20 days
CH_3I [a]	$7.28 \cdot 10^{-8}$	$6.47 \cdot 10^{-12}$	$7.28 \cdot 10^{-8}$	110 days
$CH_3CHClCH_3$ [a]	$2.12 \cdot 10^{-7}$	-	$2.12 \cdot 10^{-7}$	38 days
$CH_3CH_2CH_2Br$ [a]	$3.86 \cdot 10^{-6}$	-	$3.86 \cdot 10^{-6}$	26 days
$CH_3-C(CH_3)(Cl)-CH_3$ [a]	$3.02 \cdot 10^{-2}$	-	$3.02 \cdot 10^{-2}$	23 sec
CH_2Cl_2 [a]	$3.2 \cdot 10^{-11}$	$2.3 \cdot 10^{-15}$	$3.2 \cdot 10^{-11}$	704 yr
$CHCl_3$ [b]	-	$6.9 \cdot 10^{-12}$	$6.9 \cdot 10^{-12}$	3500 yr
$CHBr_3$ [b]	-	$3.2 \cdot 10^{-11}$	$3.2 \cdot 10^{-11}$	686 yr
CCl_4 [b,c]	-	$4.8 \cdot 10^{-7}$	$4.8 \cdot 10^{-7}$	7000 yr (1 ppm)
$C_6H_5CH_2Cl$	$1.28 \cdot 10^{-5}$	-	$1.28 \cdot 10^{-5}$	15 hr

Source: Mabey and Mill (1978)

[a] $K_h = K_N$; $K_B \ll K_N$ [b] $K_h = K_B$; $K_N \ll K_B$
[c] Rate second order with respect to $[CCl_4]$, K_h ($L\ mol^{-1}\ sec^{-1}$)

Table 10.3
Equilibrium constants for selected redox reactions

Reaction	log K, 25°C	Standard electrode potential (V), 25°C
$Na^+ + e = Na\ (s)$	-46	-2.71
$Zn^{2+} + 2e = Zn\ (s)$	-26	-0.76
$Fe^{2+} + 2e = Fe\ (s)$	-15	-0.44
$Co^{2+} + 2e = Co\ s$	-9.5	-0.28
$V^{3+} + e = V^{2+}$	-8.8	-0.26
$2H^+ + 2e = H_2\ (g)$	0.0	0.00
$S\ (s) + 2H^+ + 2e = H_2S$	+0.47	+0.14
$Cu^{2+} + e = Cu^+$	+2.7	+0.16
$AgCl\ (s) + e = Ag\ (s) + Cl^-$	+3.7	+0.22
$Cu^{2+} + 2e = Cu\ (s)$	+12.0	+0.34
$Cu^+ + e = Cu\ (s)$	+18.0	+0.52
$Fe^{3+} + e = Fe^{2+}$	+13.2	+0.77
$Ag^+ + e = Ag\ (s)$	+13.5	+0.80
$Fe(OH)_3\ (s) + 3H^+ + e = Fe^{2+} + 3H_2O$	+18.8	+1.06
$IO_3^- + 6H^+ + 5e = 1/2\ I_2\ (s) + 3H_2O$	+104	+1.23
$MnO_2\ (s) + 4H^+ + 2e = Mn^{2+} + 2H_2O$	+42	+1.23
$Cl_2\ (g) + 2e = 2Cl^-$	+46	+1.36
$Co^{3+} + e = Co^{2+}$	+31	+1.82

pe is related to E, the electrode potential, by:

$$pe = \frac{F * E}{2.3RT} \tag{10.12}$$

$$E = E° + \frac{2.3RT}{nF} \log\frac{[ox]}{[red]} \tag{10.13}$$

$$pe° = \frac{F * E°}{2.3RT} \tag{10.14}$$

This implies that pe is a measure of the free energy, ΔG:

$$pe = \frac{\Delta G}{-n * 2.3RT} \qquad pe° = \frac{\Delta G°}{-n * 2.3RT} \tag{10.15}$$

(ΔG° is the standard free energy and ΔG and ΔG° refer to the half reaction written in the form of reduction, cf. Table 10.3 and that pe is related to the equilibrium constant as follows:

$$pe = \frac{1}{n} \log K \tag{10.16}$$

The use of these equations can be illustrated by the following process:

$$Fe^{3+} + e = Fe^{2+} \tag{10.17}$$

For this process we have:

$$E° = 0.77; \quad pe° = \frac{F * 0.77}{2.3 * R * 298} \quad (t=25°C) \tag{10.18}$$

$$\log K = n * pe° = 1 * \frac{0.77}{0.059} = 13.1 \tag{10.19}$$

If in an acid solution $[Fe^{3+}] = 10^{-3}$ and $[Fe^{2+}] = 10^{-2}$ (10.20)

$$pe = pe° + \frac{1}{n} \log \frac{[Fe^{3+}]}{[Fe^{2+}]} = 13.1 + \frac{1}{1} \log 10^{-1} = 12.1 \tag{10.21}$$

Figs. 10.9 - 10.10 give examples of pertinent pe-pH diagrams of environmental interest.

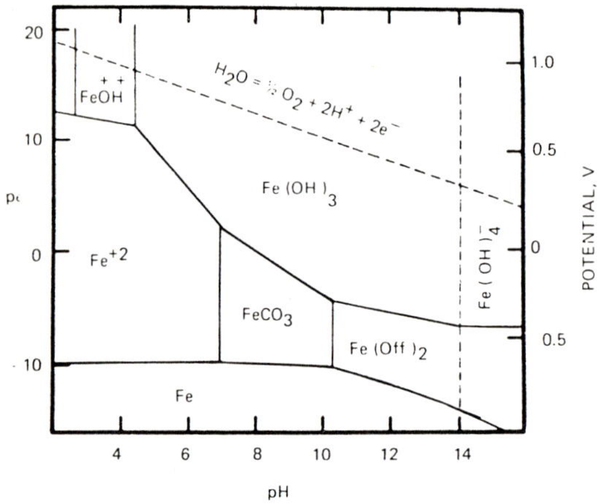

Fig. 10.9: pe-pH diagram for iron. Total concentration is $2{*}10^{-3}$ M. Ion activity from other ions 10^{-5} M.

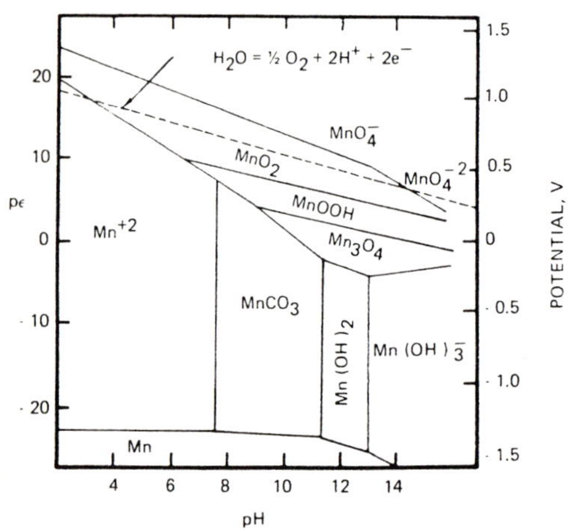

Fig. 10.10: pe-pH diagram for manganese. Total concentration is $2{*}10^{-3}$ M. Ion activity from other ions 10^{-5} M.

The importance of inorganic redox processes in environmental context can be illustrated by examples.

If FeS_2 is exposed to air e.g. by reduced water level in mines, the following processes will occur:

$$2FeS_2 + 2H_2O + 7O_2 = 2FeSO_4 + 2H_2SO_4 \tag{10.22}$$

$$4FeSO_4 + O_2 + 2H_2SO_4 = 2Fe_2(SO_4)_3 + 2H_2O \tag{10.23}$$

$$Fe_2(SO_4)_3 + 6H_2O = 2Fe(OH)_3\,(s) + 3H_2SO_4 \tag{10.24}$$

As will be seen, the formation of considerable amounts of sulfuric acid occurs, resulting in extremely low pH values, which in many cases have caused great damage to the environment.

Another example is related to the release of phosphorus from sediments in aquatic ecosystems, which may cause increased eutrophication. Phosphorus may be bound as iron(III)phosphates, but if the conditions are changed from aerobic to anaerobic, the following process will occur:

$$FePO_4 + HS^- + e = FeS + HPO_4^{2-} \tag{10.25}$$

Edgington and Callender (1970) mention a third example. Lake Michigan has ferromanganese nodules, which contain unexpectedly large concentrations of arsenic, up to 345 ppm, but averaging 180 ppm. The nodules are stable under aerobic conditions but under anaerobic conditions arsenic will be released. As arsenic is highly toxic to mammals and also carcinogenic it is obviously of great importance to formulate the redox processes in the Lake Michigan sediment in order to give predictions for the release of arsenic.

All organic matter will suffer oxidation if present in an aerobic environment for a sufficient long period of time. If reduced material is sufficiently abundant, however, the oxygen dissolved in interstitial water or the lower layers of aquatic ecosystems, will be exhausted. Oxidation of organic matter, however, will continue by denitrification and sulphate reduction. All these processes can in principle occur by pure chemical oxidation, but in general the microbiological oxidation plays a far more important role.

Aquatic systems have several oxidants such as radicals, singlet oxygen and triplet diradicals with half lifetimes of a few milliseconds as well as more stable oxidants such as peroxides, peracids and ozone. Table 10.4 reviews organic oxidation chemistry.

The oxidation rate, R_{Ox} may be written as follows:

$$R_{Ox} = k_{Ox} * [C][Ox] \tag{10.26}$$

where k_{Ox} is the specific second order rate constant for oxidation at a specific temperature and [C] and [Ox] are molar concentrations of the chemical compounds and the oxidant, respectively. The total rate of oxidation - if more oxidants work simultaneously - is the sum of the rates for reaction of each kind of oxidants:

$$R_{Ox} = (k_{Ox_1}[Ox_1] + k_{Ox_2}[Ox_2] \ldots\ldots\ldots\ldots) * [C]$$

$$= \left(\sum_{n=1}^{n=n} k_{Ox_n} * [Ox_n] \right) * [C] \tag{10.27}$$

Table 10.4
Review: Chemical oxidation of organic compounds

Oxidant	Process type
$RO^{\cdot}_n$	$RO^{\cdot}_n$ + H –C– – RO_nH + ·C– (n=1 or 2)
$RO^{\cdot}_2$	$RO^{\cdot}_2$ + C–C – RO_2 + C — C ·
HO	HO + –
$RO^{\cdot}_2$	$RO^{\cdot}_2$ + NO – $RO^{\cdot}$ + NO_2
$^1O_2$ or O_3	$^1O_2$ + C – CCH – C–C–C
$^1O_2$ or O_3	$^1O_2$ + –
$^1O_2$ or O_3	$^1O_2$ + C–C –
$^1O_2$ or O_3	$2R_2S$ + $^1O_2$ – $2R_2SO^{\cdot}$
$^1O_2$ or O_3	ArOH + $^1O_2$ – $ArO^{\cdot}$ + $HO^{\cdot}_2$

Integration between the time limits 0 and t gives:

$$\ln[C_0] / [C_t] = \sum_{n=1}^{n=n} k_{Ox_n} * [Ox_n] * t \qquad (10.28)$$

or if the half-life limit $t_{1/2}$ is used:

$$t_{1/2} = \ln 2 / (\Sigma k_{Ox_n} * [Ox_n]) \qquad (10.29)$$

Table 10.5 gives rates of oxidation by singlet oxygen in water at 25°C.

Table 10.5
Rate constants of oxidation by singlet oxygen in water at 25°C. The concentration of singlet oxygen can be estimated as 10^{-12} M

	Log k_{Ox} M^{-1} s^{-1}
Unsubstituted aliphatic carbons	3.5
Cyclic olefins	5.3
Substituted olefins	6.0
Dialkyde sulfide	6.8
Diene	7.0
Imidazole	7.6
Furan	8.2
Trialkyleneamine	8.9

10.2 MODELS OF HYDROLYSIS AND REDOX PROCESSES

Many inorganic processes proceed relatively fast compared with the usually applied intervals in environmental models. It implies that equilibrium diagrams as those presented in Figs. 10.9 - 10.10 can be applied to give equilibrium data, when pH and pe are known.

The kinetics of oxygenation of iron(II) and manganese(II) might be used as illustrations of such relatively fast processes. Iron(II) is oxidized by a rate, which is first order with respect to both iron(II) and oxygen and second order with respect to hydroxide ion:

$$\frac{d\,[Fe(II)]}{dt} = -k\,[Fe(II)]\,[OH^-]^2\,P_{O2} \qquad (10.30)$$

where $k = (8.0 \pm 2.5) * 10^{13} min^{-1} atm^{-1} mole^{-3}$ at 20°C. This differential equation can be rewritten as follows:

$$\frac{d\ln [Fe(II)]}{dt} = -k * \frac{10^{-28}}{[H^+]^2} * P_{O2} \tag{10.31}$$

The kinetics of Mn(II) oxygenation suggest an autocatalytic reaction, following the expression:

$$\frac{d\ [Mn(II)]}{dt} = -k_1\ [Mn(II)] - k_2\ [Mn(II)]\ [MnO_2] \tag{10.32}$$

where k_1 and k_2 are constants.

Both oxidation processes proceed in a matter of minutes as shown in Fig. 10.11, where the kinetics of oxidation of iron(II) and manganese(II) are shown.

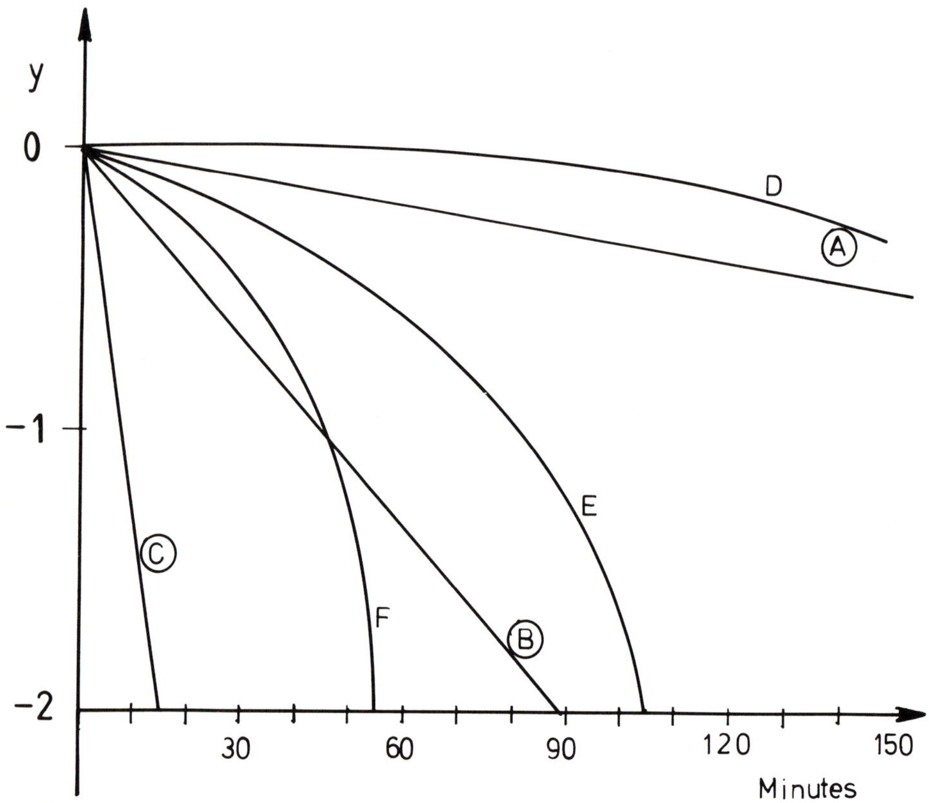

Fig. 10.11: Oxidation of Fe(II) and Mn(II) in HCO_3^- solutions. y is the logarithm of the concentration of the metal(II) ions mentioned. **A.** Fe(II) at pH=6.6. **B**. Fe(II) at pH=6.9. **C.** Fe(II) at pH=7.2. **D.** Mn(II) at pH=9.0. **E.** Mn(II) at pH=9.3 and **F.** Mn(II) at pH=9.5.

It is clear from these considerations that when intervals of half a day, or one day, are used in a model, equilibrium description will be sufficient, while intervals of a few hours, or even minutes, will require a description of the oxygenation kinetics; see equations 10.31 - 10.32.

However, in case the amount of oxygen is limited, the transfer of oxygen may be the rate-determining process. As a significant amount of oxygen may be needed for oxidation of iron(II) and manganese(II), it will under all circumstances be necessary to account for these components, when an oxygen model is set up.

In Fig. 10.12 is shown a submodel for the oxygen concentration at the sediment surface. As will be seen, this model also accounts for the oxygenation of iron and manganese compounds. Furthermore, the submodel considers the decomposition of organic matter at the surface and the diffusion of oxygen from the water column. If the oxygen transfer is the rate determining process, and the oxygen is consumed almost instantly, the submodel must describe as accurately as possible the transfer of oxygen and at the same time describe the decomposition of organic matter by anaerobic processes. The oxidation of iron and manganese compounds will proceed by a rate determined by the transfer of oxygen, but the amount of unoxidized iron and manganese compounds will require computation to determine when the aerobic conditions will occur. Obviously this will involve the submodel being coupled with other submodels - for instance, for the sedimentation of iron and manganese compounds and their addition to the sediment.

Oxidation of iron(II) is followed by hydrolysis and oxidation to insoluble iron(III) oxide, but as this process is very fast and iron(III) is incapable of further oxidiation, this process is of no interest for oxygen submodels. The amount of iron(III) compounds may, however, be important in orther relations as they have a higher adsorption capacity for manganese(II) and phosphorus compounds than the corresponding iron(II) compounds.

Hydrolysis and chemical oxidation of organic matter is usually insignificant compared with microbiological decomposition. The influence of these processes on the oxygen budget is therefore mostly unimportant. However, these processes might be of great importance for specific toxic organic compounds. Therefore, many models of toxic substances in aquatic ecosystems will have to include these processes. Organic compounds are usually decomposed by a chain of processes, of which some are chemical and others are biochemical. Here it is important to know at least the slowest steps and the type of kinetics valid for these as they will determine the overall decomposition rate. This will in many cases require quite a comprehensive chemical knowledge of environmental processes which in many cases today we do not have.

Many overall decomposition processes are described as first-order reactions, which seems to be an acceptable approximation since the slowest process actually is a first-order reaction - and this may not be known by the modeller who may use this type of kinetics because it fits with his observations.

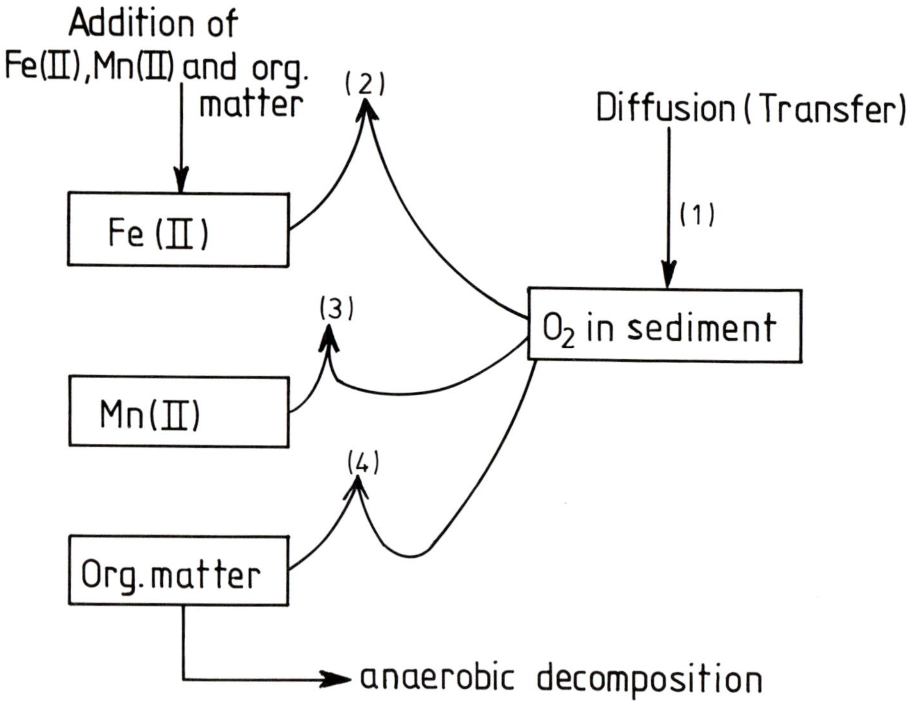

Fig. 10.12: Model of oxidations at sediment surface. The processes are: (1) diffusion (2) oxidation of iron(II) (3) oxidation of manganese(II) (4) oxidation of organic matter and (5) anaerobic decomposition.

Fig. 10.13 attempts to summarize these considerations. If hydrolysis and/or redox processes seem to be of importance for a model, several questions should be answered before the model equations are formulated. These questions are:

* What is the time step needed for the model?
* What is the rate of the processes compared with the time steps?
* Is there a chain of reactions?
* Which reaction step determines the overall rate?

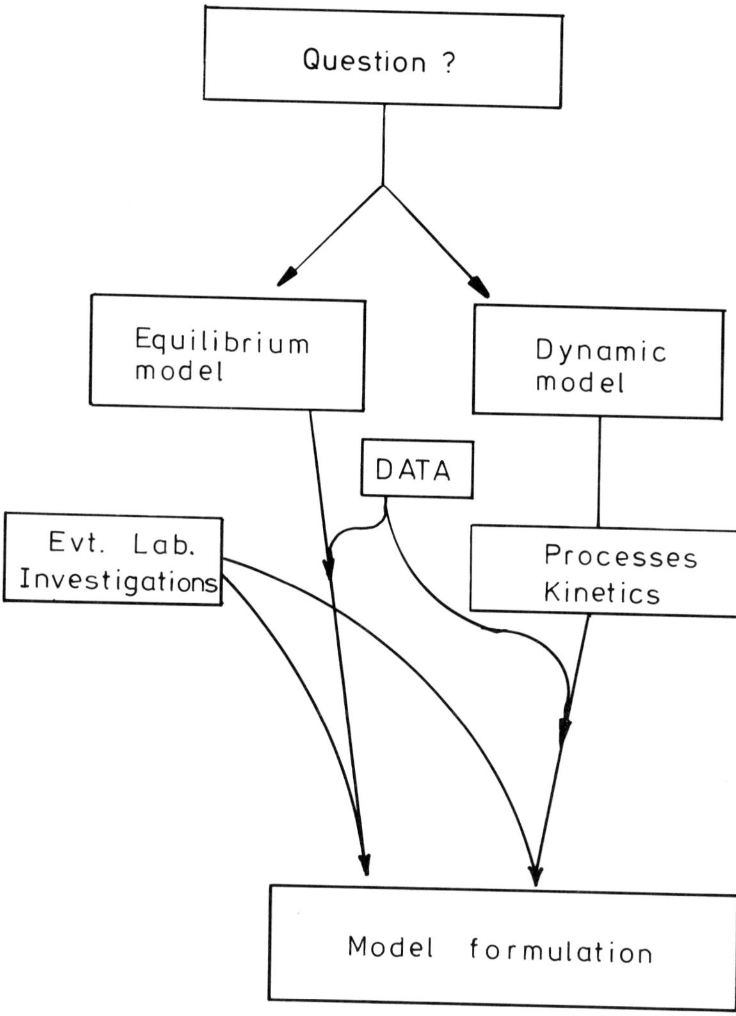

Fig. 10.13: The diagram illustrates how it is feasible stepwise to construct a model of hydrolysis and/or redox processes.

Dependent on the answers to these questions the model will be either a model of chemical equilibria or a model of the chemical kinetics. In the latter case it is of special importance to know the chain of processes and the kinetics of the slowest step(s) for the formulation of the model equations. If this knowledge cannot be provided it is necessary to test various simple kinetics and see which one(s) will best fit the observations. This will obviously raise the question of the quality of the data able to support such a test (eventually a calibration of the kinetic constants). If this is not the case, the required data must be provided by laboratory investigations.

Either equilibria or dynamic models for hydrolysis and/or redox processes have found use in the following environmental models:

1. Models of the formation of sulphuric acid by oxidation of iron sulphide; for instance, in mine drainage water. The result gives the acidity (incl. pH), which again may be translated into an environmental effect, see for instance Fig. 10.1, where the environmental effect is dissolution of aluminium ions.

2. Models of the oxygen budget in an aquatic ecosystem. The processes at the sediment surface have interest in this relation.

3. Release of phosphorus at the sediment surface, see process (10.25).

4. Models of organic toxic compounds in aquatic ecosystems. The decomposition rate is often determined by hydrolysis and oxidation processes.

10.3 PARAMETER ESTIMATION

Equilibrium data or kinetic data for a great amount of environmental hydrolysis and redox processes are available. Tables 10.2 and 10.5 illustrate which type of data it is possible to find in the literature, including handbooks of chemistry (see also Jørgensen et al. (1979)). Table 10.6 gives a more comprehensive list of standard electrode potentials, which are available for equilibrium calculations, see equations 10.12 and 10.16.

As mentioned in section 10.2 there is also a number of processes for which the required chemical data are not available. In this case observations of sufficient quality are needed, or laboratory investigations of the processes involved must be carried out to provide the required parameters (constants); see Fig. 10.13.

The influence of the temperature on the equilibrium data can be taken into account by use of thermodynamics:

$$\ln K(T) = -\Delta G^\circ (T) / RT \qquad (10.33)$$

where K(T) is the equilibrium constant as a function of T; R is the gas constant; $\Delta G°$ (T) the standard free energy as function of the temperature; and, T the absolute temperature. We have

$$\Delta G° = \Delta H° - T\Delta S° \qquad (10.34)$$

where $\Delta H°$ and $\Delta S°$ are standard enthalpy and entropy, respectively. In most cases $\Delta S°$ does not change significantly over the temperature range of interest and as the pressure is constant, we can use the Gibbs-Helmholz equation:

$$(\partial\ (\Delta G/T/\partial T)_p) = -\Delta H/T^2 \qquad (10.35)$$

and obtain:

$$\frac{d\ln K}{dT} = \Delta H°\ (T)\ /\ (RT^2) \qquad (10.36)$$

Integration of 10.36 yields

$$\ln K(T_1) = \ln K(T_2) + \int_{T_1}^{T_2} \frac{\Delta H°\ (T)\ dt}{RT^2} \qquad (10.37)$$

As

$$\Delta H°\ (T_2) = \Delta H°\ (T_1) + \int_{T_1}^{T_2} n * \Delta C_p\ (T)\ dT \qquad (10.38)$$

where ΔC_p is the change in molecular heat capacity between reactants and products, and n is the number of moles in the reaction. $\Delta H°$ (T) can be expressed by use of ΔC_p (T), which is often available, and equation (10.37) can be used to find an expression for K(T).

Chemical reaction rate constants, k, are also dependent on the temperature:

$$k = A\ \exp\ (\ (-E_a)\ /\ (RT)\ ) \qquad (10.39)$$

where A is a constant and E_a is the activation energy.

However, for rather narrow temperature ranges the following approximation may be used:

$$k_t = k_{to} * e^{(t-t_o)} \qquad (10.40)$$

where e is a constant.

Another approximation uses the so-called Q_{10} value, which is defined as:

$$k_t = k_{t-10} * Q_{10} \qquad (10.41)$$

For many chemical and biological reactions Q_{10} is in the order of 2.0.

10.4 CONCLUSIONS

In some models of environmental interest a description of hydrolysis and redox processes might be needed. It will, for instance, be the case where the oxygen consumption for oxidation of iron(II) and manganese(II) is significant for the oxygen budget, or where these processes are important for the decomposition of toxic organic compounds.

Our knowledge to these latter processes is rather limited in certain specific cases. However, as the number of toxic substance models is rapidly increasing it should be expected that more and more parameters will be available over the coming years. In addition more publications are likely to emerge about processes involving toxic substances in the environment in general, since the ecotoxicological consequences of toxic substances are of increasing concern.

Table 10.6
Standard electrode potentials at 25°C

Process			E° (volt)
$K^+ + e^-$	=	K (s)	-2.925
$Ca^{2+} + 2e^-$	=	Ca (s)	-2.87
$Na^+ + e^-$	=	Na (s)	-2.714
$Mg^{2+} + e^-$	=	Ng (s)	-2.37
$Al(OH)_4^- + 3e^-$	=	Al (s) + $4OH^-$	-2.35
$Al^{3+} + 3e^-$	=	Al (s)	-1.66
$Mn^{2+} + 2e^-$	=	Mn (s)	-1.18

Table 10.6 (continued)

Process			E° (volt)
$Cr^{2+} + 2e^-$	=	Cr (s)	-0.91
$2H_2O + 2e^-$	=	$H_2 + 2OH^-$	-0.828
$Zn^{2+} + 2e^-$	=	Zn (s)	-0.763
$2CO_2$ (g) [a] $+ 2H^+ + 2e^-$	=	$H_2C_2O_4$	-0.49
$Fe^{2+} + 2e^-$	=	Fe (s)	-0.440
$Cr^{3+} + e^-$	=	Cr^{2+}	-0.41
$Cd^{2+} + 2e^-$	=	Cd (s)	-0.403
$Co^{2+} + 2e^-$	=	Co (s)	-0.277
$Ni^{2+} + 2e^-$	=	Ni (s)	-0.24
$Sn^{2+} + 2e^-$	=	Sn (s)	-0.136
$Pb^{2+} + 2e^-$	=	Pb (s)	-0.126
$2H^+ + 2e^-$	=	H_2 (g)	0.0000
$S_4O_6^{2-} + 2e^-$	=	$S_2O_3^{2-}$	0.08
S (s) $+ 2H^+ + 2e^-$	=	H_2S	0.141
$Cu^{2+} + e^-$	=	Cu^+	0.153
$Sn^{4+} + 2e^-$	=	Sn^{2+}	0.154
$HSO_4^- + 3H^+ + 2e^-$	=	SO_2 (g) (S) $+ 2H_2O$	0.17
AgCl (s) $+ e^-$	=	Ag $+ Cl^-$	0.2222
Hg_2Cl_2 (s) $+ 2K^+ + 2e^-$	=	2Hg (l) + 2KCl (s)	0.2415
Hg_2Cl_2 (s) $+ 2e^-$	=	2Hg (l) $+ 2Cl^-$ (0.1 M KCl)	0.334
$Cu^{2+} + 2e^-$	=	Cu (s)	0.337
$Fe(CN)_6^{3-} + e^-$	=	$Fe(CN)_6^{4-}$	0.36
I_2 (s) $+ 2e^-$	=	$2I^-$	0.5345
$I_3^- + 2e^-$	=	$3I^-$	0.5355
$H_3AsO_4 + 2H^+ + 2e^-$	=	$HAsO_2 + 2H_2O$	0.559
MnO_4^-	=	MnO_4^{2-}	0.564
I_2 (aq) $+ 2e^-$	=	$2I^-$	0.62
O_2 (g) $+ 2H^+ + 2e^-$	=	H_2O_2	0.682
$Fe^{3+} + e^-$	=	Fe^{2+} (1 M HCl)	0.70
$Fe^{3+} + e^-$	=	Fe^{2+}	0.771
$Hg_2^{2+} + 2e^-$	=	2Hg (l)	0.789
$Ag^+ + e^-$	=	Ag (s)	0.7995

Table 10.6 (continued)

Process			E° (volt)
$NO_3^- + 2H^+ + e^-$	=	$NO_2 + H_2O$	0.80
$Hg^{2+} + 2e^-$	=	Hg (l)	0.854
$Cu^{2+} + I^- + e^-$	=	CuI (s)	0.86
$H_2O_2 + 2e^-$	=	$2OH^-$	0.88
$2Hg^{2+} + 2e^-$	=	Hg_2^{2+}	0.920
$NO_3^- + 3H^+ + 2e^-$	=	$HNO_2 + H_2O$	0.94
$NO_3 + 4H^+ + 3e^-$	=	$NO + 2H_2O$	0.96
$HNO_2 + H^+ + e^-$	=	$NO + H_2O$	1.00
$NO_2 + 2H^+ + 2e^-$	=	$NO + H_2O$	1.03
Br_2 (l) $+ 2e^-$	=	$2Br^-$	1.065
$NO_2 + H^+ + e^-$	=	HNO_2	1.07
Br_2 (aq) $+ 2e^-$	=	$2Br^-$	1.087
$2IO_3^- + 12H^+ + 10e^-$	=	I_2 (s) $+ 6H_2O$	1.20
O_2 (g) $+ 4H^+ + 4e^-$	=	$2H_2O$	1.229
MnO_2 (s) $+ 4H^+ + 2e^-$	=	$Mn^{2+} + 2H_2O$	1.23
$Cr_2O_7^{2-} + 14H^+ + 6e^-$	=	$2Cr^{3+} + 7H_2O$	1.33
Cl_2 (g) $+ 2e^-$	=	$2Cl^-$	1.3595
PbO_2 (s) $+ 4H^+ + 2e^-$	=	$Pb^{2+} + 2H_2O$	1.455
$2ClO_3 + 12H^+ + 10e^-$	=	$Cl_2 + 6H_2O$	1.47
$MnO_4^- + 8H^+ + 5e^-$	=	MnO_2 (s) $+ 4H_2O$	1.51
$MnO_4^- + 4H^+ + 3e^-$	=	MnO_2 (s) $+ 2H_2O$	1.695
$H_2O_2 + 2H^+ + 2e^-$	=	$2H_2O$	1.77
$Co^{3+} + e^-$	=	Co^{2+}	1.842
O_3 (g) $+ 2H^+ + 2e^-$	=	O_2 (g) $+ H_2O$	2.07
$F_2 + 2e^-$	=	$2F^-$	2.87

REFERENCES

Edgington, D.H. and E. Callender, 1970. Minor element geochemistry of Lake Michigan ferromanganese nodules. Earth Plant. Sci. Lett 8: 97-100.

Jørgensen, S.E., 1975. Do heavy metals prevent the agricultural use of municipal sludge? Water Research, 9: 163-170.

Jørgensen, S.E. and I. Johnsen, 1988. Principles of environmental science and technology. Elsevier, ISEM, Amsterdam, Copenhagen.

Jørgensen, S.E., M.B. Friis, J. Hendriksen, L.A. Jørgensen and H.F. Mejer, 1979. Handbook of environmental data and ecological Parameters (S.E. Jørgensen, ed.), ISEM, Copenhagen, p. 319.

Mabey, W.M. and T. Mill, 1978. Critical review of hydrolysis of organic compounds in water under environmental conditions. J. Phys. Chem. Ref. Data, 7: 383.

Mabey, W.M., T. Mill and D.G. Hendry, 1978. Test protocols for environmental processes: Hydrolysis. US-EPA report, Contract 69-03-2227.

Tinsley, I.J., 1979. Chemical concepts in pollutant behavior. John Wiley & Sons, New York.

Wolfe, N.L., R.G. Zepp and D.F. Paris, 1978. Use of structure reactivity relationships to estimate hydrolytic persistence of carbamate pesticides. Water Res., 12: 561.

CHAPTER 11

PHOTOCHEMICAL REACTIONS

by
George T.F. Wong

11.1 INTRODUCTION

An enormous amount of light energy can reach the hydrosphere from the sun. At low latitudes, under a cloudless sky and a summer sun, the irradiance at sea level is of the order of 1000 W/m^2 (Ochakovskii et al., 1972). An overwhelming fraction of this energy input is of course converted to heat. However, even the small fraction, about 0.04% on the average, of the light energy that enters the water and is captured by photosynthetic activities is enough to be the driving force in shaping the entire aquatic biological domain. In considering such a massive input of energy into the complex aquatic environment, the possibility of the occurrence of abiological light-induced chemical reactions and their importance in determining the chemical composition of natural waters can hardly be overlooked.

Unfortunately, until recently there was little interest in studying the chemistry of these reactions in natural waters. In the past few years, a series of review articles and a bibliography of references have been published (Zafiriou, 1977, 1983, 1984; Sundstrom and Ruzo, 1978; Zepp and Baughman, 1978; Zepp, 1980, 1982a, 1982b; Zika, 1981; Zafiriou et al., 1984) and three symposia (Miller, 1983; Newman, 1984; Zafiriou et al., 1984; Anonymous, 1985) devoted largely to this subject area were held to call attention to the importance of photochemical reactions in the chemistry of natural waters. Some chemical species in natural waters may absorb light energy directly and undergo direct photochemical reactions. Examples of this type include the photolysis of nitrate, nitrite and methyl iodide in seawater (Zafiriou and True, 1979a, 1979b; Zafiriou and McFarland, 1981; Zika et al., 1984). In some cases, the products of these direct photochemical reactions such as free radicals and excited organic chromophores may themeselves be reactive (Zafiriou, 1974; Zafiriou and True, 1979c; Mill et al., 1980; Zafiriou et al., 1980; Zepp et al., 1981a, 1981b, 1985; Zepp and Schlotzhauer, 1981a). They may react with compounds that are not photoreactive resulting in indirect or sensitized photochemical reactions. If the products are still reactive, such as singlet oxygen and hydrogen peroxide, further reactions may occur (Van Baalen and Marler, 1966; Joussot-Dubien and Kadiri, 1970; Zepp et al., 1977; Baxter and Carey, 1982; Haag et al., 1984a, 1984b).

Since the major emphases of the recent reviews and symposia were on identifying directly or indirectly photochemically reactive chemical species in natural waters, on proposing the mechanisms and the rates of these reactions, and on speculating on the fate of the resulting products, their findings will not be re-iterated here. The primary purpose of this article is to consider what may be needed in order to incorporate photochemical reactions into aquatic geochemical models.

11.2 THEORETICAL BASIS FOR THE MODELLING OF PHOTOCHEMICAL REACTIONS

Only light that is absorbed by a molecule can produce chemical changes. This fundamental law of photochemistry has been recognized since the early nineteenth century by Grotthus and Draper. The absorption of light by a molecule is governed by the Lambert-Beer law. The fraction of light that is absorbed as it passes through an absorbing medium is given by:

$$I_a = I_o (1 - 10^{eCd}) \qquad (11.1)$$

where I_a is the intensity of the radiation absorbed, I_o is the intensity of the incident light, C is the concentration of the chromophore, d is the pathlength and e is the extinction coefficient which is a characteristic of the chromophore and is dependent on the wavelength of the radiation. Thus, the initiation of photochemical reactions will depend on the spectral characteristics and the intensity of the incident light and the ability of a given chemical species to absorb light at various wavelengths.

The molecule which has absorbed light becomes 'energy-rich' or 'excited'. **The Stark-Einstein law** states that if a species absorbs radiation, then one particle is excited for *each quantum* of radiation absorbed. The energy that can be imparted on the chromophore is of course limited by the energy of the photon absorbed. This energy is given by:

$$E = h v \qquad (11.2)$$

where E is the energy of the photon, h is the Planck's constant and v is the frequency of the photon. Thus, depending on the nature of the incident light and the property of the chromophore, different excited states of the chromophore may be formed. Although it is not exclusively true, in general light absorption, which results in photochemical reactions, causes electronic excitation of the chromophore. Excitation of the vibrations and rotations of the absorbing species in its ground electronic state is less likely to induce chemical changes. The excess energy in the excited species may be dissipated by a variety of processes as shown in Fig 11.1 (Owen, 1971; Wayne, 1976). These are called primary reactions. The chromphore may be consumed and new species formed by dissociation (process 1), spontaneous isomerization (process 2) or direct reactions with another species Z (process 3). If a chemical bond is ruptured following the absorption of light, then, photolysis is said to have occured. In other cases, the excited form is reverted back to the stable ground state through intermolecular energy transfer (process 5), physical quenching (process 6) or luminescence (process 7). If the products produced in these primary reactions are chemically reactive, further, secondary reactions, may occur. In the process, other new chemical species may be formed. These primary and secondary reactions are not equivalent to the direct and indirect photochemical reactions as discussed previously. A photochemical reaction is said to have occurred only if a new chemical species is formed. Thus, in processes 1 through 4, the species XY has undergone direct photochemical reaction. Species W, Z and any other species which reacts with any of the reactive products produced in the primary reactions have undergone indirect photochemical reactions.

The effectiveness of photochemical reactions in the consumption or production of a given chemical species is given by **the overall quantum yield**, ß, of that process. The overall quantum yield is the number of molecules of a given reactant consumed for each photon absorbed. It may be further sub-divided into that which is caused by primary reactions, or primary quantum yield, and that which is caused by secondary reactions.

Since more than one primary reaction may be involved in a photochemical reaction, the primary quantum yield may be defined with respect to a specific primary reaction. Moreover, the sum of the quantum yields of all primary reactions, including de-activation, must be unity. The overall quantum yield, which is the quantity most often determined, on the other hand, does not distinguish between the effects of primary and secondary reactions and it may exceed unity. If multiple reaction pathways leading to the formation of a multitude of products are possible, the overall quantum yield may be defined with respect to the removal of the reactant or the formation of a particular product. Thus, whether a particular photochemical reaction occurs and to what extent does it occur are linked to the manner in which the excited chemical species may dispose of the excess energy that it has acquired through the absorption of photons. It is often difficult to identify unequivocally and quantify precisely all the pathways of the primary reactions of a photochemical reaction. It is even more difficult to identify and quantify each secondary reaction. As a result, overall quantum yield of the disappearance or the formation of a particular chemical species is more often determined.

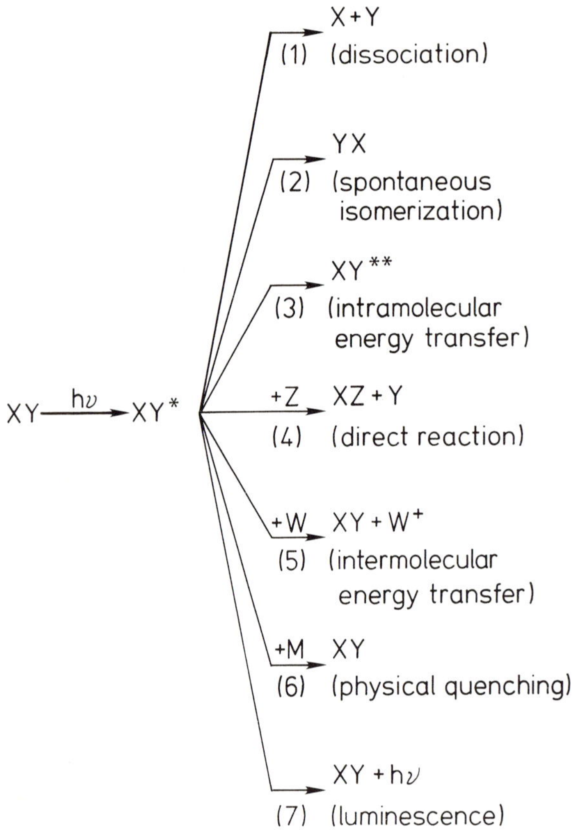

Fig. 11.1: The different pathways of primary reactions for the dissipation of light energy absorbed by a chemical species XY.
The symbols *, ** and + indicate electronically excited species. The products in processes (1), (2) and (4) may also be excited.

11.3 MODELLING PHOTOCHEMICAL REACTIONS IN NATURAL WATERS

Modelling photochemical reactions in natural waters is still in its infancy. No proven general model,that is applicable to every kind of natural water is presently available for estimating the rates of consumption or production of chemical species through light-induced reaction pathways. A number of pieces of fundamental and essential information for constructing a precise and realistic model are still not available. For example, the chemical composition of natural waters and their temporal and spatial variabilities are still inadequately known. Thus, the reactants that may participate directly and indirectly in photochemical reactions in a competitive, synergistic or catalytic manner cannot yet be fully identified. The rate at which a particular photochemical reaction may proceed in this poorly defined medium and to what extent will it proceed *cannot* be quantified *without* extensive empirical data. Moreover, as stated previously, even in a well defined medium, the detailed reaction pathways of photochemical reactions can still rarely be identified definitively.

Despite of these limitations, several investigators have attempted to construct models for estimating the rates of direct photolysis, a specific type of photochemical reaction, in natural waters under restrictive conditions (Zepp and Cline, 1977; Mancini, 1978; Zepp and Baughman, 1978; Zepp, 1978, 1982a, 1982b). If the overall rate of consumption of a chemical species, A, by direct photolysis with light at a given wavelength, l, is proportional to the ability of the species to absorb photons at that wavelength and the quantum yield of the reaction, then, the rate law governing the reaction can be written as:

$$-(d[C_A]/dt) = k_{al} ß_l [C_A] \qquad (11.3)$$

where t is time, ß is the quantum yield of the reaction, k_a is the specific rate of absorption of photons and $[C_A]$ is the concentration of A. For an incident beam of light with a certain spectral characteristic, if the quantum yield is not strongly wavelength-dependent (Turro, 1965), the complete rate law for direct photolysis will be:

$$-(d[C_A]/dt) = k_a [C_A] \qquad (11.4)$$

where k_a is the sum of the k_{al} values of all the wavelengths represented by the photons in the incident light. Thus,

$$k_a = \Sigma k_{al} \qquad (11.5)$$

If k_a and ß are constants, then the solution of equation (11.4) is

$$\ln[C_A] = \ln[C_A]_0 - k_a ß t \qquad (11.6)$$

where $[C_A]_0$ is the concentration of A at t = 0. Thus, at

$$[C_A] = 0.5 [C_A]_0$$

$$t = t_{1/2} = \ln 2/k_a\, \beta = 0.693/k_a\, \beta \qquad (11.7)$$

where $t_{1/2}$ is the half/life of A with respect to direct photolytic decomposition.

For *photosensitized* reactions, the light energy absorbed by the photosensitizer must first be transferred to the species of interest before indirect photolysis of the species can occur. The rate law for the indirect photolysis of a species B becomes:

$$-d[C_B]/dt = k_a[C_S]\, \beta_s\, [C_B] \qquad (11.8)$$

where k_a is the specific rate of absorption of photons by the photosensitizer S, C_S and C_B are the concentrations of S and B respectively and β_s is the quantum yield of the photosensitized reactions (Zepp, 1982a, 1982b).

Once the reaction rates of a photochemical reaction at various locations and times in a body of natural water are estimated, they must be integrated over space and/or time to evaluate the overall rate of the photochemical reaction. The integrated results may then be incoporated into geochemical models.

11.4 ESTIMATION OF MODEL PARAMETERS

In order to apply these models to natural water systems, the rate of light absorption of a given chemical species, k_a, and its quantum yield, β, must be estimated. The rate of light absorption is dependent on the intensity of the incident light, the extinction coefficient of the chromophore and its concentration. However, this rate as deduced from the Beer-Lambert law (equation (11.1)) is valid only for a pure solution of the chromophore at a defined concentration, in a solvent that does not interact with the chromophore through secondary light-induced reactions, and under an incident beam of light which has a constant intensity and invariant spectral characteristics. None of these conditions is met in the natural environment.

The intensity and spectral characteristics of sunlight at a given depth in a body of water are variable as they are affected by a number of temporally and spatially variable processes. Initially, at the top of the Earth's atmosphere, the spectral characteristics of extraterrestrial sunlight are well defined and they depend on the temperature of the sun. If the sun is considered as a *black body* (Williams, 1970; Coulson, 1975), then, the radiant flux at a wavelength l, F_l, is

$$F_l = \frac{2\pi hc^2}{l^5} \left[\frac{1}{\exp (hc/kT) - 1} \right] \qquad (11.9)$$

and the **total radiant flux**, F, is

$$F = \int_0^{\infty} F_l d\, l \qquad (11.10)$$

where h is Planck's constant, c is the speed of light, k is Boltzmann's constant and T is the absolute temperature of the sun. However, the radiant flux may vary for several percents with the sidereal date since the distance between the Earth and the sun changes during the year. The average radiant flux per unit area normal to the incident sunlight at the top of the atmosphere at the mean Earth-sun distance is the solar constant, F'. It has a value of about 2 $cal/cm^2/min$ (Williams, 1970). The radiant flux, F at sidereal date D is

$$F = F' \; [\, 1 + 0.0335 \; \sin(360 \; D/365) \,] \qquad (11.11)$$

As sunlight passes through the atmosphere, it is attenuated as a significant fraction may be absorbed or scattered. The atmosphere, as a result of the presence of water vapor, absorbs photons with long wavelengths in the red end of the spectrum more effectively. On the other hand, a larger fraction of the photons with short wavelengths in the blue end of the spectrum is scattered back into space. Photons may be scattered by Rayleigh scattering centres, such as the molecules of oxygen, nitrogen, the noble gases and ozone, and by Mie scattering centres which are conposed of the various kinds of aerosol particles. In Rayleigh scattering, the resulting light field is isotropic. In Mie scattering, the resulting light field is anisotropic. Thus, as sunlight passes through the atmosphere, it is separated into two components: direct sunlight that is transmitted through the atmosphere without alteration, and the diffuse skylight resulted from *scattering* processes. **The total solar radiation that reaches the Earth's surface is the sum of these two kinds of light.** The intensity and spectral characteristics of sunlight at the Earth's surface will then be affected by the concentrations and the photochemical properties of the gases and the aerosol particles and the thickness of the atmosphere, or the pathlength, through which sunlight has to pass to reach the Earth. The concentrations of some of the gases, such as ozone, the chemical compositions and concentrations of the aerosols and the amount of cloud cover of the sky are of course temporally and spatially variable. **The pathlength** is related to the time of the day, the elevation at a given geographical location and the solar zenith angle which is a function of latitude and the sidereal date. Even on a clear day, only about 55% of the energy at a wavelength of 350 nm can be transmitted through the atmosphere while 95% of the energy in the red end of the spectrum may reach the Earth's surface (Forsythe, 1964; Williams, 1970). However, the more energetic photons with the shorter wavelengths are more likely to cause photochemical reactions.

At the atmospheric-hydrospheric interface, a fraction of the light is reflected back to the atmosphere. Only the remaining fraction enters a body of water. For direct sunlight, the reflectance, which is the ratio of the reflected radiant flux to the incident radiant flux, at the water surface is dependent on the angle of incidence and refraction and the roughness of the water surface. The angle of incidence is defined by the zenith angle which may vary as described previously. Reflectance increases with increasing angle of incidence reaching total reflection, or 100% reflectance, at an angle of incidence of 90°. Roughness of the water surface will be affected by surface wave conditions. The reflections of the diffuse skylight is more difficult to quantify. The reflectance of skylight has been estimated to be about 6.6% for a smooth water surface (Burt, 1954). Thus, the total reflectance, ¶, for a smooth water surface is

$$\P = (I_r/I) = \P_s (1 - n) - \P_d n \qquad (11.12)$$

where I and I_r are the incident and reflected irradiance respectively and n is the ratio of sky radiation to total radiation, and $\P_s$ and $\P_d$ are the reflectance of direct sunlight and skylight, and

$$\P_s = 1/2 \left[\frac{\sin^2 (i - j)}{\sin^2 (i + j)} + \frac{\tan^2 (i - j)}{\tan^2 (i + j)} \right] \qquad (11.13)$$

where i and j are the angles of incidence and refraction respectively (Jerlov, 1968). The irradiance, I_i, that enters a body of water will then be

$$I_i = I (1 - \P) \qquad (11.14)$$

The light that enters a body of water is refracted and again attenuated as it passes through the water. Refraction is governed by Snell's Law, so that, for a given medium,

$$n = \sin i / \sin j \qquad (11.15)$$

where n is the refractive index which is an inherent property of the medium into which a beam of light enters. Refraction thus alters the direction of the incident light, and thus, the distance a beam of light must travel, or the pathlength, to reach a given depth in a body of water. LIght can be attenuated in a body of water by scattering and absorption (Smith and Tyler, 1976). Both forward and backward scatterings may occur.

Scattering re-distribute the incident light. Water molecules and suspended particles are the major scatterers. Scattering caused by water is varied by changes in temperature and pressure although the variations within the normal ranges of temperature and pressure in natural waters are small. Scattering caused by particles is much more variable since the concentration and the physical and chemical nature of suspended particles may exhibit large spatial and temporal variations (Gibbs, 1974; Sackett, 1978). Miller and Zepp (1979) reported that photolysis rates can actually be enhanced by light-scattering caused by suspended sediments, and they attributed this enhancement to the increased diffuseness and thus availability of light for absorption as a result of scattering. On the other hand, Oliver et al. (1979) suggested that suspended sediments may decrease the rate of photolysis of pollutants by shielding them from the available light. Obviously, once a given amount of light energy has entered a body of water, the amount of energy available for photochemical reactions is fixed. Scattering will not affect the total effect of photochemical reactions in the body of water unless the absorption coefficient or the quantum yield is changed in the process. However, the rate of photochemical reactions at a given point of the body of water may well be altered as a result of changes in the availability of light.

A number of constitutents in natural waters are known to absorb light. Water molecules themselves absorb light and its absorption spectrum exhibits strong wavelength-dependence. A transmittance maximum occurs at around 470 nm. The transmittance drops off at longer and shorter wavelengths reaching 10% of the maximum value at about 570 nm and 330 nm (Williams, 1970). Among the common inorganic salts in natural waters, bromide and nitrate may contribute significantly to the absorption of light at short wavelengths in the ultraviolet region (Armstrong and Boalch, 1961; Ogura and Hanya, 1966, 1967). The concentration of nitrate can be quite variable in natural waters as it is involved in a variety of anthropogenic and biological processes. Large quantities of bromide can be found in seawater. However, its concentration is well known and it bears a simple and well defined relationship to salinity (Wilson, 1975). Absorption of visible light by the common inorganic salts in natural waters is small. Dissolved organic matter absorbs light in both the visible and ultraviolet regions and the absorption increases with decreasing wavelengths (Armstrong and Boalch, 1961; Ogura and Hanya, 1966, 1967; Foster and Morris, 1974). The contribution of dissolved organic matter to light absorption is quite variable and cannot yet be easily predicted since both the concentrations and the chemical nature of dissolved organic matter in natural waters can be temporally and spatially variable and are still poorly known (Kerr and Quinn, 1975, 1980; Williams, 1975; Head, 1976).

Suspended particulate matter also absorbs light. The amount absorbed is again variable as the concentrations and the physical and chemical nature of these particles in natural waters are also variable and poorly known. Particulate materials in natural waters include biogenic organic and inorganic particles, mineral grains, coatings of metal oxides and/or hydroxides, materials co-precipitated or co-crystallized with these particles and chemi- cal species absorbed onto the particulate surface (Sackett, 1978). Biogenic particles such as algal cells have been shown to be an important light ab- sorber in many aquatic environments (Baker and Smith, 1982; Zepp and Schlotzhauer, 1983). Adsorption of a species and the manner in which it is adsorbed onto a particulate surface may also affect the photochemical behavior of that species (Zepp and Schlotzhauer, 1981). The light energy absorbed by the particles may lead to heterogeneous direct or indirect photochemical reactions (Zepp et al., 1983; Waite and Morel, 1984; Funden et al., 1984). These competing species which can absorb light with varying efficiencies at different wavelengths will effectively reduce the amount of light that is available to a photochemical reaction of interest. The resulting combined effects of refraction, scattering and absorption will determine the irradiance and the spectral characteristics of sunlight at a given depth in a body of water.

Supposedly, one can initiate a theoretical estimation of the intensity and spectral characteristics of sunlight at a given depth of a body of water at any point in the path of the light as it travels from the sun to the point of interest on Earth. In this approach, obviously, the farther is the starting point from the point of interest, the more difficult the task and the less certain the results will be. The irradiance at the Earth's surface has been estimated theoretically by this approach by a number of investigators (Green et al., 1974a, 1974b, 1980; Mo and Green, 1974; Johnson et al., 1976; Willis, 1978; Baker et al., 1982). However, only *limited* success was achieved. Recently, some field measurements have become available (Smith and Baker, 1979; Baker et al., 1980, 1982; Baker and Smith, 1982a) for calibrating the various theoretical models. The results indicate that the theoretically estimated irradiance is model-dependent. Models have also been proposed for estimating the spectral irradiance in non-terrestrial waters, where the attenuation is caused primarily by biogenic material, as a function of the total concentrations of chlorophyll-like pigments and dissolved organic matter (Smith and Baker, 1978a, 1978b; Baker and Smith, 1982b) The existing models may be sufficient for estimating a first order approximation of the underwater light fields. A highly accurate general model for estimating the light fields at various depths in different kinds of natural water has yet to be devised.

Once **the light field** is defined, one still needs to know the **extinction**

coefficient and the concentration of the chromophore in order to estimate the rate of light absorption, k_a, at a given depth. Extinction coefficients as a function of the wavelengths of photons can be determined readily in the laboratory. The temporal and spatial distributions of the chromophore are controlled by biological processes, other geochemical processes as well as hydrodynamics. There is no proven model that can be used to predict the concentrations and the distribution patterns of a chromophore in natural waters. They will have to be determined by actual field observations. Thus, at the present time, for accurate results, many of the parameters needed for estimating k_a must be obtained empirically by actual field and/or laboratory measurements.

The quantum yield, ß, of a given chemical species at various wavelengths can be determined experimentally in the laboratory in well-defined media. However, natural water is not a well-defined medium with a uniform and homogeneous composition. In the primary reactions (fig. 11.1), chemical species that can react with the photochemically excited form of the species of interest as represented by Z and W in processes (11.4) and (11.5) in figure 1 will change the overall quantum yield. If coupled secondary reactions occur so that other light sensitive chemical species may form products that can react with the species of interest, the overall quantum yield will again be altered. Thus, for accurate results, the value of ß to be used in the model may also have to be determined empirically in the water of interest.

In most cases, extensive empirical data bases are not available. Thus, several investigators (Zepp and Cline, 1977; Mancini, 1978; Zepp, 1978, 1982a, 1982b) have proposed the estimation these parameters by using some combinations of theoretical and empirical approaches. However, the assumptions that must be invoke in these models place severe restrictions on their general applicability to natural water systems. For example, Zepp and Cline (1977) had to assume that (a) there is exposure to the sun and whole sky at sea level without cloud cover; (b) there is little daily variation in the concentrations of atmospheric ozone; (c) there is no secondary light-induced reactions; (d) the effect of light scattering in water bodies can be ignored; and (e) the chemical species of interest is uniformly distributed in the body of water and is exposed to the same amount of light during a given period of time. Under these asumptions, the model is more likely to be valid for estimating the statistical average reaction rate over a relatively long period of time, so that short term variations in the compositions of the atmosphere and the body of water may be damped out; in shallow water, so that the body of water may be well mixed; and with low concentrations of organic matter and suspended matter, so that variations in the attenuation of light in the water column may be minimized. Not too many bodies of natural water can meet these criterion. It woud be especially difficult to use these models to predict short term phenomenon, such as a pulse input of a highly photochemically reactive species, in coastal and terrestrial waters where the composition is variable and the concentrations of organic matter and suspended matter are high.

11.5 INCORPORATION OF PHOTOCHEMICAL REACTIONS INTO GEOCHEMICAL MODELS

In geochemical models, aside from the consumption or production of a chemical species by photochemical reactions, one must also consider hydrodynamics, consumption and production by biological as well as other physico-chemical reactions. The importance, or lack of it, of photochemical reactions in controlling the concentrations and/or speciations of a certain element in a body of natural water must be evaluated relative to the contributions from other sources and/or sinks.

Thus, photochemical reactions become one term among a large array of terms in a mathematical equation. Furthermore, the number of boundary conditions for the physics, chemistry and biology of the system are usually quite limited. As a result, in order to solve such equations readily, the functionality of each term must be kept relatively simple and must involve few variables.

Two types of models have been used most extensively in recent studies. The first kind is the 'box model' (Keeling and Bolin, 1967; Imboden and Lerman, 1978; Veronis, 1978; Officer, 1980). In this model, a body of water is subdivided into conceptual reservoirs which are well mixed. Then, for each reservoir,

$$\frac{d\,[D_A]_a}{dt} = \sum_p (Q_A)_p + \sum_q (R_A)_q - \sum_r (J_A)_r \qquad (11.16)$$

where $[C_A]_a$ is the concentration of a species A in the ath reservoir, Q_A is an influx or eflux of A into or out of the a'th reservoir as a result of hydrodynamic processes, R_A and J_A are the production and consumption of A within the a'th reservoir. Each term in the right hand side of equation (11.16) may be made up of a number of sub-terms since A may be transported into or out of the reservoir through various routes and a variety of biological and physico-chemical reactions may occur within the reservoir. Thus, for a particular photochemical reaction, the net effect, that is net consumption or production, will be only one of these sub-terms of R_A or J_A. If the individual effects of production and consumption are to be considered, then, the photochemical reaction will appear as a sub-term of R_A and a sub-term of J_A.

The objective of the modelling exercise is often to quantify each sub-term in the right hand side of the equation in order to evaluate the relative quantitative importance of the various processes that may affect the concentration of A in the reservoir and to predict the rate of change of concentration with time. If the reservoir is at a steady state relative to the concentration of A, then, equation (11.16) becomes:

$$\frac{d\,[D_A]_a}{dt} = 0 = \sum_p (Q_A)_p + \sum_q (R_A)_q - \sum_r (J_A)_r \qquad (11.17)$$

In this case, one of the *sub-terms* in the right hand side of the equation, most likely the one representing the process that is most difficult to quantify, may be estimated if all the other sub-terms are known. If photochemical reactions were to be included in this model, it must either be estimated independently or be treated as the only unknown in the equation to be evaluated. In the former case, its effects integrated over time and space for that particular reservoir must be estimated as discussed in the previous sections. In the latter case, if photochemical reactions were to be estimated as an unknown, the result would include the cumulative uncertainties of all the other sub-terms in the equation. Moreover, the result is a net statistical average over the entire reservoir and, by itself, does not give information on the mechanisms of the reactions involved.

The second kind of model that has been used extensively is the advection-diffusion model (Riley, 1951; Craig, 1969; Veronis, 1975, 1977; Csanady, 1978) whereby

$$\frac{d\,[D_A]_a}{dt} = \frac{\partial K_x \partial\,[C_A]}{\partial x^2} + \frac{\partial K_y \partial\,[C_A]}{\partial y^2} + \frac{\partial K_z \partial\,[C_A]}{\partial z^2} - u\frac{\partial\,[C_A]}{\partial x} \qquad (11.18)$$

$$- v\frac{\partial\,[C_A]}{\partial y} - w\frac{\partial\,[C_A]}{\partial z} + \sum_q (R_A)_q - \sum_r (J_A)_r$$

where $[C_A]$ is the concentration of A at a given point in a body of water; K_x, K_y and K_z are the coefficients of eddy diffusivity; and u, v and w are the advective velocities in the x, y and z directions respectively; and R_A and J_A are production and consumption terms. Thus, in the right hand side of the equation, the first three terms describe transport by eddy diffusion, the second three terms describe transport by advection, and the last two terms sum up the effects of biological and physico-chemical processes. Again, the effects of photochemical reactions will be a sub-term in the last two terms of the equation. Since, frequently, neither the physics of the body of water, nor the chemistry, nor the biology, nor the temporal and spatial variations of the concentrations of the chemical species of interest are known in details, the equation can rarely be solved analytically. Even solving the equation numerically is difficult (Riley, 1951). Drastic simplifications must be made. The assumptions that are most often invoke are **a.** a steady state in the concentration of A has been reached; **b.** the transport processes in a certain direction dominates; **c.** the coefficient of eddy diffusivity and advective velocity at a given direction are independent of distance in that direction, and the chemical nature or concentration of the species involved; **d.** the dependence of the production and consumption terms on C_A and distance in a given direction must be defined and are usually arbitrarily considered to be zero-th order or first order functions or to vary exponentially with distance. The resulting model is usually one dimensional (e.g., Craig, 1969, 1974; Spencer and Brewer, 1971) and occasionally two dimensional (e.g., Imboden and Emerson, 1978; Spencer et al., 1981). Thus, for a **vertical advection-diffusion model**, the equation becomes:

$$\frac{d\,[C_A]}{dt} = K_z \frac{\partial^2\,[C_A]}{\partial\,z^2} - w \frac{\partial [C_A]}{\partial\,z} + \sum_q (R_A)_q - \sum_r (J_A)_r \qquad (11.19)$$

and for a **horizontal advection-diffusion model**, the equation is:

$$\frac{d\,[C_A]}{dt} = K_x \frac{\partial^2\,[C_A]}{\partial\,x^2} - u \frac{\partial [C_A]}{\partial\,x} + \sum_q (R_A)_q - \sum_r (J_A)_r \qquad (11.20)$$

and J and R may have the forms of

$$J_A = J_0, \text{ and } R_A = R_0 \qquad (11.21)$$

where J_0 and R_0 are constants (e.g., Craig, 1969; Craig and Weiss, 1970; Craigh, 1971), or

$$J_A = K_{J1}\,[C_A], \text{ and } R_A = K_{R1}\,[C_A] \qquad (11.22)$$

where K_{J1} and K_{R1} are constants (e.g., Craig, 1974), or

$$J_A = J_0 e^{-k' z}, \text{ and } R_A = R_0 e^{-k'' z} \qquad (11.23)$$

or $$J_A = J_0 e^{-k' x}, \text{ and } R_A = R_0 e^{-k'' x} \qquad (11.24)$$

where J_0, R_0, k' and k" are constants (e.g. Wyrtki, 1962; Craig and Weiss, 1970; Spencer and Brewer, 1971; Brewer and Spencer, 1975). Frequently in the modelling exercise, the observed distribution of a conservative tracer such as temperature or salinity is used to define the physics, or K_z and w, of the system. Only one unknown in the sub-terms in the produciton or consumption term is allowed. Then, the value of this unknown sub-term in J_A or R_A will be optimized in order to obtain the best agreement between the model-generated and the observed distributions of the species of interest. In the process, this particular consumption or production sub-term is quantified. Once again, if a *photochemical* reaction is to be included in the model, it must be present either as an empirically defined term or as the only unknown among all the processes that may affect the concentration of A. If it is treated as an unknown, the result again cannot be related to any particular reaction mechanism without collaborating information.

11.6 CONCLUSIONS AND RESEARCH NEEDS

Photochemical reactions may have a significant impact on the geochemistry of natural waters. However, at the present time, the overall picture of this field of study is one of immense complexity and lack of information, partially as a result of the state of affairs of aquatic geochemistry in general. Several levels of modelling efforts may be identified. These levels present increasing degree of complexity, increasing resemblance to reality and increasing usefulness in estimating the effects of photochemical reactions in the geochemistry of natural water. In the first level, the models, which include most of the studies done to date, attempt to describe the rate of a given light-induced reaction in a defined medium under a defined light-regime in the laboratory. In the next level, the model may describe photochemical reactions in various media and light-regimes as found in the natural environments. This will be followed with a model that can integrate the total effects of photochemical reactions in the entire water column of different bodies of natural water ranging from highly coloured to relatively clear water, eutrophic to oligotrophic water, freshwater to brine, and waters with high to low particulate loads and high to low concentrations of dissolved organic matter. Finally, a model should include physical, biological as well as other geochemical processes that may affect the chemical composition of natural waters. Thus, advances in modelling photochemical reactions in natural waters will be closely linked to a better appreciation of the geochemistry, physics and biology of the aquatic environment as a whole.

Several pieces of fundamental and yet presently unavailable information have been referred to repeatedly by various investigators.

(1) **Natural water as a chemical medium needs to be better characterized.** The specifications of elements in both the dissolved and particulate phases have to be known in much greater details. This is especially important in studying photochemical reactions since chemical kinetics rather than thermodynamics are often the controlling factor. A temporally and spatially variable sub-environment that is not at chemical equilibrium with the environment in general is created in the presence of light, and chemical species that are not at equilibrium with the prevailing chemical conditions of the body of water are formed in photochemical reactions. The possible competitive, synergistic or catalytic effects of unknown chromophores and other chemical species that undergo indirect photochemical reactions are often eluded to, but these effects cannot even begin to be identified and quantified until these unknown species are identified and quantified. The inorganic specification of a number of elements in the dissolved phase is reasonably, although by no means exhaustively, understood (Stumm and Brauner, 1975). However, we know little about the compostion of dissolved organic material and particulate matter in natural waters (Kerr and Quinn, 1975, 1980; Williams, 1975; Head, 1976; Sackett, 1978). Chemical species that involve interactions among inorganic and organic species, such as metal ions associated with organic chelators, and dissolved and particulate phases, such as adsorbed species, are also poorly known.

(2) **The ability to quantify the underwater light field at various depths is essential.** The present understanding of the physics and physico-chemical processes that affect the propagation of light in natural waters may be sufficient to allow one to construct theoretical models for making crude estimations of the underwater light field. However, if a higher degree of precision is needed, further improvements in these models will be needed.

(3) **In geochemical modelling, the physics of the system, namely, advective and diffusive processes, must be defined in order that the effects of geochemical processes may be isolated from the observed total signal.** However, in many cases, the physics of the body of water is a major unknown.

Unfortunately, these essential pieces of information are difficult to obtain and they have been long sought for. Breakthroughs in these areas are clearly not in sight in the foreseeable future. Under these limitations, extensive efforts in formulating a general model applicable to photochemical reactions in every kind of natural water may be premature. In the mean-time, modelling efforts may have to be concentrated on specific sites of interest. Input parameters for the modelling effort may then be obtained directly by field measurements and laboratory observations using samples collected from the particular site. As a result, the conclusions will, out of necessity, be restrictive in scope, and may be applicable to a particular site for a particular period of time. This situation is not unlike the one that aquatic biologists, who study primary production in natural waters, are in. Photosynthesis, as a first approximation, may be viewed as a special kind of photochemical reaction. It has been studied extensively for a long period of time and the mechanism of the process, down to the molecular level, is known. However, even though there have been attempts in estimating primary production by modelling, it is still estimated for a given location basically by field measurements (Strickland, 1958, 1965; Ryther, 1959; Jamart et al., 1977; Morel, 1978). Photochemists probably will have to use the same approach in the foreseenable future. Any broad brush modelling effort will yield results that have high inherent uncertainties. However, in such a modelling effort, since processes, other than photochemical reactions, that may affect the concentrations of a given chemcial species in natural waters also often cannot be precisely quantified, a first-order approximation of the effects of photochemical reactions may be all that is warranted for evaluating the relative importance of the various processes in controlling the concentrations of that species. A more precise estimation of the effects of photochemical reactions may require much additional work but may not yield discernible improvement in the overall understanding of the geochemistry of that species.

ACKNOWLEDGMENTS

I wish to thank Dr. O. Zafiriou for enlightening discussions while I prepared this manuscript and for reviewing it after it was complete.

REFERENCES

Anonymous (1985). Final program. 189th ACS national meeting. Chem. Engineer. News, 63(11), 58.

Armstrong, F.A.J. and G.T. Boalch (1961). The ultra-violet absorption of seawater. J. Mar. Biol. Assoc. U.K., 41, 591-597.

Baker, K.S. and R.C. Smith (1982a). Sprectral irradiance penetration in natural waters. In: The Role of Solar Ultraviolet Radiation in Marine

Ecosystem. J. Calkins Ed. Plenum Press. New York. pp. 233-246.
Baker, K.S. and R.C. Smith (1982b). Bio-optical classification and model of natural waters. 2. Limnol. Oceanogr., 27, 500-509.
Baker, K.S., R.C. Smith and A.E.S. Green (1980). Middle ultraviolet radiation reaching the ocean surface. Photochem. Photobiol., 32, 367-374.
Baker, K.S., R.C. Smith and A.E.S. Green (1982). Middle ultraviolet irradiance at the ocean surface: Measurements and models. In: The Role of Solar Ultraviolet Radiation in Marine Ecosystems. J. Calkins Ed. Plenum Press. pp. 79-91.
Baxter, R.M. and J.H. Carey (1982). Reactions of singlet oxygen in humic waters. Freshwater Biol., 12, 285-292.
Brewer, P.G. and D.W. Spencer (1975). Trace element models in coastal waters. In: Marine Chemistry in the Coastal Environment. T.M. Church Ed. ACS Symp. Ser. No. 18. Americal Chemical Society. Washington D.C. pp. 80-96.
Burt, W.V. (1954). Albedo over wind-roughened water. J. Meterorol., 11, 283-290.

Cooper, W.J. and R.G. Zika (1983). Photochemical formation of hydrogen peroxide in surface and ground water exposed to sunlight. Science, 290, 711-712.
Coulson, K.L. (1975). Solar and Terrestrial Radiation: Methods and Measurements. Academic Press. New York. 322 pp.
Craig, H. (1969). Abyssal carbon and radiocarbon in the Pacific. J. Geophys. Res., 74, 5491-5506.
Craig, H. (1971). The deep metabolism: oxygen consumption in abyssal ocean water. J. Geophys. Res., 76, 5078-5086.
Craig, H. (1974). A scavenging model for trace elements in the deep sea. Earth Plan. Sci. Lett., 23, 149-159.
Craig, H. and R.F. Weiss (1970). The Geosecs 1969 intercalibration station: introduction, hydrographic features, and total CO_2-O_2 relationships. J. Geophys. Res., 75, 7641-7647.
Csanady, G.T. (1978). Water circulation and dispersal mechanisms. In: Lakes. Chemistry Geology Physics. A. Lerman Ed. Springer-Verlag, New York. pp. 21-64.

Finden, D.A.S., E. Tipping, G.H.M. Jaworski and C.S. Reynolds (1984). Light-induced reduction of natural iron (III) oxide and its relevance to phytoplankton. Nature, 309, 783-784.
Forsythe, W.E. (1964). Smithsonian physical tables. 9th ed. Publication 4169. Smithsonian Institution. Washington D.C.
Foster, P. and A.W. Morris (1974). Ultra-violet absorption characteristics of natural waters. Water Res., 8, 137-142.

Gibbs, R.J. (1974). Principles of studying suspended materials in water. In: Suspended Solids in Water. R.J. Gibbs ed. Plenum, New York. pp. 3-15.
Green, A.E.S., T. Sawada and E.P. Chettle (1974a). The middle ultraviolet reaching the ground. Photochem. Photobiol., 19, 251-259.
Green, A.E.S., T. Mo and J. Miller (1974b). A study of solar erythema radiation doses. Photochem. Photobiol., 20, 473-483.
Green, A.E.S., K.R. Cross and L.A. Smith (1980). Improved analytical characterization of ultraviolet skylight. Photochem. Photobiol., 31, 59-65.

Haag, W.R., J. Hoigne, E. Gassmann and A.M. Braun (1984a). Singlet oxygen in surface waters -- Part I: Furfuryl alcohol as a trapping agent. Chemosphere, 13, 631-640.
Haag, W.R., J. Hoigne, F. Gassmann and A.M. Braun (1984b). Singlet oxygen in

surface water -- Part II: Quantum yields of its production by some natural humic materials as a function of wavelength. Chemosphere, 13, 641-650.
Head, P.C. (1976). Organic processes in estuaries. In: Estuarine Chemistry. J.D. Burton and P.S. Liss Eds. Academic Press, New York. pp. 54-92.

Imboden, D.M. and S. Emerson (1978). Natural radon and phosphorus as limnologic tracers: Horizontal and vertical eddy diffusion in Greifensee. Limnol. Oceanogr., 23, 77-90.
Imboden, D.M. and A. Lerman (1978). Chemical models of lakes. In: Lakes. Chemistry Geology Physics. A. Lerman Ed. Springer-Verlag, New York, pp. 341-356.

Jamart, B.M., D.F. Winter, K. Banse, G.C. Anderson and R.K. Lam (1977). A theoretical study of phytoplankton growth and nutrient distribution in the Pacific Ocean off the northwestern U.S. coast. Deep-Sea Res., 24, 753-773.
Jerlov, N.G. (1968). Optical Oceanography. Elsevier. New York. 194 pp.
Johnson, F.S., T. Mo and A.E.S. Green (1976). Average latitudinal variation in ultraviolet radiation at the Earth's surface. Photochem. Photobiol., 23, 179-188.
Joussot-Dubien, J. and A. Kadiri (1970). Photosensitized oxidation of ammonia by singlet oxygen in aqueoua solution and in seawater. Nature, 227, 700-701.

Keeling, C.D. and B. Bolin (1967). The simultaneous use of chemical tracers in oceanic studies. I. General theory of reservoir models. Tellus, 19, 566-581.
Kerr, R.A. and J.G. Quinn (1975). Chemical studies in the dissolved organic matter in seawater. Isolation and fractionation. Deep-Sea Rea., 22, 107-116.
Kerr. R.A. and J.G. Quinn (1980). Chemical compostion of dissolved organic matter isolated from different oceanic environments. Mar. Chem., 8, 217-229.

Mancini, J.L. (1978). Analysis framework of photodecomposition in waters. Environ. Sci. Technol., 12, 1274-1276.
Mill, T., D.G. Hendry and H. Richardson (1980). Free-radical oxidants in natural waters. Science, 207, 886-887.
Mill, T., J.S. Winterle, A. Fischer, W.M. Mabey, D. Tse and A. Baraze (1984). Indirect photolysis and excited states of humic acid. In: Gas-Liquid Chemistry of Natural Waters, Vol. 1. L. Newman Ed. Brookhaven National Laboratory Publication BNL 51757. Upton, New York. pp. 25.1 - 25.11.
Miller, G.C. and R.G. Zepp (1979). Effects of suspended sediments on photolysis rates of dissolved pollutants. Water Res., 13, 453-459.
Miller, S. (1983). Photochemistry of natural water systems. Environ. Sci. Technol., 17, 568A-570A.
Mo, T. and A.E.S. Green (1974). A climatology of solar erythema dose. Photochem. Photobiol., 20, 483-496.
Morel, A. (1978). Available, usable, and stored radiant energy in relation to marine photosynthesis. Deep-Sea Res., 25, 673-688.

Newman, L. Ed. (1984). Gas-Liquid Chemistry of Natural Waters. Brookhaven National Laboratory Publication BNL 51757. Vols. 1 and 2. Brookhaven National Laboratory. Upton, New York.

Ochakovskii, Y.Y., O.V. Kopelvich and V.I. Voyton (1972). Light in the Sea. NTIS Report No. FTD-HT-23-0105-72. 207 pp.

Officer, C.B. (1980). Box models revisited. In: Estuarine and Wetland Processes. P. Hamilton and R.B. Macdonald Eds. Plenum Press, New York. pp. 65-114.

Ogura, N. and T. Hanya (1966). Nature of ultra-violet absorption of seawater. Nature, 212, 758.

Ogura, N. and T. Hanya (1967). Ultraviolet absorption of the seawater, in relation to organic and inorganic matters. Internat. J. Oceanol. Limnol., 1, 91-102.

Oliver, B.G., E.G. Cosgrove and J.H. Carey (1979). Effect of suspended sediments on the photolysis of organics in water. Environ. Sci. Technol., 13, 1075-1077.

Owen, E.D. (1971). Principles of photochemical reactions in aqueous solution. In: Organic Compounds in Aquatic Environments. S.J. Faust and J.V. Hunter Eds. Marcel Dekker, New York. pp. 387-423.

Riley, G.A. (1951). Oxygen, phosphate and nitrate in the Atlantic Ocean. Bull. Bingham. Oceanogr. Coll., 13, 1-126.

Ryther, J.H. (1959). Potential productivity of the sea. Science, 130, 602-608.

Sackett, W.M. (1978). Suspended matter in sea-water. In: Chemical Oceanography, 2nd ed., Vol. 7, J.P. Riley and R. Chester Eds. Academic Press, New York. pp. 127-172.

Smith, R.C. and K.S. Baker (1978a). The bio-optical state of ocean waters and remote sensing. Limnol. Oceanogr., 23, 247-259.

Smith, R.C. and K.S. Baker (1978b). Optical classification of natural waters. Limnol. Oceanogr. 23, 260-267.

Smith, R.C. and K.S. Baker (1979). Penetration of UV-B and biologically effective dose-rates in natural waters. Photochem. Photobiol., 29, 311-323.

Smith, R.C. and J.E. Tyler (1976). Transmission of solar radiation into natural waters. Photochem. Photobiol. Rev., 1, 117-155.

Spencer, D.W. and P.G. Brewer (1971). Vertical advection diffusion and redox potential as controls on the distribution of manganese and other trace metals dissolved in waters of the Black Sea. J. Geophys. Res., 76, 5877-5892.

Spencer, D.W., M.P. Bacon and P.G. Brewer (1981). Models of the distribution of ^{210}Pb in a section across the North Equatorial Atlantic Ocean. J. Mar. Res., 39, 119-138.

Strickland, J.D.H. (1958). Solar radiation penetrating the ocean. A review of requirements, data and methods of measurement, with particular reference to photosynthetic productivity. J. Fish. Res. Bd. Can., 15, 453-493.

Strickland, J.D.H. (1965). Phytoplankton and marine primary production. Ann. Rev. Microbiol., 19, 127-162.

Stumm, W. and P.A. Brauner (1975). Chemical speciation. In: Chemical Oceanography, 2nd ed., Vol. 1. J.P. Riley and G. Skirrow Eds. Academic Press, New York. pp. 173-240.

Sundstrom, G. and L.O. Ruzo (1978). Photochemical transformation of pollutants in water. In: Aquatic Pollutants: Transformation and Biological Effects. O. Hutzinger, I.H. vann Lelyveld and B.C.J. Zoeteman, Eds. Pergamon Press, New York, pp. 205-222.

Turro, N.J. (1965). Molecular Photochemistry. Benjamin, New York. pp. 18-19.

Van Baalen, C. and J.E. Marler (1966). Occurrence of hydrogen peroxide in sea water. Nature, 211, 951.

Veronis, G. (1975). The role of models in tracer studies. In: Numerical Models of Ocean Circulation. National Academy of Sciences, Washington D.C., pp.

133-146.
Veronis, G. (1977). Use of tracers in circulation studies. In: The Sea: Ideas and Observations on Progress in the Study of the Seas. Vol. 7. E.D. Goldberg Ed. Wiley, New York. pp. 169-188.

Waite, T.D. and F.M.M. Morel (1984). Photoreductive dissolution of colloidal iron oxide: effect of citrate. J. Colloid. Interface Sci., 102, 121-137.
Wayne, R.P. (1970). Photochemistry. Elsevier, New York. 263 pp.
Williams, J. (1970). Optical Properties of the Sea. U.S. Naval Institute. Annapolis, Maryland. 123 pp.
Williams, P.J. le B. (1975). Biological and chemical aspects of dissolved organic material in sea water. In: Chemical Oceanography, 2nd ed. Vol. 2, J.P. Riley and G. Skirrow Eds. Academic Press, New York. pp. 301-364.
Willis, W.M. (1978). The photosynthetic action spectra of the phytoplankton and role in governing spatial and temporal distribution: a numerical modeling approach. Ph. D. Thesis. Old Dominion University. Norfolk, VA. 98 pp.
Wilson, T.R.S. (1975). Salinity and the major elements of sea water. In: Chemical Oceanography, 2nd ed. Vol. 1, J.P. Riley and G. Skirrow Eds. Academic Press, New York. pp. 365-414.
Wyrtki, K. (1962). The oxygen minimum in relationship to ocean circulation. Deep Sea Res., 9, 11-23.

Zafiriou, O.C. (1974). Sources and reactions of OH and daughter radicals in seawater. J. Geophys. Res., 79, 4491-4497.
Zafiriou, O.C. (1977). Marine organic photochemistry previewed. Mar. Chem., 5, 497-522.
Zafiriou, O.C. (1983). Natural water photochemistry. In: Chemical Oceanography, 2nd edition. J.P. Riley and R. Chester Eds. Vol. 8. Academic Press, United Kingdom. pp. 339-376.
Zafiriou, O.C. (1984). A Bibliography of References in Natural Water Photochemistry. Woods Hole Oceanographic Institution Tech. Memo. WHOI-2-84. Woods Hole, MA. 51 pp.
Zafiriou, O.C. and M. McFarland (1981). Nitric oxide from nitrite photolysis in the central equatorial Pacific. J. Geophys. Res., 86, 3173-3182.
Zafiriou, O.C. and M.B. True (1979a). Nitrite photolysis in seawater by sunlight. Mar. Chem., 8, 9-32.
Zafiriou, O.C. and M.B. True (1979b). Nitrate photolysis in seawater by sunlight. Mar. Chem., 8, 33-42.
Zafiriou, O.C. and M.B. True (1979c). Nitrite photolysis as a source of free radicals in productive surface water. Geophys. Res. Lett., 6, 81-84.
Zafiriou, O.C., J. Joussot-Dubien, R.G. Zepp and R.G. Zika (1984). Photochemistry of natural waters. Environ. Sci. Technol., 18, 358A-371A.
Zafiriou, O.C., M. McFarland and R.H. Bromund (1980). Nitric oxide in seawater. Science, 207, 637-639.
Zepp, P.G. (1978). Quantum yields for reaction of pollutants in dilute aqueous system. Environ. Sci. Technol., 12, 327-329.
Zepp, P.G. (1980). Assessing the photochemistry of organic pollutants in aquatic environments. In: Dynamics, Exposure and Hazard Assessment of Toxic Chemicals. R. Haque Ed. Ann Arbor Science. An Arbor, MI. pp. 69-110.
Zepp, P.G. (1982a). Photochemical transformations induced by solar ultraviolet radiation in marine ecosystems. In: The Role of Solar Ultraviolet Radiation in Marine Ecosystems. J. Calkins, Ed. Plenum Press, New York. pp. 293-307.
Zepp, P.G. (1982b). Experimental approaches to environmental photochemistry. In: The Handbook of Environmental Chemistry. O. Hutzinger Ed. Vol. 2/ Part B. Springer-Verlag. Berlin. pp. 19-40.

Zepp, P.G. and G.L. Baughman (1978). Prediction of photochemical transformation of pollutants in the aquatic environment. In: Aquatic Pollutants: Transformation and Biological Effects. O. Hutzinger, I.H. van Lelyveld and B.C.J. Zoeteman, Eds. Pergamon Press, New York. pp. 237-263.

Zepp, P.G. and D.M. Cline (1977). Rates of direct photolysis in aquatic environment. Environ. Sci. Technol., 11, 329-366.

Zepp, P.G. and P.F. Schlotzhauer (1981a). Comparison of photochemical behavior of various humic substances in water: III. Spectroscopic properties of humic substances. Chemosphere, 10, 479-486.

Zepp, P.G. and P.F. Schlotzhauer (1981b). Effects of equilibration time on photoreactivity of the pollutant DDE sorbed on natural sediments. Chemosphere, 10, 453-460.

Zepp, P.G. and P.F. Schlotzhauer (1983). Influence of algae on photolysis rates of chemicals in water. Environm. Sci. Technol., 17, 462-468.

Zepp, P.G., G.L. Baugham and P. F. Schlotzhauer (1981a). Comparison of photochemical behavior of various humic substances in water: I. Sunlight induced reactions of aquatic pollutants photosensitized by humic substances. Chemosphere, 10, 109-117.

Zepp, P.G., G.L. Baugham and P. F. Schlotzhauer (1981b). Comparison of photochemical behavior of various humic substances in water: II. Photosensitized oxygenation. Chemosphere, 10, 119-126.

Zepp, P.G., P.F. Schlotzhauer and R.M. Sink (1985). Photosensitized transformation involving electronic energy transfer in natural waters: Role of humic substances. Environ. Sci. Technol., 19, 74-81.

Zepp, P.G., N.L. Wolfe, G.L. Baugham and R.C. Hollis (1977). Singlet oxygen in natural waters. Nature, 267, 421-423.

Zika, R.G. (1981). Marine organic photochemistry. In: Marine Organic Chemistry. E.K. Duursma and R. Dawson Eds. Elsevier Oceanography Series, No. 31. Elsevier. New York. pp. 299-325.

Zika, R.G., L.T. Gidel and D.D. Davis (1984). A comparison of photolysis and substitution decomposition rates of methyl iodide in the ocean. J. Geophys. Res., 11, 353-356.

CHAPTER 12

MICROBIAL DECOMPOSITION

by
Bruno Tassin and Daniel R. Thévenot

12.1 INTRODUCTION

Role of microbial decomposition in biodegradation processes of organic pollutants.

Within aquatic ecosystems, microbial decomposition of organic matter plays a prominent role in the energy and mass transformation processes (Jørgensen, 1983a). Organic waste water discharge in a river results in dissolved oxygen uptake, either directly by chemical oxidation of the reducing pollutants (see Chap. 10), or by their metabolism by micro-organisms, i.e. their biodegradation. Raising the temperature of rivers reduces oxygen solubility and accelerates the kinetics of chemical and microbiological oxygen and polluting load reduction.

Several terms are used to describe globally these processes. **Self-purification** is a neutral expression, meaning elimination of dissolved or particulate matter with "polluting" properties, irrespective of the mechanism involved (Wuhrmann, 1972). **Biodegradation** is a more commonly accepted term which describes alterations of a compound sufficiently important to change its identity (Dart and Stretton, 1980). All biodegradable compounds are ultimately converted into one of the very few key intermediates such as acetyl-coenzyme A or one of the tricarboxylic acid cycle intermediates before total oxidation occurs. As such compounds are also used for the resynthesis of larger molecules, only a portion of biodegradable compounds are totally oxidized to inorganic constituents.

Water, soil, sewage and mud are extremely complex environments biologically, chemically and physically (Dart and Stretton, 1980). Large numbers of organisms from different species and genera may be present displaying a wide range of degradative abilities. The organisms may show a spectrum of interactions, ranging from predation through commensalism to synergism. The presence or absence of predators, parasites or hosts may cause microbial populations to fluctuate over a wide range. This chapter will not describe such microbial decomposition at a biological level, but will present mathematical and experimental submodels used to simulate this wide range of interacting processes.

A large part of waste water disposal regulations refers to the **ability of various organic wastes to be biodegraded** in the ecosystem. Various procedures have been developed for the measurement of biodegradability of chemical compounds, either in the innoculated die-away tests where water, salts and biomass is incubated during 5 to 40 days with the tested substance, or with activated sludge test methods where large amounts of biomass and additional organic substrates are continously fed with air or oxygen for better adaptation ot large amounts of substrates (Blok, 1975). Such tests were proposed for monitoring either carbon dioxide formation or oxygen consumption in oxygen regulated "Sapromat" respirometric cells (Remiche and Van der Wielen, 1975). In the former case, an alternative to barium carbonate measurement is the direct monitoring of $^{14}CO_2$ formation from traces of ^{14}C radiolabelled organic compounds, such as amines (Boethling and Alexander, 1979). The simultaneous monitoring of carbon dioxide and oxygen allows discrimination between microbial decomposition and chemical oxidation.

Determination of parameters related to biodegradation: dissolved oxygen (DO), organic matter (OM) and biomass (B).

Since organic pollution control has been the first concern in water quality studies and regulations parameters relating to biodegradation are relatively well defined.

Dissolved oxygen (DO) is easily determined either with absolute chemical methods such as the Winkler volumetric titration using manganous and iodide salts in basic media, or with relative electrochemical methods using Clark-type electrode, i.e. cathodic reduction of oxygen diffusing through an hydrophobic membrane.

If DO may easily and rapidly be monitored with a precision better then 1-2%, **organic matter** (OM) represents a large variety of compounds and may not be determined without ambiguity. Several methods have appeared, each of them corresponding to a given goal. The Biochemical Oxygen Demand test (BOD) intends to assess the pollutional potential of a waste water which contains an available organic carbon source for aerobic organotrophic microorganisms, by measuring the amount of oxygen utilized during growth of the organisms in a sample of waste water (Gaudy, 1972). Standard methods for measuring BOD have been described and compared. DO variations are usually measured in diluted samples after 5 or 21 days incubation at 20°C (Stack, 1972 and 1973); alternatively, automatic respirometers enable the direct recording of oxygen consumption of undiluted samples. In the Warburg method, gas volume variations are monitored, whether in Marais or

"Sapromat" methods, and oxygen is supplied to the sample so that its concentration remains constant throughout the incubation test (Mytelka and Brenner, 1969; Remiche and Van der Wielen, 1975; Hung and Eckenfeld, 1976; Illic, 1978). The term 'ultimate BOD' is usually referred to in 21-days incubation tests - this test duration being necessary for total biodegradation of nitrogen-containing compounds (Stack, 1972). Nevertheless, as nitrogeneous compounds are often specifically measured and their microbial decomposition kinetics taken into account (see § 2.2.1), carbonaceous OM is then estimated by 5 days BOD, called BOD_5. The amount of OM may alternatively be determined either by chemical oxidation in standard conditions, such as 2 hr ebullition with excess bichromate in sulphuric acid (Chemical Oxygen Demand, COD) (Green et al., 1981), or by total oxidation by oxygen at 900°C, or by persulphate in the presence of heavy UV radiation (Total Oxygen Demand, TOD) (Clark, 1974); finally the carbon dioxide released during such total oxidation procedures is monitored in various ways. Carbonaceous compounds present in organic pollutants may be determined by their Total Organic Carbon (TOC). Since the OM content in surface water and in waters generally is of great economic and ecological importance, BOD, COD, TOD and TOC have been thoroughly studied and "mixed" methods with the accuracy of chemical titration have been proposed (Stack, 1973).

Biomass concentration (B) is even more difficult to determine than OM. Numerous methods have been developed, but none of them is generally accepted by the scientific community. Adenosine Triphosphate (ATP) is frequently chosen for evaluation of biomass. Indeed it is a fundamental constituent of all living cells, and only living ones, since it is destroyed very rapidly after cell death. It has been shown that the amount of ATP is proportional to the amount of cell organic carbon, which is the usual indicator of biomass: 0.30 to 0.39 g ATP is contained in 100 g cell organic carbon of algae and bacteria of different sizes and species (Holm-Hansen, 1970). Furthermore, trace amounts of ATP can be easily determined by bioluminescence, after cell filtration and extraction. When mixed with luciferine and luciferase, ATP extracts yield light emission in which the intensity maximum is proportional to the ATP amount, with the lower detection limit being as low as 10^{-12} mol (Tobin et al., 1978; Apoteker, 1981). But an alternative to ATP determination has been proposed - Deoxyribonucleic acid (DNA) content, though this rarely used for biomass estimation since it is known to retain its concentration level even after cell death. Finally, when phytoplanktonic biomass is to be determined, chlorophyl s are measured after cell filtration and extraction.

12.2 MATHEMATICAL MODELS OF MICROBIAL DECOMPOSITION

Different types of kinetic models of microbial decomposition.

Numerous kinetic models of water quality have been proposed to describe DO and BOD variations along a stream (Gromiec et al., 1983; Gromiec, 1983) in lakes or reservoirs (Orlob, 1983, Jørgensen, 1983b) and in waste water treatment processes. The first, and most widely used model, was proposed by Streeter and Phelps (1925) for the Ohio river. It states that oxygen uptake is equal to BOD uptake, and that both kinetics are first-order. Since then several modifications have been proposed to take into account the effect of dispersion (O'Connor, 1961, 1962, 1967), sedimentation and adsorption (Dobbins, 1964), photosynthesis and respiration (O'Connor and Di Torro, 1970), thermal exchanges (Lin et al., 1973a, b, c), bental demand (Shelton et al., 1978) and nitrogeneous biodegradation (Shelton et al., 1978; Knowles and Wakeford, 1978). Nevertheless, it has been frequently observed that first-order curves do not fit well the experimental variations of BOD, so that *second-order* kinetics were proposed for DO and BOD uptakes (Woodward, 1953; Revelle et al., 1965; Young and Clark, 1965; Zanoni, 1967; Edeline and Lambert, 1979). Their success in describing actual river evolution was not generally accepted.

Beside these field studies, Flegal and Schroeder (1976), Rauwel and Thévenot (1976) and Apoteker and Thévenot (1983) conducted laboratory experiments without renewal of DO: after organic discharges. All these authors observed lag-phases with no oxygen evolution, followed by exponential phases, similar to those encountered in experiments on cellular growth.

First-order kinetic models: Streeter-Phelps, O'Connor, Dobbins, Lin and Shelton models; examples and discussion.

Streeter and Phelps (1925) kinetic model of DO and BOD evolution in streams is based upon two main assumptions: firstly, both microbial decomposition of OM and reoxygenation of water by the atmosphere follow first-order kinetics; second, microbial DO and BOD uptakes are equal whatever the stage of OM decomposition and the type of OM. Taking these two assumptions as valid, two kinetic equations may be written:

$$dL / dt = -K_1 L \qquad (12.1)$$

$$dC / dt = -K_1 L + K_2 (C_s - C) \qquad (12.2)$$

where
L is the OM load, usually estimated as ultimate BOD, i.e. 21 days BOD ($M\ L^{-3}$),
C is the DO concentration ($M\ L^{-3}$) at time t (T),
C_s is the saturated DO concentration ($M\ L^{-3}$) at the water temperature T,
K_1 is the first-order biodegradation kinetic constant, also called deoxygenation rate constant (T^{-1}),
K_2 is the first-order oxygen transfer or reaeration kinetic constant (T^{-1}).

As presented later in this paragraph and in § 3.1, K_1 frequently does not take into account BOD excretion by nitrogeneous compounds and by sediments.

As DO concentration is frequently replaced by oxygen saturation deficit, $D = C_s - C$, equation (12.2) becomes:

$$dD / dt = K_1 L - K_2 D \tag{12.3}$$

The solutions to (12.1) and (12.3) are:

$$L = L_0 \exp(-K_1 t) \tag{12.4}$$

$$D_t = \frac{K_1 L_0}{K_2 - K_1} (\exp(-K_1 t) - \exp(-K_2 t)) + D_0 \exp(-K_2 t) \tag{12.5}$$

where
D_t is the dissolved oxygen saturation deficit after time t,
L_0 is the BOD at an initial reference time (t = 0),
D_0 is the dissolved oxygen deficit at t = 0.

Kinetic constants K_1 and K_2 are sometimes replaced by

$\underline{k}_1 = 0.434\ K_1$ and $\underline{k}_2 = 0.434\ K_2$

when exponential functions of (12.5) are replaced by power of 10 functions. Equation 12.5 is the Streeter-Phelps "oxygen sag formula" and a profile of DO along a stream is referred to as a "dissolved oxygen sag curve" (Figure 12.1). The critical point of this curve corresponds to the minimum of DO or the maximum of saturation deficit D_c, reached at critical time t_c. At this point $dD / dt = 0$, and the critical deficit is

$$D_c = \frac{K_1}{K_2} L = \frac{K_1}{K_2} L_0 \exp(-K_1 t_c) \tag{12.6}$$

and the critical time equals

$$t_c = \frac{1}{K_2 - K_1} \mathrm{Ln} \left(\frac{K_2}{K_1} \left(1 - \frac{D_0 (K_2 - K_1)}{L_0 K_1} \right) \right) \tag{12.7}$$

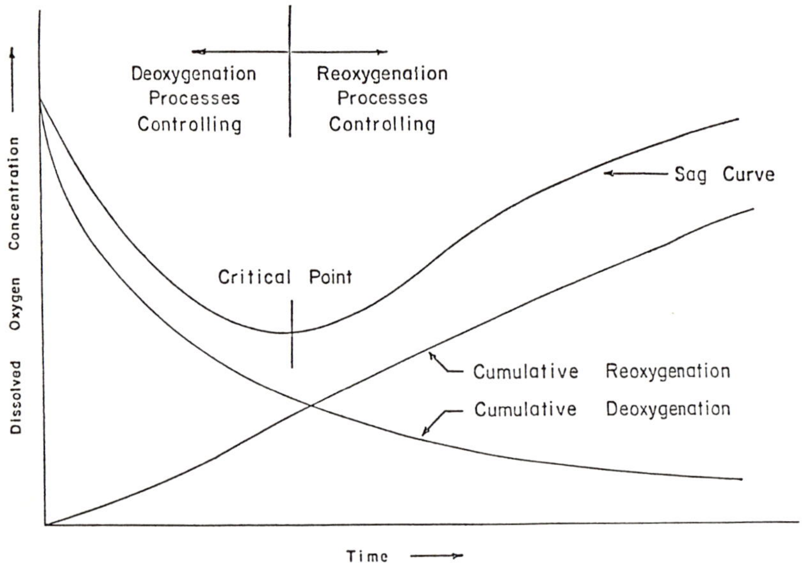

CLASSICAL OXYGEN SAG CURVE WITH DEOXYGENATION AND REOXYGENATION

Fig. 12.1: Dissolved oxygen sag curve using first-order kinetics. After Gates (1971).

Equations 12.6 and 12.7 are sometimes simplified, using the Fair coefficient $f = K_2 / K_1$, also called the self-purification factor:

$$t_c = \frac{1}{K_1 (f - 1)} \mathrm{Ln} \left(f \left(1 - (f - 1) \frac{D_0}{L_0} \right) \right) \tag{12.8}$$

$$D_c = \frac{L_0}{f} \left(f \left(1 - (f - 1) \frac{D_0}{L_0} \right) \right)^{1/(1-f)} \tag{12.9}$$

Arababi et al. (1974, 1975) have given a linear approximation of D_c in relation to the initial oxygen deficit and pollution load

$$D_c = R L_0 + A D_0 \tag{12.10}$$

with

$$R = f^{f/(1-f)} \quad \text{and} \quad A = 1 - f^{1/(1-f)}$$

this approximation, better than 6.2%, being very usefull for water quality assessment studies.

Besides the two previously mentioned assumptions, the Streeter-Phelps model is based on a constant pollution load discharged at a single given point along a stream having a constant flow rate Q and uniform cross section. The lateral and vertical DO and BOD concentrations are assumed uniform throughout any cross section. Within lakes and reservoirs such assumptions are obviously erroneous. Water stratification, related to temperature gradients, play frequently a prominent role in lake water quality models. Thus equations 12.4 to 12.9 are mainly used for stream water quality models.

Since 1925 when the Streeter-Phelps kinetic model was presented, numerous modifications and improvements have been incorporated without changing the first-order kinetic basic assumptions.

Nitrogeneous compounds biodegradation.

When BOD tests are performed with waste water samples containing nitrogeneous compounds such as organic nitrogen (amines, amides, proteins) or ammonium salts, oxygen evolution curves, recorded on Sapromat or respiration equipments, frequently present two steps (Figure 12.2). The second step, significant after 7-10 days, is attributed to nitrogen containing compound from microbial decomposition, i.e. organic nitrogen mineralisation and ammonium nitrification.

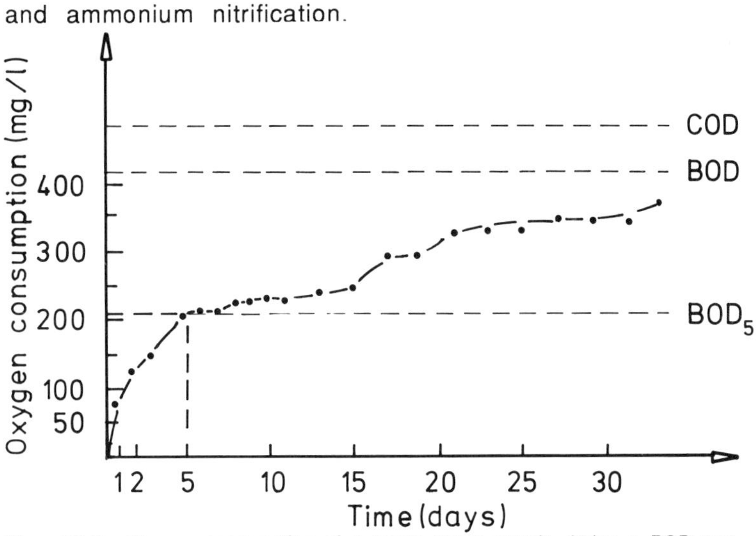

Fig. 12.2: Oxygen consumption of a waste water sample during a BOD test. After Wilderer et al. (1970).

As microbial decomposition of nitrogen-containing compounds is known to occur more slowly than carbonaceous biodegradation, the Streeter-Phelps kinetic models may incorporate in OM and DO uptake kinetic equations 12.1 and 12.2 specific terms for each form of nitrogen (O'Connor and Di Toro, 1970; Shelton et al., 1978):

$$d\,(NH_4\text{-}N)\,/\,dt = K_{1,NH_4}\,(NH_4\text{-}N) \tag{12.11}$$

$$d\,(NO_2\text{-}N)\,/\,dt = K_{1,NO_2}\,(NO_2\text{-}N) \tag{12.12}$$

$$dD/dt = K_1\,L + 4.57\,K_{1,NH_4}\,(NH_4\text{-}N) + 1.14\,K_{1,NO_2}\,(NO_2\text{-}N) - K_2\,D \tag{12.13}$$

the numerical factors 4.57 and 1.14 being equal to the number of grams of oxygen necessary for the transformation of 1 g of NH_4-N or NO_2-N into nitrate. In equation 12.13 L relates to carbonaceous OM and is usually estimated by BOD_5.

When such models have been applied to OM and DO variations in stream water, first-order kinetic constants were found to vary significantly from stream quality: K_1 ranges between 0.09 and 0.64 day^{-1} (the lowest values being obtained in non-polluted rivers) (Eckenfelder, 1970; Hung et al., 1978) and $K_{1,NH4}$ ranges between 0.07 and 0.14 day^{-1} (Lefort, 1971). In fact when values of DO, OM, NH_4-N and NO_2-N are measured along a stream section of homogeneous properties at initial and 't' time, K_1, K_2, $K_{1,NH4}$ and $K_{1,NO2}$ are determined from best fit of the data (Rinaldi and Soncini-Sessa, 1977).

Besides this specific introduction of nitrogenous compound biodegradation, the Streeter-Phelps kinetic model has received several sets of modifications and improvements which nevertheless do not change its first-order character.

Dispersion and transport.

O'Connor (1961, 1962, 1967) introduced Fick's diffusion law in BOD and DO profiles:

$$\frac{dL}{dt} + \frac{v\,dL}{dx} - \frac{E\,d^2L}{dx^2} = -K_1\,L \tag{12.14}$$

$$\frac{dC}{dt} + \frac{v\,dC}{dx} - \frac{E\,d^2C}{dx^2} = -K_1\,L + K_2\,(C_s - C) \tag{12.15}$$

where

v is the stream velocity $(L\ T^{-1})$

E is the turbulent transport (longitudal mixing) coefficient $(L^2\ T^{-1})$

Sediments

Dobbins (1964) introduced the BOD removal by sedimentation and/or adsorption with a first-order kinetic constant $K_{1,s}$ and the removal of DO by benthic processes and by plant respiration/photosynthesis:

$$\frac{dL}{dt} + \frac{v\,dL}{dx} - \frac{E\,d^2L}{dx^2} = -(K_1 + K_{1,s})\,L \qquad (12.16)$$

$$\frac{dC}{dt} + \frac{v\,dC}{dx} - \frac{E\,d^2C}{dx^2} = -K_1 L + K_2 (C_s - C) - D_b \qquad (12.17)$$

where

$K_{1,s}$ is the kinetic constant of BOD evolution on sediments (T^{-1})

D_b is the rate of removal of DO caused by benthic demand and plants $(M\ L^{-3}\ T^{-1})$

The water exertion rate of BOD in streams, including all sinks such as sediments, is usually refered to as $K_{1,r}$. Thus one may replace in equation (12.16) $K_1 + K_{1,s}$ by $K_{1,r}$ if other terms such as resuspension or volatilisation of OM are neglected.

Photosynthesis and respiration

O'Connor and Di Toro (1970) adjusted the temporal form of the photosynthetic oxygen source in streams by a half-cycle sine wave. The daily rate of photosynthetic oxygen production P as a function of time was expressed as:

$$P = P_m \sin\,(pi/s)\,(t - t_s)) \quad \text{when } t_s \le t \le t_s + s$$
$$P = 0 \quad \text{when } t_s + s \le t \le t_s + 1 \qquad (12.18)$$

where

P_m is the maximum oxygen photosynthesis, in the stream $(M\ L^{-3}\ T^{-1})$

t_s is the time, expressed as a fraction of the day, at which the source becomes active (T)

s is the fraction of the day during which the source is active (period of sunlight) (T).

Taking into account the benthic and algal respiration rate and the algal photosynthesis, the DO equation becomes:

$$\frac{v\,dC}{dx} - \frac{E\,d^2C}{dx^2} = -K_1 L + K_2 (C_s - C) - S - R + P \qquad (12.19)$$

where

S = a function of distance x, is the benthic respiration rate (sink)($M\ L^{-3}\ T^{-1}$)

R = a function of distance x, is the algal respiration rate (sink) ($M\ L^{-3}\ T^{-1}$)

P = a function of time t and distance x, is the algal photosynthetic oxygen production rate (source) ($M\ L^{-3}\ T^{-1}$).

Alfi et al. (1972) have determined in situ the above mentioned S, R and P parameters by implanting cylinder containers into a stream and recording in each of them the DO evolution. These closed cylinders were either transparent (with bottom open to sediments for benthic demand or bottom closed) or black (bottom closed) to prevent K_2, S and P influence. K_2 was determined using an unconfined oxygen probe situated in the vicinity.

Thermal exchanges

To the Dobbins model, Lin et al. (1973a, b, c) introduced a third equation representing water temperature T evolution:

$$\frac{dT}{dt} - \frac{E\,d^2T}{dx^2} + \frac{v\,dT}{dx} - Q_T = 0 \qquad (12.20)$$

where Q_T represents the air-water energy transfer rate, function of air and water temperature, specific mass and calorific capacity as well as wind velocity and radiation flux (Jacquet, 1983) (see Chap. 5).

Waste water discharge

Lin et al. (1973b) introduced also a continuous discharge of wastes along the stream and modified equation (12.14):

$$\frac{dL}{dt} + \frac{v\,dL}{dx} - \frac{E\,d^2L}{dx^2} = -K_1 L + L_a \qquad (12.21)$$

where L_a is the rate of BOD release along a river length unit ($M\ L^{-4}$).

Discussion

Whatever the improvements brought to the original Streeter-Phelps kinetic equations (12.1) and (12.2), all these models suffer from the two basic assumptions mentioned above. If one summarizes microbial decomposition processes by a biomass catalysed overall reaction

$$OM + O_2 \longrightarrow \text{products} \qquad (12.22)$$

then, assuming a first-order kinetic of OM evolution means that reaction (12.22) is independent of DO and biomass concentration:. This is obviously erroneous, and in contradiction to any experimental microbial reaction kinetics. Both DO and B levels play a prominent role in aerobic biodegradation reactions (Gates, 1971; Wuhrmann, 1972; Edeline and Buet, 1978; Edeline and Lambert, 1979; Apoteker, 1981; Apoteker and Thévenot, 1983). Furthermore microbial metabolism as well as cell culture experiments demonstrate that the substrates and oxygen demands by microorganism are not proportional to but depend upon the cell development state.

As literature contains numerous reports of field studies in which the Streeter-Phelps kinetic models are used with reasonable success, there is an apparent contradiction between field and laboratory studies. Validity of experimental simulation methods for self-purification studies (Gates, 1971; Edeline and Lambert, 1974) and of cell culture kinetics does not seem to be questionable. Thus the apparent validity of first-order models may be related to the lack of precision of field determinations of concentration parameters such as C, L or T and even more dramatically to the very approximate evaluation of waste fluxes and quality, i.e. L_o and/or L_a.

Within lakes and reservoirs the main processes taken into account for water quality modelling are usually eutrophication reactions related to phosphorus and nitrogen cycles. In such systems the absence of any kinetic relation to biomass evolution is obviously erroneous. On the other hand, most stream kinetic models used in environmental assessment studies neglect eutrophication processes and biomass evolution:. Field data seem reasonably well fitted with previously mentioned Streeter-Phelps first-order kinetic models.

Zero-order kinetic models; examples and discussion

When Seine River water DO and BOD was modelled by Lesouef (1981), it was found that, although first-order kinetic models provided a good fit above the Paris general sewage plant at Achères, below that point zero-order BOD kinetics seem to fit much better. Equation (12.13) becomes:

$$\frac{dD}{dt} = K_1 + 4.57\,K_{1,NH4}\,(NH_4\text{-}N) - K_2\,D \qquad (12.23)$$

One should point out that for very slow biodegradation processes there is no large difference between first- and zero-order kinetics and this seems to be the case in the Seine River below Achères general sewage plant of Paris.

Furthermore, Lesouef (1979) demonstrated that the ammonium uptake first-order kinetic constant $K_{NR,NH4}$ was frequently two times larger than the ammonium oxidation (nitrification) kinetic constant $K_{1,NH4}$. Indeed ammonium uptake correspond to nitrification but also to assimilation or even ammoniac evaporation reactions. For this equations (12.11) (12.23) and (12.5) become:

$$\frac{d\,(NH_4\text{-}N)}{dt} = -\,K_{NR,NH4}\,(NH_4\text{-}N) \qquad (12.24)$$

$$\frac{dD}{dt} = K_1 + 4.57\,K_{1,NH4}\,(NH_4\text{-}N)_o\,\exp(-K_{NR,NH4}\,t) - K_2\,D \qquad (12.25)$$

$$D_t = \frac{4.51\,K_{1,NH4\text{-}N}\,(NH_4\text{-}N)_o}{K_2 - K_{NR,NH4\text{-}N}}\,(\exp(-K_{NR,NH4\text{-}N}\,t) - \exp(-K_2\,t))$$

$$+ \frac{K_1}{K_2}\,(1 - \exp(-K_2\,t)) + D_o\,\exp(-K_2\,t) \qquad (12.26)$$

with K_1 at 20°C equal to 1.4 mg $l^{-1}day^{-1}$ and $K_{NR,NH4}$ equal to 0.11 and 0.06 day^{-1} in spring-summer and in autumn respectively.

When first-order kinetics are used for DO, BOD and NH_4 evolution, as well as for ammonium nitrification, equations (12.23), (12.25) and (12.26) become:

$$\frac{dD}{dt} = K_1 L + 4.57 K_{1,NH4} (NH_4\text{-}N) - K_2 D \qquad (12.27)$$

$$\frac{dD}{dt} = K_1 L_o \exp(-K_1 t) + 4.57 K_{1,NH4} (NH_4\text{-}N)_o \exp(-K_{NR,NH4} t) - K_2 D \qquad (12.28)$$

$$D_t = \frac{4.51 K_{1,NH4\text{-}N} (NH_4\text{-}N)_o}{K_2 - K_{NR,NH4\text{-}N}} (\exp(-K_{NR,NH4\text{-}N} t) - \exp(-K_2 t))$$

$$+ \frac{K_1 L_o}{K_2 - K_1} (\exp(-K_1 t) - \exp(-K_2 t)) + D_o \exp(-K_2 t) \qquad (12.29)$$

with K_1 ranging between 0.4 and 0.6 day^{-1}.

Such zero-order kinetic models for DO evolution in highly polluted streams seem to be only related to attempts to fit correctly sets of field data. They are not based on chemical or microbiological kinetic assumptions.

12.3 PARAMETERS ESTIMATION

Determination of the reaeration kinetic constant K_2

Although this point has been already considered in the Chapter 3, the most recent methods of calculation of the reaeration kinetic constant K_2 will be presented here. It will also be assumed that the reoxygenation will occur at the interface air-water and follow a first order kinetic law:

$$\frac{dR_o}{dt} = K_2 (C_s - C) \qquad (12.30)$$

where

dR_o/dt is the rate of reoxygenation ($M\ L^{-3}\ T^{-1}$)

C_S is the concentration of oxygen at saturation ($M\ L^{-3}$)

C is the concentration of oxygen in the river ($M\ L^{-3}$)

K_2 is the reaeration kinetic constant (T^{-1}).

The reaeration kinetic constant is influenced by many factors. The most important to be considered are: (see Chapter 3 for further details)

a. the water temperature T
b. the river flow characteristics (flow velocity v, depth of the river H)
c. the meteorological characteristics (air temperature and wind velocity)

a. The water-temperature T dependence of the reaeration coeffient K_2 is usually assumed to follow an Arrhenius-type law:

$$K_2 = (K_2)_{20} \quad theta^{(T-20)} \tag{12.31}$$

where
T is the temperature of the water (°C),
$(K_2)_{20}$ is the value of K_2 at 20°C (T^{-1}), and
theta is a dimensionless parameter varying from 1.008 (Truesdale et ai., 1958) to 1.046 (Streeter et al., 1936).

b. The influence of two parameters characteristic of the river flow (stream velocity and depth) on K_2 was investigated by Covar, 1976.

A graph has been plotted of the data collected by O'Connors et al. (1958; Churchill et al. (1968) and Owens et al. (1964), as presented in Chapter 3.

Table 12.1 presents some reaeration coefficient values for rivers and streams of different depth and velocities.

Determination of the biodegradation kinetic constant K_1

Numerous methods have been developed and used to determine the biodegradation kinetic constant rate, K_1, of the first-order Streeter-Phelps kinetic model. Three different rates can be found in the literature (Bowie et al., 1985):

K_1 water deoxygenation rate due to BOD exertion (T^{-1})
$K_{1,l}$ laboratory deoxygenation rate due to BOD exertion (T^{-1})
$K_{1,r}$ water BOD exertion rate, including all sinks (T^{-1})

The significance of these three parameters is very different and has to be taken into account when a model is built up.

Table 12.1
Reaeration coefficient $\underline{K}_2$ for different rivers and streams. After Hydroscience (1971)

River name	Category	Depth (m)	Area (m^2)	Flow (m^3/s)	Velocity (m/s)	K_2 20°C (l/day)	Reference
Grand River (Michigan)	Shallow	0.579	29.729	8.336	0.280	4.500	O'Connor et al. (1970)
Clinton River (Michigan)	Shallow	0.482	4.143	0.909	0.219	5.900	O'Connor et al. (1970)
Truckee River (Nevada)	Shallow	0.509	13.935	7.688	0.552	6.600	O'Connor et al. (1970)
Jackson River (Virginia)	Shallow	0.914	33.910	2.791	0.082	4.100	Hydroscience (1967)
Ivel River (GB)	Shallow	0.369	3.200	0.138	0.043	2.350	Owens et al. (1962)
Mohawk River (New York)	Inter-mediate	4.572	353.032	22.597	0.064	0.07-4.0	O'Connor et al. (1968)
New River (Virginia)	Inter-mediate	1.524	159.793	34.093	0.213	1.040	Hydroscience (1966)
Clinch River (T.V.)	Inter-mediate	2.185	208.594	167.918	0.805	1.130	Churchill et al. (1962)
Holston (T.V.)	Inter-mediate	3.478	330.416	294.070	0.890	0.280	Churchill et al. (1962)
Wautasa (T.V.)	Inter-mediate	1.042	57.82	88.122	1.524	5.600	Churchill et al. (1962)
Ohio River	Deep	9.754	3994.831	170.467	0.043	0.060	Streeter et al. (1925)
Upper Hudson	Deep	6.401	627.096	286.708	0.457	0.165	Hydroscience (1965)
Upper James	Deep	4.724	1393.546	55.218	0.040	0.220	Hydroscience (1964)

The $K_{1,r}$ coefficient includes all sinks of BOD and especially the physical processes like sedimentation, floculation, resuspension and mixing, and special chemical processes like volatilization. Thus the use of $K_{1,r}$, or of K_1, in the model depends only on the system analysis of the system which is modelled. If just BOD and DO are taken into account, then $K_{1,r}$ has to be used; if the model deals specifically with nitrogen-containing species, then K_1 is prefered.

Relation between K_1 and $K_{1,l}$

The difference between $K_{1,l}$ and K_1 is related to the difference of flow characteristics in the river and in a BOD bottle. Many relations have been developed between these parameters. We will present two of these methods.

The first one (Nejedly, 1973) relates $K_{1,l}$ and K_1 to the longitudinal dispersion coefficient, E, of the river. Indeed, the dispersion increases significantly the rate of BOD exertion. Thus for a given temperature, the following relation is used:

$$K_1 = K_{1,l}\, a^{E*b} \qquad (12.32)$$

where
a and b are adjustable dimensionless parameters
E is the longitudinal dispersion ($L^2\ T^{-1}$)
E can be calculated with empirical relations or determined experimentally on the river (using dyes such as rhodamine, for example).

The second formulation (Bosko, 1966; Novtony, 1969) replaces in equation (12.32) the multiplicative correction by an additive one. Thus the BOD exertion is more activated by the river, but caused partially by the river itself.

$$K_1 = K_{1,l} + A \qquad (12.33)$$

where A is a kinetic coefficient describing absorption by biological slimes attached to a solid bottom or by dispersed biological flocs (Velz et al., 1962). Bosko proposed for coefficient A:

$$A = n\,(v/H) \qquad (12.34)$$

where
v is the river velocity ($L\ T^{-1}$),
H is the river depth (L), and
n is a dimensionless coefficient of bed activity.
The last mentioned coefficient takes into account hydraulic parameters and chemical properties of the bed of the river. It is taken as a step function of the slope of the bed of the river (Table 12.2).

An alternative relation between coefficient A and stream flow characteristics was proposed by Novtony (1969):

$$A = w * \frac{S_e^{1/6}}{H^{0.75}} \qquad (12.35)$$

where

w is a coefficient varying from 0.1 (streams with moving bottoms) to 3.0 (rocky streams)

S_e is the slope of the river (m/km).

Table 12.2
Stream slope dependance of coefficient of bed activity n. After Bowie et al. (1985) $K_1 = K_{1,l} + n\ (v/H)$

Slope of the river (km/m)	1.22	2.45	4.90	12.26
n	0.10	0.15	0.25	0.40

Novotny et al. (1975) compared equations (12.34) and (12.35) for different types of rivers, assuming that K_1 took into account a benthal oxygen demand (Table 12.3).

Two possibilities are available for determining a biodegradation kinetic constant: either calculate $K_{1,l}$ through laboratory experiment and evaluate K_1 with one of the previously mentioned equations (12.34) or (12.35), or calculate directly K_1 with a numerical optimisation technique or an statistical-experimental law.

Table 12.3
Comparison of experimental and calculated values of the A coefficient, according to Bosco and Novotny, for Holston River, Tenessee. After Novotny et al. (1974)

Character of bottom	Velocity (m/s)	Depth (m)	Slope (m/km)	A Bosko	A Novotny	Measured K_1-$K_{1,l}$
Gravel	0.19	0.76	0.62	0.30	1.08	1.0
Sand	0.13	0.76	0.62	0.10	0.43	0.25
Grav. rock	0.52	0.58	0.50	1.34	1.90	2.12
Grav. rock	0.32	0.73	0.80	0.65	1.79	1.60
Grav. rock	0.29	0.73	0.234	0.59	1.34	1.14
Grav. low poll.	0.42	0.70	0.234	0.41	0.55	0.55
Grav. low poll.	0.27	1.86	0.234	0.12	0.22	0.15

Calculation of $K_{1,l}$

a. Experimental study of BOD exertion.

The results of a BOD test in standard conditions (see paragraph 12.2) provides a DO versus time curve similar to that presented in Figure 12.2 (Wilderer et al., 1970). This curve frequently presents two plateaus, the first between 5 and 15 days, and the second one after 25 days. The first step with a quite marked slope corresponds to the carbonaceous degradation while the second corresponds to the nitrification stage of the biochemical degradation. Thus we see that the hypothesis of a first order reaction with a constant parameter $K_{1,l}$ is really an approximation. However, a first-order kinetic may be reasonably fitted during the first step of the experiment (before the curve reaches the first plateau): $K_{1,l}$ can be determined following DO uptake during 5 or 7 days, this last value being frequently preferred for practical considerations.

b. Graphical and mathematical procedures for determining $K_{1,l}$.

Before the general use of computers - before c.1965 - many authors developed graphical methods to determine the constant K_1. These all suppose that the biodegradation follows the first-order Streeter-Phelps kinetic law:

$$C_o - C_t = L_o \ (1 - \exp(K_{1,l} * t)) \tag{12.36}$$

where

$C_o - C_t$ is the oxygen consumption at time t ($M\ L^{-3}$)

L_o is the ultimate BOD at time zero ($M\ L^{-3}$)

Hewitt et al. (1975) compared seven of these methods with 20 sets of data (Table 12.4). It is beyond the scope of this book to detail all of these methods, which have now been overtaken by other methods. The different approaches used by each author depend on the limitations inherent in the local conditions of the study (Bansal, 1975). The variation of the results lies between 0 and 50% depending to the method. However, according to Marske et al. (1972) who compared four of the previous methods, the Reed-Theriault method seems to give the best results.

Table 12.14
Graphical and mathematical K_1 procedures compared by Hewitt et al. (1975)

Name of the author	Name of the method	Date	Formulation
Reed & Theriault	Theriault	1927-31	integral
Fair	Log-differences	1936	integral
Thomas	Slope	1937	differential
Rhame	Two point	1957	integral
Sheehy	Slide Rule	1960	integral
Moore	Moments	1960	integral
Navone	Simplified slope	1960	differential

c. Numerical methods for determining $K_{1,l}$.

The advent of the computer now allows the use of numerical optimization techniques for determining $K_{1,l}$. More efficient than the graphical techniques, these methods allow checking of the validity of the results through confidence contour determination, for example. The most used method is a modified least-square method for non-linear problems (Marske et al., 1972; Berthouex, 1971; Balland, 1982). In the following, we note:

$(C_o - C_t)_{obs}$ the observed value of DO consumed at time t,

$\underline{L}$ the calculated value of BOD exerted at time t, with the relation $\underline{L} = L_o$ (1 - exp($K_{1,l}$ * t),

S = sigma $((C_o - C_t)_{obs} - \underline{L}$), over 1 the number of observations.

Because of the non-linearity of the problem a search is conducted with an iterative algorithm for L_o and $K_{1,l}$ which minimize S. If S remains high even when these L_o and $K_{1,l}$ values represnet the best fit for the observed data - calculated data, the first-order law does not represent the observed phenomena well. The confidence contour around the best estimate are those for which S (L_o, $K_{1,l}$) = constant. Although this problem is non-linear, a confidence level: 100 (1 - q) % associated to a confidence contour can be defined:

$$S_q = S(L_o, K_{1,l})(1 + (p/l\text{-}p)\ F(p,l\text{-}p,1\text{-}q)) \qquad (12.37)$$

with p the number of parameters for the model,

1 the number of observations
q the confidence level

For $p = 2$; $F(2,1-2,1-q) = l-2/2\ (q^{-2/l-2} - 1)$ (12.38)

$S_q = S(L_o \text{ and } K_{1,l}) * q^{-2/l-2}$ (12.39)

Since this method gives the best parameters $K_{1,l}$ and L_o or $K_{1,l}$ and BOD_5 for each sample, it is possible to improve it and then to simplify the $K_{1,l}$ calculation, searching a regression relation between the two parameters (linear or log-log). Thus with this relation, valid at a certain location, it becomes possible to calculate $K_{1,l}$ with a simple measure of BOD_5 (Balland, 1982).

Marske et al. (1972) compared this technique with three of the previously mentioned graphical methods, and plotted their results as the percentage of the distance from the best fit of $K_{1,l}$ vs ultimate BOD (Figure 12.3). It will be noticed that the graphical Reed-Theriault method gives as good results as the numerical least-squares method.

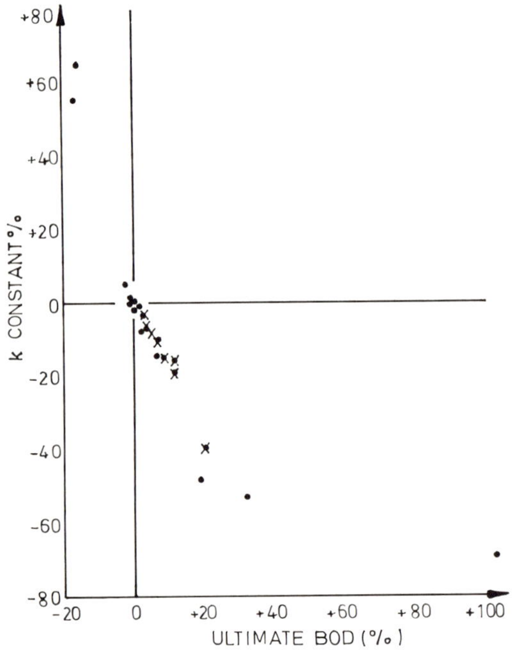

Fig. 12.3: Biodegradation kinetic constant determination: deviation from the best estimate. After Marske et al., 1972.
o: Method of Moments, x: Thomas Slope Method, •: Log-difference Method.

d. Order of magnitude of K_1

Table 12.5 presents some deoxygenation rate constants K_1 observed for different locations and temperatures, in the United States. From these values it is clear that K_1 may vary over more than one order of magnitude.

Table 12.5
Deoxygenation rate constants K_1. After Bansal (1975)

	Discharge (m^3/s)	Section (m^2)	Top width (m)	Temperature (°C)	K_1 (base e) (l/day)
Kansas River at Bonner Springs	430.416	399.483	234.696	25.000	0.020
	61.164	111.464	153.924	28.000	0.120
Kansas River at Topeke	86.083	134.709	142.646	22.000	0.080
	76.172	119.380	142.037	18.000	0.370
Big Blue River at Tuttle Creek	30.016	97.548	59.131	22.000	0.210
	27.212	92.903	59.131	8.000	0.600
Republican River below Milford	7.306	17.094	59.741	24.000	0.180
	17.243	36.418	75.795	24.000	0.070
Smoky Hill River at Enterprise	10.562	19.603	39.929	24.000	0.160
	4.446	8.175	29.566	14.000	0.240
Flint River. Michigan	3.794	19.510	30.480	20.000	0.560
	4.927	37.161	46.878	20.000	0.690
	4.927	18.581	23.439	2.000	0.630
Upper Hudson, Troy, New York	84.951	557.416	104.546	20.000	0.125
	127.426	627.096	97.841	20.000	0.165

Determination of K_1 or $K_{1,r}$ by direct methods

a. Statistical experimental method

The aim of this method is to determine K_1 without use of any relation between $K_{1,l}$ and K_1 . As the river flow may be characterized by Froude (Fr) and Reynolds (Re) numbers, Bansal (1975), tried to relate, on the basis of data gathered by the E.P.A., K_1 , Re and Fr. The following relation was found with a correlation coefficient of 0.93 and a standard deviation of 0.41.

$$\text{Log}\,(K_1 * H^2 / nu) = 3.606 + \text{Log}\,(Re/Fr) \qquad (12.40)$$

with

$$Re = v * H / nu \tag{12.41}$$

$$Fr = v / (g * H)^{1/2} \tag{12.42}$$

where
nu is the kinematic viscosity coeffcient (L^2 T^{-1})
H is the depth of the river (L)
g is the acceleration of gravity (L T^{-2})

The temperature influence is included into the variation of nu with temperature. It is also possible to use this relation at 20°C and then use Arrhenius law. The major advantage of this method is its simplicity. It may be possible to use a Streeter-Phelps model with good accuracy, without any knowledge of the location of the river or the pollution. This is very useful for fast diagnosis of the biodegradation in a river.

b. Determination of K_1 or $K_{1,r}$ with manual or automatic calibration

These methods are no longer based on laboratory BOD test. Now the computer is considered as a numerical laboratory and after seting the model description (simple with Streeter-Phelps first-order kinetics), parameter values are determined for achieving the best fit between the observed and calculated data along the river. The fact that either K_1 or $K_{1,r}$ are determined depends on the previous analysis of the system which is being modelled. If just the BOD exertion is taken into account, then K_1 will be found. If BOD exertion and N-NH_4^+ nitrification in water and on the sediments, and the respiration and photosynthesis are taken into account, then $K_{1,r}$ will be found.

b.1 Manual methods

The simplest way, for a simple Streeter-Phelps model, with just two parameters, is to simulate the behaviour of the river for different sets of parameters until a good adjustment between experimental and simulated values is observed (Lesouef, 1980). Although this method is especially rough, it can be very helpfull and it provides good results if the model remains simple.

b.2 Numerical methods

If the model is a more complicated one, the difficulty of manual parameter estimation increases. The only way which can be used is an automatic calibration procedure. Two strategies are available for a numerical calculation of the estimate (Rinaldi et al., 1980): the recursive and the accumulative estimation. Accumulative procedures are designed for calculating estimates from a fixed set of data, while recursive schemes are suited for estimates which are to be updated as new information becomes available. While the first method is used when the parameters are not supposed to vary in time, the second one is very usefull for real time state estimation. Since BOD measurements need a minimum of 5 days, the utilization of such a method in biodegradation remains rare.

b.2.1 Accumulative procedures

These procedures are based on an error estimator function to minimize , for example, a least squares estimator. Various numerical methods are available to solve a minimization problem, such as the *quasi-linearization* technique (Stehfest, 1973), which has the advantage of a fast convergence rate. These accumulative techniques are now much used (Shastry et al., 1973; Rinaldi, 1979; Radhakrishnan et al., 1974), and allow the building of sophisticated models with many parameters. After a successful calibration the set of parameters found has to be validated on another set of data.

b.2.2 Recursive procedures

The recursive procedures are based on stochastic mechanistic models. The problem is to determine the state estimation for the time interval, t_S, from the input - output measurement for the previous intervals $(0 < t < t_S)$. Parameter estimation can be done with the well known numerical technique of Kalman filtering, or another method which appears to provide an improvement in statistical efficiency - the multivariate extension of the Instrumental Variable Approximate Maximum Likehood (IVAML) method of time series analysis (Whitehead et al., 1979). This procedure has been successfully applied to the real-time monitoring scheme of the River Ouse at Bedford on a modified Streeter-Phelps model (DO, $N\text{-}NH_4^+$) (Whitehead, 1983).

12.4 FUTURE RESEARCH NEEDS

Difficulties encountered in modelling microbial decomposition of organic matter are possibilities of descriptive or interpretative mathematical models and relation between the model structure and its applications.

Experimental and field research needs for microbial decomposition models: differences between intrinsic parameters and parameters fitted from field data, difficulties arising from non-linear kinetics.

ACKNOWLEDGMENTS

The help of Mrs Gamrasni, from l'Association Française de l'Etude des Eaux, Paris, is greatly acknowledged.

NOTATIONS

A	is an additional coefficient describing the augmentation of $K_{1,l}$ in natural sites (T^{-1})
ATP	adenosine triphosphate
B	biomass concentration ($M\ L^{-3}$)
BOD	biochemical oxygen demand ($M\ L^{-3}$)
C	dissolved oxygen concentration ($M\ L^{-3}$)
C_s	saturated dissolved oxygen concentration ($M\ L^{-3}$)
COD	chemical oxygen demand ($M\ L^{-3}$)
D	dissolved oxygen deficit, i.e. $C_s - C$ ($M\ L^{-3}$)
D_b	rate of removal of DO caused by benthic demand and plants ($M\ L^{-3}$)
D_c	critical dissolved oxygen deficit, i.e. maximum of D ($M\ L^{-3}$)
DNA	deoxyribonucleic acid
DO	dissolved oxygen
dR_o/dt	rate of reoxygenation ($M\ L^{-3}\ T^{-1}$)
E	turbulent transport (longitudinal mixing) coefficient ($L^2\ T^{-1}$)
F	source or drain of organic pollutant ($M\ L^{-3}$)
Fr	Froude number (dimensionless)
f	Fair coefficient, also called self-purification factor, i.e. K_2 / K_1 (dimensionless)
g	acceleration of gravity ($L\ T^{-2}$)
H	depth (L)
K_1	biodegradation kinetic constant in first-order kinetic models, i.e. water deoxygenation rate due to BOD exertion (T^{-1})
$K_{1,l}$	laboratory deoxygenation rate due to BOD exertion (T^{-1})
$K_{1,r}$	water exertion rate of BOD, includes all sinks of BOD (T^{-1})
$K_{1,s}$	kinetic constant of BOD evolution on sediments (T^{-1})
$K_{1,NH4}$	ammonia biodegradation (oxidation) kinetic constant in first-order nitrification kinetic models (T^{-1})
$K_{NR,NH4}$	ammonium uptake kinetic constant in first-order kinetic models (T^{-1})
$K_{1,NO2}$	nitrite biodegradation (oxidation) kinetic constant in first-order

	kinetic models (T^{-1})
K_2	oxygen transfer kinetic constant, also called reaeration kinetic constant (T^{-1})
$(K_2)_{20}$	value of K_2 at 20°C (T^{-1})
L	concentration of organic matter, generally expressed in ultimate BOD and frequently related to carbonaceous compounds ($M\ L^{-3}$)
L_o	initial value of ultimate BOD ($M\ L^{-3}$)
L_a	rate of BOD release along a river length unit ($M\ L^{-4}$)
l	number of observations (dimensionless)
NH_4-N	ammonium nitrogen concentration ($M\ L^{-3}$)
NO_2-N	nitrite nitrogen concentration ($M\ L^{-3}$)
n	coefficient of bed activity (Bosko, 1966) (dimensionless)
OM	organic matter
P	algal photosynthetic oxygen production rate (source) ($M\ L^{-3}\ T^{-1}$)
P_m	maximum algal photosynthetic oxygen production rate (source) ($M\ L^{-3}\ T^{-1}$)
P	number of parameters of the model
Q	flow rate (downstream) ($L^3\ T^{-1}$)
Q_T	thermal air-water energy transfer rate
q	confidence level
R	algal respiration rate (oxygen sink) ($M\ L^{-3}\ T^{-1}$)
Re	Reynolds number (dimensionless)
S	benthic respiration rate (oxygen sink) ($M\ L^{-3}\ T^{-1}$)
S_e	slope of the river (dimensionless or m/km)
s	period of sunlight, i.e. fraction of the day at which photosynthesis is active (T)
T	temperature of the water (°C)
t	time (T)
t_s	time, expressed as a fraction of the day, at which oxygen photosynthesis becomes active (T)
TOC	total organic carbon ($M\ L^{-3}$)
TOD	total oxygen demand ($M\ L^{-3}$)
V	volume (L^3)
v	velocity (downstream) ($L\ T^{-1}$)
w	coefficient of bed activity (Novotny, 1969)
x	distance (downstream) (L)
Y	biomass yield from organic matter (dimensionless)
mu	bacterial growth rate (T^{-1})
mu_M	maximal growth rate (T^{-1})
nu	kinematic viscosity coefficient ($L^2\ T^{-1}$)
ro	water specific mass ($M\ L^{-3}$)

pi 3.1416

theta temperature coefficient of K_1 (dimensionless)

phi temperature coefficient of K_2 (dimensionless)

REFERENCES

Alfi S., Argaman Y. and Shelef G. (1972). Mathematical model for the prediction of dissolved oxygen levels in the Alexander Stream (Israel), Proc. 6th Int. Water Pollution Research (17) C 8/17/1-8.

Apoteker A. (1981). Réactions de biodégradation en rivière. Evolution de l'oxygène dessous, de la charge polluante et de la biomasse, Thèse de 3e Cycle, Université Paris-Val de Marne, Créteil, France, 212 p.

Apoteker A. and Thévenot D.R.)1983). Experimental simulation of biodegradation in rivers. Oxygen, organic matter and biomass concentration changes, Water Res., 17 (10) 1267-1274.

Arbabi M., Elzinga J. and Revelle C. (1974). The oxygen sag equation: new properties and a linear equation for the critical deficit, Water Res., 10 (5) 921-929.

Arbabi M. and Elzinga J. (1975). A general linear approach to stream water quality modeling, Water Res., 11 (2) 191-195.

Blok J. (1975). Methods for measurement of biodegradability of chemical compounds. Int. Biodetn. Bull., 11 (3) 78-84.

Balland P. (1982). Méthode d'estimation numérique du coefficient K_1 de dégradation biochimique de le DBO, T.S.M. l'eau, 3 157-161.

Bansal M.K. (1975). Deoxygenation in natural streams. Water Resources Bulletin, 11 (3) 491-504.

Berthouex P.M. and Hunter W.G. (1971). Problems associated with BOD planning experiments, Jour. San. Eng. Div. proc ASCE, 97 (SA3) 333-343.

Boethling R.S. and Alexander M. (1979). Microbial degradation of organic compounds at trace levels. Envir. Sci. Technol., 13 (8) 989-991.

Bosko K. (1966). Discussion to paper I-1, Advances in water pollution research, Pergamon Press, Oxford, 3

Bowie G.B., Mill W.B., Porcella D.B., Campbell C.L., Pagenkopf J.R., Rupp G.L., Johnson K.M., Chan P.W.H., Gherini S.A. and Chamberlin C.E. (1985). Rates, constants and kinetics formulations in surface water quality modeling. Second edition, Tetra-Tech, Lafayette, California, EPA/600/3-85/040.

Churchill M.A., Elmore H. and Buckingham R. (1962). The prediction of stream reaeration rates, Jour. San. Eng. Div. proc. ASCE, 88 (SA4) 1-46.

Clark D.W. (1974). BOD: a re-evaluation, Water and Sewage Works, 5, 68-69.

Covar A.P. (1976). Selecting the proper reaeration coefficient for use in water quality models, Presented at the U.S. EPA Conference on Environmental Simulation and Modeling, April, 19-22.

Dart R.K. and Stretton R.J. (1980). In: Microbiological aspects of pollution control, Chap. 11. Biodegradation, 2nd ed., Elsevier, Amsterdam, pp. 216-256.

Dobbins W.E. (1964). BOD and oxygen relationships in streams, J. Sanit. Engng. Div. Am. Soc. civ. Engrs 90, SA3, 53-78.

Eckenfelder W.W.Jr. (1970). Water quality engineering for practicing engineers, Barnes and Noble, New York.

Edeline F. and Buet J.L. (1978). $$$$$$$$$, La Tribune du Cebedeau, 415-416, 273-280.

Edeline F. and Lambert G. (1974). A simple simulation method for river self-purification studies, Water Res., 8, 297-306.

Edeline F. and Lambert G. (1979). Self purification of rivers: a two phase model for bacterial degradation, Water Res., 13, 473-480.

Fair G.M. and Gordon M. (1936). The Log-difference method of estimating the constants of the first BOD stage curve, Sewage Works Journal, 8 (3) 430-434.

Flegal T.M. and Schroeder E.D. (1976). Temperature effects on BOD stoichiometry and oxygen uptake rate, J. Wat. Pollut. Control Fed., 48, 2700-2707.

Gates W.E. (1971). Laboratory and mathematical simulation of oxygen balances effected in streams, School Civil Engng. Envir. Res. Center, Georgia Inst. of Technology, Atlanta, Georgia, ERC-0171, 52 p.

Gaudy A.F. (1972). In: Water pollution microbiology, Chap. 13, Biochemical oxygen demand, R. Mitchell ed., Wiley Intersciences, New York, pp. 305-332.

Green M., Shelef G. and Moraine R. (1981). Chemical and biochemical oxygen demands as indicators of biodegradable substrate concentration, Water Pollut. Control Fed., 80 (5) 655-658.

Gromiec M.J., Loucks D.P. and Orlob G.T. (1983). Stream quality modeling, In: Mathematical modeling of water quality: streams, lakes and reservoirs, G.T. Orlob ed., International Series on Applied Systems Analysis, 12, Wiley Interscience, New York, pp. 176-226.

Gromiec, M.J., 1983. Biochemical oxygen demand - dissolved oxygen: river models, In: Application of Ecological Modelling in Environmental Management, Part A, S.E. Jørgensen, ed. Elsevier Scientific Publishing Company, Amsterdam, pp. 131-225.

Hewitt J.P. and Hunter J.V. (1975). A comparison of methods used to calculate first-order BOD equations constants, Water Research, 9, 683-687.

Holm-Hansen O. (1970). ATP levels in algal cells as influenced by environmental conditions, Pl. Cell Physiol., 11, 689-700.

Hung Y.-T. and Eckenfeld W.W. (1976). Comparison of Warburg and Marais methods for determining oxygen uptake rate constants, Water Pollution Control, 114 (2) 6-25.

Hung Y.-T., Eckenfelder W.W. and Reynolds T.D. (1978). Effect of treatment and flow on the deoxygenation rate constant, Water Air Soil Pollut., 10 (2) 175-185.

Hydroscience (1964). Assimilation capacity of the upper James River, State Water Control Board Commonwealth of Virginia.

Hydroscience (1965). Pollution analysis of the Upper Hudson River estuary, Malcolm Pirnie Engineers.

Hydroscience (1966). Water quality studies New River Celanese Corp. of America, Celco Plant, Narrows, Virginia.

Hydroscience (1967). Water quality analysis of the Jackson River, West Virginia Pulp and Paper Company, Covington.

Illic P. (1978). Biochemischer Sauerstoffbedarf. Die Rolle des CO_2 bei der Respirometrischen Bestimmung, Wasser Luft Betr., 22 (1-2), 20-21.

Jacquet J. (1983). Simulation of the thermal regime of rivers, In: Mathematical modeling of water quality: streams, lakes and reservoirs,

G.T. Orlob ed., International Series on Applied Systems Analysis, 12, Wiley Interscience, New York, pp. 150-175.
Jørgensen S.E. (1983a). Modeling the ecological processes, In: Mathematical modeling of water quality: streams, lakes and reservoirs, G.T. Orlob ed., International Series on Applied Systems Analysis, 12, Wiley Interscience, New York, pp. 116-149.
Jørgensen S.E. (1983b). Ecological modeling of lakes, In: Mathematical modeling of water quality: streams, lakes and reservoirs, G.T. Orlob ed., International Series on Applied Systems Analysis, 12, Wiley Interscience, New York, pp. 337-394.

Knowles G. and Wakeford A.C. (1978). A mathematical deterministic river quality model. Part I: Formulation and description, Water Res., 12, 1149-1153.

Lefort D. (1971). Modèles mathématiques de pollution de riviére, La Tribune du Cebedeau, 335, 392-398.
Lesouef A. (1979). Modèles d'auto-épuration de l'azote, in Les problèmes de l'azote dans les eaux, Martin ed., Techniques et Documentation, Paris.
Lesouef A. (1981). Proposition d'un modèle 1980 de qualité de la Seine de Montereau à Poses, Agence Financère de Bassin Seine Normandie, Sous-direction Pollution, 40 p.
Lin S.H., Fan L.T. and Ericson L.E. (1973a). Transient temperature effects on BOD and DO concentrations along a stream, Water Res., 7, 1861-1873.
Lin S.H., Fan L.T. and Hwang C.L. (1973b). Digital simulation of the effect of thermal discharge on stream water quality, Water Resour. Bull., 9, 689-702.
Lin S.H., Fan L.T. and Hwang C.L. (1973c). Water quality control by artificial aeration of stream receiving thermal and organic discharge, Water Resour. Bull., 9, 874-883.

Marske D.M. and Polkowski L.B. (1972). Evaluation of methods for estimating biochemical demand parameters, J.W.P.C.F. 44 (10) 1987-2000.
Moore E.W., Thomas H.A. and Snow W.B. (1950). Simplified method for the analysis of BOD data, Sewage and Industrial Wastes 22 (12), 1343-1353.
Mytelka A.I. and Brenner W.E. (1969). An automatic recording respirometer, Water Quality Instrumentation, 1, 142-150.

Navone R. (1966). A new method for calculating K and L for sewage, Sewage Works Journal 107 285.
Nejedly A. (1973). The effect of longitudinal dispersion on the deoxygenation rate in polluted streams, J.W.P.C.F., 45 (7) pp. 1601-1605.
Novotny V. (1969). Boundary layer effect on the course of the self-purification of small streams, Advances in water pollution research, Pergamon Press, Oxford.
Novotny V. and Krenkel P.A. (1975). A waste assimilative capacity model for a shallow, turbulent stream, 9, 233-241.

O'Connor D.J. and Dobbins W.E. (1958). Mechanisms of reaeration in natural streams, A.S.C.E. Transactions Biotechnology, Bioengineering, 123 (2934) 641.
O'Connor D.J. (1961). Oxygen balance of an estuary, Trans. Am. Soc. Civ. Engrs., 126, part III, 556.
O'Connor D.J. (1962). The effect of stream flow on waste assimilation capacity, Proceedings of 17th Industrial Waste Conference, Purdue University Lafayette, 1N, 1-3 May, Engng. Extension Series N° 112, pp. 608-629.
O'Connor D.J. (1967). The temporal and spatial distribution of dissolved

oxygen in streams, Water Resour. Bull., 3, 65-79.
O'Connor D.J. and Di Toro D.M. (1970). Photosynthesis and oxygen balance in streams, J. Sanit. Engng. Div. Soc. Civ. Engrs., 96 (SA2) 547-571.
Orlob G.T. (1983). One-dimensional models for simulation of water quality in lakes and reservoirs, In: Mathematical modeling of water quality: streams, lakes and reservoirs, G.T. Orlob ed., International Series on Applied Systems Analysis, 12, Wiley Interscience, New York, pp. 227-273.
Owens M., Edwards R.W. and Gibbs J.W. (1964). Some reaeration studies in streams, Int. Jour. Air and Water Poll., 8 469.

Radhakrishnan K.P., Lizcano J.J., Erickson L.E. and Fan L.T. (1974). Evaluation of methods for estimating stream water quality parameters, in a transient model from stochastic data, Water Resources Bulletin, 10 (5) 899-913.
Rauwel F. and Thévenot D.R. (1976). Comparison of dissolved oxygen evolution after thermal and pollution discharge. Final contract report for Agence Financière de Bassin Seine Normandie, 50 p.
Reed L.J. and Theriault E.J. (1931). The statistical treatment of reaction-velocity data I,II, Jour. Phys. Chem. 35 673, 950.
Remiche C. and Van Der Wielen (1975). Quelques considérations concernant l'utilisation du "Sapromat" pour la mesure de la toxicité. Trib. Cebedeau, 381-382 (8-9) 314-319.
Revelle C.S., Lynn W.R. and Riviera M.A. (1965). Biooxidation kinetics and a second order equation describing the BOD reaction, J. Wat. Pollut. Control Fed., 37, 1679.
Rhame G.A. (1956). A two point method for estimating the first stage BOD, Sewage and Industrial wastes 28 (9) 1987-1093.
Rinaldi S. and Soncini-Sessa R. (1977). Sensitivity analysis of Streeter-Phelps models, Intern. Inst. Applied Syst. Anal., Laxenburg, Austria, RR 77-1, 18 p.
Rinaldi S., Soncini-Sessa R., Stehfest H. and Tamura H. (1979). Modeling and control river quality, Mac-Graw Hill, New York, 380 p.

Shastry J.S., Fan L.T. and Erickson L.E. (1973). Non-linear estimation in water-quality modeling. Jour. Env. Eng. Div. proc. ASCE, 99 (ee3) 315-331.
Sheeny J.P. (1960). Rapid method for solving first-order equations, J.W.P.C.F. 32 (6) 646-652.
Shelton S.P., Burdick J.C. III and Drewry W.A. (1978). Water quality modeling in a low flow stream, J. Wat. Pollut. Control Fed., 50 2289-2306.
Stack V.T. (1972). In: Water and water pollution handbook, Vol. 3, Chap. 15, Biochemical oxygen demand measurement, L.L. Ciaccio ed., Dekker, New York, pp. 801-829.
Stack V.T. (1973). Stabilization oxygen demand, in Biological Methods for the assessment of Water Quality, ASTM STP 528, Amer. Soc. for Testing and Materials, 209-220.
Stehfest H. (1973). Mathematical modelling of self-purification of rivers (in German), report KFK 1654 UF, Kernforschungzentrum, Karlsruhe.
Streeter H.W. and Phelps E.B. (1925). A study of the pollution and natural purification of the Ohio River, Public Health Bulletin 146, Washington D.C.
Streeter H.W., Wright C.T. and Kehr R.W. (1936). Measures of natural oxidants in polluted streams II, Sewage Works Journal, 8 282.

Theriault E.T. The oxygen demand of polluted waters, Public Health Service Bulletin, U.S. Public Health Service, Washington D.C., 173.
Thomas H.S. Jr. (1937). The slope method of evaluating the constants of the first stage BOD curve, Sewage Works Journal, 9 (3) 425-430.

Truesdale G.A. and Van Dyke K.G. (1958). The effect of temperature on the aeration of flowing waters, Water and Waste Treatment Journal 7 9.
Tobin R.S., Ryan J.F. and Afghan B.K. (1978). An improved method for the determination of adenosine triphosphate in environmental samples, Water Res., 12, 783-792.

Velz C.J. and Gannon J.J. (1962). Biological extraction and accumulation in stream self-purification, Advances in Water pollution research, Pergamon Press, Oxford.

Whitehead P.G. (1983). Modeling and forecasting water quality in non-tidal rivers: the Bedford Ouse study, In: Uncertainty and forecasting of water quality, Beck M.B. and Van Straten G. eds., Springer-Verlag, Berlin, 321-337.
Whitehead P.G. and Young P.C. (1979). Water quality in rivers systems: Monte-Carlo analysis, Water Resources Research, 15 (2) 451-459.
Wilderer P., Hartmann L. and Janeckova J. (1970). Objections against the use of long-term BOD for characterizing raw waste water (in German). Wasser- und Abwasser-Forschung, 1 7-12.
Woodward R.L. (1953). Deoxygenation of sewage. A discussion, Water Res., 12, 783-792.
Wuhrmann K. (1972). In: Water pollution Microbiology, Chap. 6. Stream purification, Mitchell R. ed., Wiley-Interscience, New York, pp. 119-151.

Young J.C. and Clark J.W. (1965). Second order equation for BOD, J. Sanit. Engng. Div. Am. Soc. civ. Engrs. 91, SA1, 43.

Zanoni A.E. (1967). Waste water deoxygenation at different temperatures, Water Res., 1, 543.

CHAPTER 13

NITRIFICATION

by
Shinichiro Ohgaki and Chalermraj Wantawin

13.1 INTRODUCTION

Many mathematical or physical submodels have been investigated for the prediction of biogeochemical phenomena in aquatic ecosystems. These models have been useful for the improvement and control of water quality in natural water bodies such as rivers, lakes, bays and estuaries. It is important, however, that the developer of a model has the ability to determine the parameters in the submodel and to confirm the validity of substantial assumptions. The most substantial assumption in modelling of the nitrification process is in the growth model of the nitrifying microorganisms. Thus reliability of the nitrification submodel depends on the growth model.

Because of the slow growth rate of *nitrifying bacteria,* it is expected that nitrification would be significant only in water bodies with a long detention time. However, nitrification can also occur from the very beginning below the outfall of waste water, by virtue of micro-organisms attached to the stream bed. The system in natural water bodies is quite sophisticated due to the large number of physical, chemical and biological processes linked together. A model applicable to the characteristics of one type of water body may be meaningless for others. Therefore, the basic knowledge on nitrification and clear understanding of the correlation between characteristics of water flow and nitrifying bacteria growth will help in developing and applying an effective model for any natural water body. The nitrifying bacteria growth models applied to treatment processes also have to be reviewed as these serve as references for part of the biological processes in modelling nitrification in aquatic ecosystems.

13.2 NITRIFYING MICRO-ORGANISMS AND FACTORS AFFECTING NITRIFICATION

The metabolism of autotrophic nitrifying micro-organisms is quite specific. They derive their energy from inorganic nitrogen-oxidation and obtain carbon for cell-synthesis from inorganic carbon substances. The apparent inability of organic utilization has been investigated and there is good evidence that *Nitrobacter* can utilize organic carbon as energy source and, in general, is to be classified as a facultative autotroph.

In contrast with many heterotrophs, the growth of nitrifiers is very slow and the yield of cells per unit of energy source oxidized is low. Like other micro-organisms, nitrifiers can grow at their maximum rate with optimum environmental factors and in an environment without any toxic substances.

The nitrification process consists of two successive reactions. These are the oxidation of ammonia to nitrite, and of nitrate to nitrate. The initial conversion of ammonia to nitrite by an ammonia oxidizer has traditionally been regarded as the rate-limiting step for nitrification. However, *Nitrosomonas;* ammonia oxidizers, are believed to be capable of out-competing *Nitrobacter* under stressful conditions e.g. limiting oxygen, low temperature condition, etc. Thus even operating with the same loading, nitrification can cease under environmental stress, due to the accumulation of nitrite built up to a level which is toxic to both nitrifiers.

Organic matter generally is not inhibitory as was once thought. However, the presence of non-toxic organic substances may indirectly inhibit nitrification through the *activity of heterotrophs.* During oxidization of organic compounds by heterotrophs, inhibiting intermediates may be produced or the available ammonia-nitrogen and dissolved oxygen successfully competed for by their cell synthesis and respiration, respectively.

Figure 13.1 shows the schematic representation of parameters affecting the activity of both nitrifiers. The growth kinetics and the quantitative values of inhibiting factors will be reviewed and summarized in the following sections.

13.3 GROWTH KINETICS OF NITRIFYING MICRO-ORGANISMS

Monod Model

For the modelling of heterotrophic growth, the empirical microbial growth equation expressed after Monod, analogous to Michaelis-Menten (single enzyme, single substrate) kinetics, has been popularly employed to describe the growth of nitrifying bacterias as follows:

$$\frac{1}{X_n} * \frac{dX_n}{dt} = \mu_n = \mu_{max,n} \frac{S_n}{K_{s,n} + S_n}$$

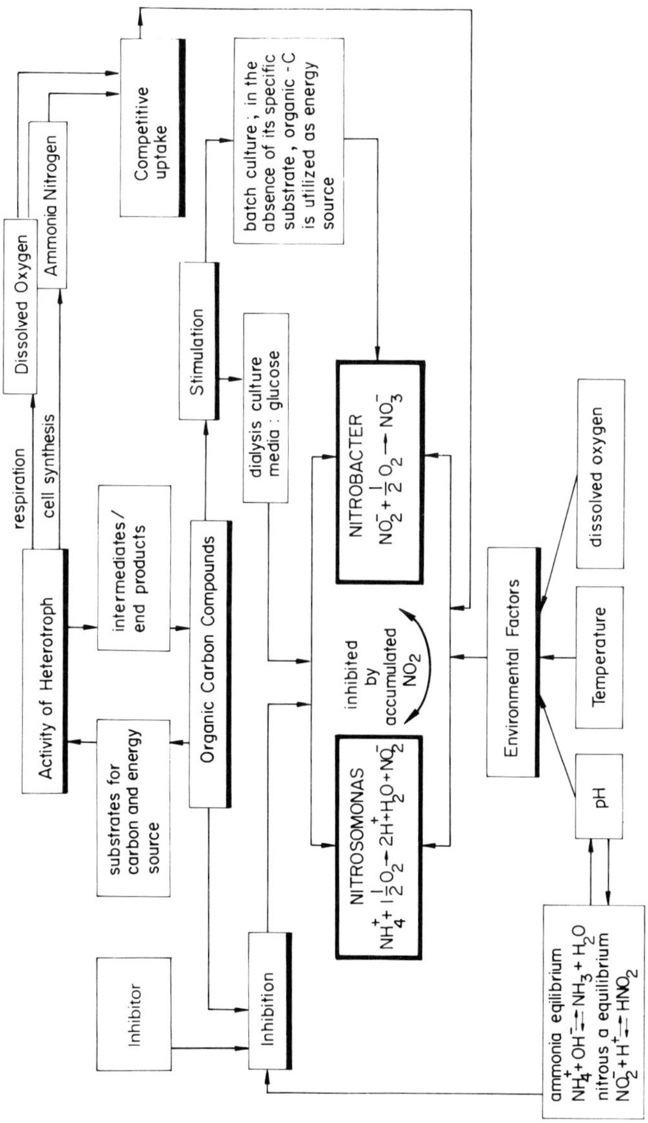

Fig. 13.1: Schematic Representation of Parameters Affecting Nitrification.

The ammonia removal rate derived from the Monod equation is

$$\frac{d\,S_n}{dt} = -\frac{1}{Y_n} * \frac{d\,X_n}{dt} = -\frac{K_n\,S_n\,X_n}{K_{s,n} + S_n}$$

where

K_n = $\mu_{max,n}$ / Y_n

X_n = nitrifying mass cell concentration

μ_n = specific growth rate of nitrifiers

$\mu_{max,n}$ = maximum specific growth rate of nitrifiers

S_n = ammonia nitrogen concentration

$K_{s,n}$ = saturation concentration

Y_n = yield coefficient

Painter (1970) pointed out that Downing et al. (1964) were the first group of workers to elucidate nitrification from the kinetics viewpoint in the waste water treatment system. The Monod equation was proposed for nitrifying growth rate in their study. Gujer and Jenkins (1975), Ericsson (1975), Shieh and Lamotte (1979), Charley et al. (1980) etc. used Monod expression or Michalis Menten kinetics to explain the mechanism of ammonia or nitrite utilization in activated sludge. Saunder and Bazin (1973), Atkinson and Fowler (1974), Williamson and McCarty (1976) etc. applied the Monod model as the biochemical reaction rate in nitrifying biofilm processes. Knowles et al. (1965) demonstrated the method to complete constants in Michaelis Menten kinetics for nitrifying organism growth in water from Thames Estuary. It has also been employed by Stratton and McCarty (1967, 1969) and Garland (1978) as an expression for the rate of nitrifying bacterial growth in natural streams and its application has been demonstrated in predicting ammonia concentration in the Clinton River and Trent River, respectively.

The Monod expression provided for a continuous transition between *first* and *zero order kinetics* based on substrate concentrations. Typical ranges of kinetics constants for nitrification follow the Monod model which mostly were reviewed by Painter (1977) and Sharma and Ahlert (1977), are shown in Table 13.1.

The saturation constant, $K_{s,n}$, in Monod kinetics for nitrification is much smaller than the concentration of ammonia or nitrite found in the effluent (typical ammonia concentration of 20 mg/l as N), therefore zero-order reaction rate usually was demonstrated as the rate of nitrification in suspended growth processes (Jenkins, 1969; Wild et al.,

1971; Wong-Chong and Loehr, 1975; Srinath et al., 1976; Hall and Murphy, 1980). Engel and Alexander (1958), Nicholas and Jones (1960) also applied zero-order nitrification for pure culture. Lamotta (1976), Watanabe et al. (1980) also applied a zero-order reaction as the intrinsic growth rate of nitrifying organisms in their models which involved the effects of diffusion coefficients of substrates on the reaction order of overall nitrification rate for biofilm processes.

TABLE 13.1
Typical values of kinetics constant for nitrifying organism

Organism	Max. Spec. Growth Rate $\mu m\ d^{-1}$	Cell Yield $\frac{\text{wt. of Cell}}{\text{wt. of N}}$	Saturation Constant: Substrate $K_{s,n}$ mg/l	Saturation Constant: Oxygen $K_{s,o}$ mg/l	Reference
Nitrosomonas	0.46-1.86 (30°C)	0.06	10 (30°C) 3.5 (25°C) 1.2 (20°C)	0.5 (30°C) 0.3 (20°C)	Painter H.A. (1977)
	0.46-2.20	0.03-0.13	0.06-5.6	0.3-1.3	Sharma & Ahlert (1977)
Nitrobacter	1.39 (32°C)	0.02	8 (32°C) 5 (25°C)	1.0 (30°C) 0.5 (32°C) 0.25 (18°C)	Painter H.A. (1977)
	0.28-1.44	0.02-0.08	0.06-8.4	0.25-1.3	Sharma & Ahlert (1977)

Monod Model for Nonsteady State

The Monod model for microbial growth is useful for steady state cultures. However, several researchers have shown its limitation applied to predicting the dynamic behavior of mixed heterogeneous populations in chemostats (Storer and Gaudy, 1969; Chu, Erickson and Fan, 1973). A lag between microbial growth and substrate utilization during the transient stage has been observed. Bungay and co-workers (Bungay, 1967; Young et al., 1970) argue a systems engineering approach was useful in predicting the dynamic behaviour of a chemostat and used a time dependent equation for the specific growth rate.

Murphy and Sutton (1977) demonstrated that parameter values derived from steady state data (Sutton, Murphy, Jank and Monaghan, 1975) were unable to describe the dynamic behavior of nitrifying processes. They utilized time series analysis as a technique to develop empirical transfer function models for nitrification under non-steady state conditions.

However, the Monod model has been applied to the growth rate during the transient state in which impulse and step loads (Poudska and Andrew, 1975) and diurnal variation loads (Gujer, 1977) have been the inputs in nitrifying suspended growth systems. The dynamic models developed by both investigators explained well the dynamic responses of nitrification in their experiments.

Double Substrate Limiting Model and Effect of dissolved Oxygen.

Most mathematical models *for biological growth* take into account only one substrate, such as the Monod model, since experimental studies are usually performed with all other nutrients in excess. However, Megee et al. (1972) have proposed the growth model for their studies of intermicrobial symbiosis under double substrate limitation - phenomenon by replacing the single substrate Michaelis-Menten expression with a double substrate expression. Ryder and Sinclair (1972) and Sykes (1973) have investigated the possibilities of double substrate-limiting model but not simultaneous limiting kinetics. Both investigators showed that at one operating point substrate A was limiting, while at another operating point substrate B might be limiting. They proposed the determination of the maximum growth rate by evaluating the microbial yields associated with the two growth rates and then selecting the expression which was limiting (Stenstrom and Poduska, 1980). Sinclair and Ryder (1975) applied their resulting model to predict the behavior of a pure culture of *Candida utilis* under both oxygen- and carbon-limiting conditions.

Bader et al. (1975) and Bader (1978) have discussed that the approach of Ryder and Sinclair (1972) and of Sykes (1973) on substrate limitation. Showing that it is not a general one. Bader (1978) proposed simultaneous double substrate limiting kinetics analogous to a two substrate enzyme model where an enzyme requires two different substrates for producing a single product. Following Bader's (1978) analysis, Stenstrom and Poduska (1980) used the double substrate limiting kinetics modified from the Monod single substrate model to describe the effects of dissolved oxygen simultaneously with those of ammonia nitrogen on the growth rate as shown in the following equation, although no experimental work was done.

$$\mu = \mu_{max} \left(\frac{S_n}{K_{s,n} + S_n} \right) * \left(\frac{DO}{K_{s,o} + DO} \right) - K_d \qquad (13.3)$$

where

μ = specific growth rate (s^{-1})
μ_{max} = maximum specific growth rate (s^{-1})
S_n = ammonia concentration ($M\ l^{-3}$)
DO = dissolved oxygen concentration ($M\ l^{-3}$)
$K_{s,n}$ = half saturation constant for ammonia nitrogen ($M\ l^{-3}$)
$K_{s,o}$ = half saturation constant for dissolved oxygen ($M\ l^{-3}$)
K_d = decay or maintenance coefficient (s^{-1})

The *double substrate limiting kinetics* have been of interest to some researchers investigating attached micro-organism processes, since substrates diffusing through biofilms will result in the limitation of either electron donors or electron acceptors in the biochemical reaction. Muller, Paquin and Famularo (1980) proposed double Monod kinetics for intrinsic growth rates in their mass transfer equations to evaluate nitrification in rotating biological contactors. Tanaka and Dunn (1982) used these kinetics in their diffusion reaction model applied on a fluidized bed nitrification reactor. Kissel, McCarty and Street (1984) and Wanner and Gujer (1984) developed the biofilm model studying the competition between *heterotrophs* and *nitrifying organisms* by utilizing the *double Monod function* for growth rates of both microorganisms.

Typical values of the half saturation constant for dissolved oxygen ($K_{s,o}$) have been shown in Table 13.1. There is no explanation for the methods used to analyze these $K_{s,o}$ values in the literature reviews of Painter (1977) or Sharma and Ahlert (1977).

Results from a number of studies on the effect of dissolved oxygen are summarized in Table 13.2.

It seems that the activity of *Nitrobacter* is suppressed under low dissolved oxygen concentrations more than that of *Nitrosomonas.* Evidence to support this is provided by the dissolved oxygen limits shown in Table 13.2 (Knowles et al., 1965) and half-saturation oxygen coefficients $K_{s,o}$ (0.25 mg/l for *Nitrosomonas,* 1.84 mg/l for *Nitrobacter* - Peeters et al., 1969), where those of *Nitrosomonas,* are lower than those of *Nitrobacter.* Other values of limiting dissolved oxygen in Table 13.2 and half-saturation oxygen coefficients in Table 13.1 cannot be used for comparison due to differences in the techniques used, as well as in environmental conditions. The accumulation of nitrite observed recently by researchers, cited in Alleman (1984), also imply the lesser activity of *Nitrobacter* than *Nitrosomonas* under limiting dissolved oxygen concentrations. Painter (1977) noted that the presence of organic matter can directly inhibit nitrifiers by virtue of

TABLE 13.2: Effects of dissolved oxygen

Dissolved oxygen mg/l	Observation	Conditions	Reference
0.2-0.7	Retarding but not inhibiting nitrification	Activated sludge: wea nitrogen in the absence of sewage	Ingols (1941), cited in Heukelekian (1942)
Lack of O_2	Reactive to its original value in a few minutes after re-aeration began	Anaerobic condition 4 hr. in activated sludge, re-aeration again	Downing and Scragg (1958) cited in Downing et al., (1964)
< 0.2	Nitrification ceased	Activated sludge	" "
> 1.0	Zero-order	Pure culture of *Nitrosomonas*	Schoberl and Engel (1964), cited in Stenstrom and Poduska (1980)
0.3-1.0	Zero-order	Activated sludge	Downing et al. (1964)
2.0	Zero-order	*Nitrosomonas* growth - water from Thames Estuary	Knowles et al. (1965)
4.0	Zero-order	*Nitrobactor* growth - water from Thames Estuary	" "
> 1-1.5	Zero-order	Activated sludge	Wuhrman (1968), cited in Sharma and Ahlert (1977)
0.5	Critical zero-order	Activated sludge	Jenkin (1969)
20% saturation	Zero-order	Batch Tests	Kiff (1972)
> 2.0	Preventing denitrification in final settling tanks	Activated sludge	Beckman et al. (1972)
38	No inhibition after acclimatization	Activated sludge at 30°C	Charley et al. (1980)
up to 60	No inhibition; no increase in ammonia oxidation	Submerged filtered receiving preoxygenated feed	Haug and McCarty (1972)

zero-order: Minimum DO concentration for zero order reaction
critical: Minimum DO concentration for nitrification to occur

heterotrophs oxidizing the compounds and successfully competing for the available dissolved oxygen if this is at a fairly low concentration since the $K_{s,o}$ for heterotrophs is generally lower than that for nitrifiers (Sharma and

Ahlert, 1977).

The experimental verification of the simultaneous **double substrate limitation model** is needed.

13.4 SENSITIVITY OF GROWTH PARAMETERS TO NITRIFICATION PROCESS

The nitrification process consists of the ammonium oxidizing process and the nitrite-oxidizing process. If the growth of each oxidizer follows the Monod type, the differential equations, excluding the decay terms on the whole nitrification process, are expressed as follows:

$$\frac{dX_1}{dt} = \mu_{1,m} * \frac{S_1}{K_1 + S_1} * X_1$$

$$\frac{dX_2}{dt} = \mu_{1,m} * \frac{S_2}{K_2 + S_2} * X_2$$

$$\frac{dX_1}{dt} = - Y_1 * \frac{dS_1}{dt}$$

$$\frac{dX_2}{dt} = Y_2 * \left(\frac{dS_3}{dt} + A_2 * \frac{dX_2}{dt}\right)$$

$$\frac{dS_2}{dt} = - \frac{dS_1}{dt} - A_1 * \frac{dX_1}{dt} - \frac{1}{Y_2} * \frac{dX_2}{dt}$$

where

X_1 = concentration of ammonia oxidizing bacteria (mg l^{-1})

X_2 = concentration of nitrite oxidizing bacteria (mg l^{-1})

S_1 = concentration of ammonia nitrogen (mg l^{-1})

S_2 = concentration of nitrite nitrogen (mg l^{-1})

S_3 = concentration of nitrate nitrogen (mg l^{-1})

$\mu_{1,m}$ = maximum specific growth rate of ammonia oxidizing bacteria (h^{-1})

$\mu_{2,m}$ = maximum specific growth rate of nitrite oxidizing bacteria (h^{-1})

K_1 = saturation constant of NH_4-N (mg l^{-1})

K_2 = saturation constant of NO_2-N (mg l^{-1})

Y_1 = yield constant of ammonia oxidizing bacteria (mg dry weight of cell/mg nitrogen)

Y_2 = yield constant of nitrite oxidizing bacteria (mg dry weight of cell/mg nitrogen)

A_1 = nitrogen content of ammonia oxidizing bacteria (mg nitrogen/mg dry weight of cell)

A_2 = nitrogen content of nitrite oxidizing bacteria (mg nitrogen/mg dry weight of cell)

By these differential equations, the relationships between different terms, such as substrate concentration and biomass concentration, are expressed but not in terms of the perspective of whole nitrification. Akai, Miki and Ohgaki (1983) has shown that perspective by the method of sensitivity analysis on parameters.

Computer simulation was used to explore the behavior of the model. Figure 13.2 shows temporal changes of NH_4-N, NO_2-N and NO_3-N concentrations as initial concentrations of ammonia oxidizing bacteria ($X_{1.0}$) were changed and other parameters remained constant. As the initial concentration of ammonia oxidizing bacteria was increased, ammonia utilization was accelerated so that the ammonia depletion time decreased and the maximum concentration of accumulated nitrite increased.

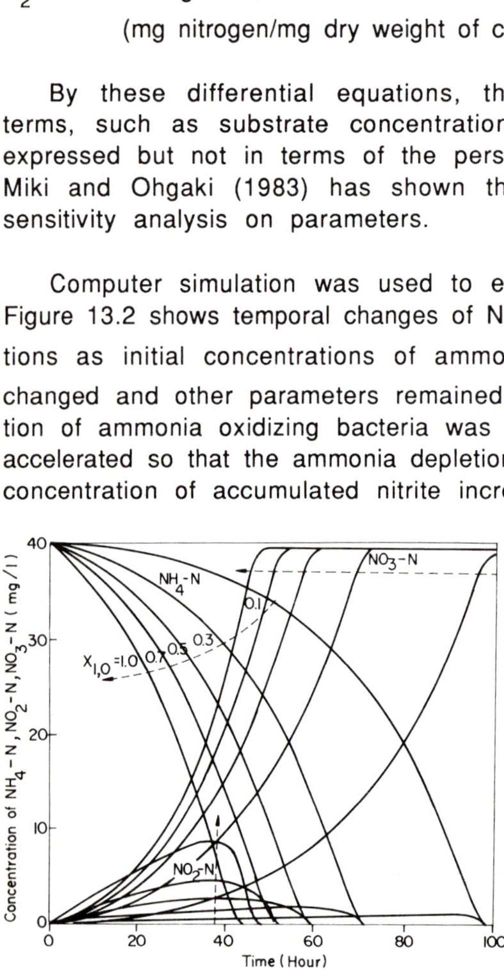

Fig. 13.2: Time related changes in nitrification time-courses as the initial concentrations of ammonia bacteria are increased. $\mu_{1,m}$ = 0.04; $\mu_{2,m}$ = 0.06; $K_1 = K_2$ = 1.0; $Y_1 = Y_2$ = 0.1; $A_1 = A_2$ = 0.1; $X_{2,0}$ = 0.5; $X_{1,0}$ = 0.1, 0.3, 0.5, 0.7, 1.0 (Akai, Miki and Ohgaki, 1983)

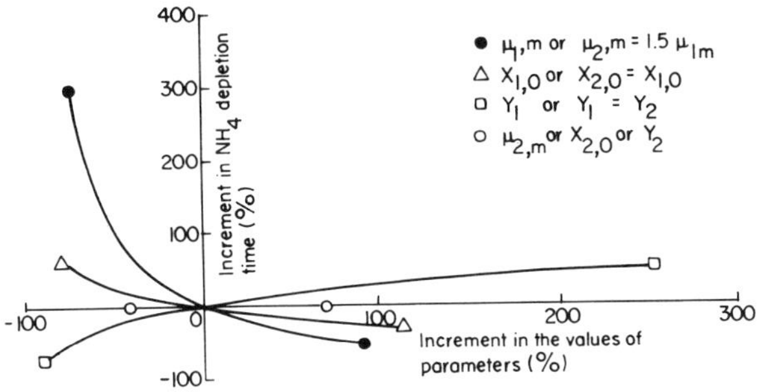

Fig. 13.3: Sensitivity of parameters to ammonia depletion time. The basic values are $\mu_{1,m} = 0.04$; $\mu_{2,m} = 0.06$; $X_{1,0} = 0.5$; $X_{2,0} = 0.5$; $Y_1 = 0.1$; $Y_2 = 0.1$; $K_1 = K_2 = 1.0$; $A_1 = A_2 = 0.1$ (Akai, Miki and Ohgaki, 1983)

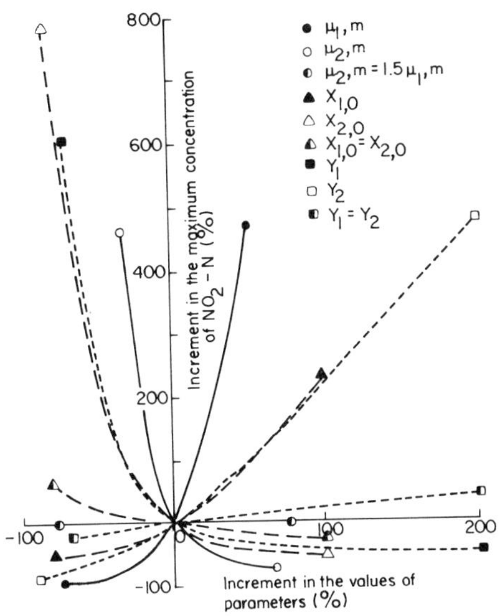

Fig. 13.4: Sensitivity of parameters to the maximum concentration of accumulated NO_2-N. The basic values are the same as those in Fig. 13.3. (Akai, Miki and Ohgaki, 1983)

Figs. 13.3 and 13.4 show the results of a sensitivity analysis of the parameters affecting the ammonia depletion time and the maximum concentration of accumulated nitrite respectively. While each of the parameters was changed stepwise from its basic value, the other paramters were assumed constant. The basic values of changing parameters in the sensitivity analysis were as follows: $X_{1.0} = 0.5$, $X_{2.0} = 0.5$, $\mu_{1,m} = 0.04$, $\mu_{2,m} = 0.06$, $Y_1 = 0.1$, $Y_2 = 0.1$ and the values $K_1 = K_2 = 1.0$ (Knowles et al., 1965),

$A_1 = A_2 = 0.1$ were set constant.

The depletion time of ammonia was rather sensitive to $\mu_{1,m}$ and $X_{1.0}$ and Y_1 but remained unchanged with $\mu_{2,m}$, $X_{z,o}$ and Y_2, while maximum concentrations of nitrite were highly sensitive to all the parameters: $\mu_{1,m}$, $\mu_{2,m}$, $X_{1.0}$, $X_{2.0}$, Y_1 and Y_2. If both of the corresponding parameters of ammonia-oxidizing bacteria and nitrite oxidizing bacteria (e.g., both $\mu_{1,m}$, $\mu_{2,m}$) were changed simultaneously, assuming $\mu_{2,m} = 1.5 \quad \mu_{1,m}$ (Knowles et al., 1965) or $X_{1.0} = X_{2.0}$ or $Y_1 = Y_2$, the variation in maximum concentration of accumulated nitrite was small due to a cancelling effect of the two corresponding parameters.

It is important to note that the yield constants and the initial concentrations of bacteria have great effect on the nitrification process as well as the maximum specific growth rates.

13.5 KINETICS EXPRESSIONS FOR INHIBITION

In the previous section, the sensitivity analysis showed the great effect of maximum specific growth rates on the whole nitrification process. The reduction of maximum specific growth rates which resulted from the effect of environmental parameters on enzyme reaction can be expressed by the enzyme inhibition model.

An enzyme inhibitor is a compound which acts to reduce the rate of enzymatically catalyzed reaction by binding with either the free enzyme (E) and/or with the enzyme-substrate complex (ES). Types of enzyme inhibition can be classified into five groups (Grady and Lim, 1980) as:

1. **Competitive inhibition**
   An inhibitor which is classed as competitive tries to bind at the same active sites as the substrate and therefore competes with the substrate for a spot on the enzyme.

2. **Uncompetitive inhibition**
   An uncompetitive inhibitor binds with the enzyme-substrate complex to form an inactive enzyme substrate-inhibitor complex which cannot undergo further reaction to yield the product.

3. **Noncompetitive inhibition**
A noncompetitive inhibitor can combine with both the free enzyme and the enzyme substrate complex.

4. **Substrate inhibition**
When their concentrations are very high, some substrates will bind with the enzyme substrate complex as well as with the free enzyme.

5. **Product inhibition**
The product may bind with the enzyme substrate complex, forming an unreactive enzyme substrate product complex (ESP).

The hypothesized mechanisms and inhibition-model of these different types are shown in Table 13.3 and Figure 13.5.

Furthermore, there are several linear equations transforming the Michaelis-Menten expression and the plots of these transforming equations make it easier to quantify the various parameters. As shown in Fig. 13.5 the effects of the inhibitor on μ_m and K_s can be easily distinguished and categorized as each type of inhibition. The linear-form equations of the plots in Fig. 13.5 are as follows (Grady and Lim, 1980).

$$\frac{1}{\mu} = \frac{1}{\mu_m} + \frac{K_s}{\mu_m} * \frac{1}{S} \quad : \text{Lineweaver-Burke equation}$$

$$\mu = \mu_m - K_s \frac{\mu}{S} \quad : \text{Hofstee equation}$$

$$\frac{S}{\mu} = \frac{S}{\mu_m} + \frac{K_s}{\mu_m} \quad : \text{Hanes equation}$$

The specific substrate utilization rate (V) is more convenient for experimental quantification of parameters than the specific growth rate (μ). Since the definitions of μ and V are:

$$\mu = \frac{dX/dt}{X}, \quad \mu_m = \frac{(dX/dt)\ \text{maximum}}{X},$$

$$V = \frac{dS/dt}{X}, \quad V_m = \frac{(dS/dt)\ \text{maximum}}{X},$$

and

$$\frac{dX}{dt} = -Y\frac{dS}{dt}$$

where

dX/dt is the rate of increase in microbial concentration in a culture ($mg\ l^{-1}\ h^{-1}$),

dS/dt is the rate of substrate removal ($mg\ l^{-1}\ h^{-1}$),

Y is the yield coefficient ($mg * mg^{-1}$)

Lineweaver - Burke equation for V is obtained as follows:

$$\frac{1}{V} = \frac{1}{V_m} + \frac{K_s}{V_m} * \frac{1}{S}$$

The K_s and V_m values can be determined by the plot of (1/V) against (1/S).

For example, Krittiya (1984) has used the Lineweaver-Burke plot for demonstrating the effect of sodium ion on the nitrite oxidizing bacteria as shown in Fig. 13.6. Her batch experimental results showed that the sodium ion inhibition on the nitrite-oxidizing process was categorized as a non-competitive type and the inhibition constant (K_i) was 2.0 $g\ l^{-1}$ as Na^+. The value of 2.0 $g\ l^{-1}$ as Na^+, is roughly equivalent to 20% of the sodium content in natural sea water. Visut (1985) has investigated the effect of sodium ions on the ammonium-oxidizing process. The ammonium-oxidizing bacteria in Visut's experiment and nitrite oxidizing bacteria in Kittiaya's experiment were isolated from the fresh water taken from the same oxidation pond. His experimental result (Fig. 13.7) showed also that inhibition was of the noncompetitive type, and the ammonium-oxidizing bacterial growth submodel with an inhibitory effect of sodium ion was expressed as follows:

$$\mu = \frac{\mu_m * S}{(K_s + S)\ (1 + I/K_i)} - K_d$$

where

μ = specific growth rate, h^{-1}

S = ammonium concentration, $mg\ l^{-1}$ as N

I = inhibitor concentration, $g\ l^{-1}$ as Na^+

μ_m = maximum specific growth rate (= 0.0313 h^{-1})

K_s = saturation constant (= 11.6 ~ 13.5 mg l^{-1} as N)

K_i = inhibition constant (= 6.64 g l^{-1} as Na^+)

K_d = decay rate (= 3.1 * 10^{-3} h^{-1})

Other researchers also applied the rate expression for enzyme inhibition in their studies of effects of some inhibitors on nitrification. (Hockenbury and Grady, 1977; Beg et al., 1982; Akai et al., 1983).

TABLE 13.3: Different type of inhibition model

Types of inhibition	Mechanism	Rate expression	Michaelis-Menten form
Competitive	$E + S \underset{k_{-1}}{\overset{k_1}{\rightleftharpoons}} ES \overset{k_2}{\rightarrow} E + P$ $E + I \underset{k_{-3}}{\overset{k_3}{\rightleftharpoons}} EI$	$\mu = \dfrac{\mu_m S}{(K_s(1+I/k_I)+S)}$ where $K_I = k_3/k_{-3}$	$\mu'_m = \mu_m$ $K'_s = K_s(1+I/K_I)$
Un-competitive	$E + S \underset{k_{-1}}{\overset{k_1}{\rightleftharpoons}} ES \overset{k_2}{\rightarrow} E + P$ $E + I \underset{k_{-3}}{\overset{k_3}{\rightleftharpoons}} EI$	$\mu = \dfrac{\mu_m S}{(K_s + S\ (1+I/k_I))}$ where $K_I = k_3/k_{-3}$	$\mu'_m = \dfrac{\mu_m}{1 + I/K_I}$ $K'_s = \dfrac{K_s}{1+I/K_I}$
Non-competitive	$E + S \underset{k_{-1}}{\overset{k_1}{\rightleftharpoons}} ES \overset{k_2}{\rightarrow} E + P$ $E + I \underset{k_{-3}}{\overset{k_3}{\rightleftharpoons}} EI$ $ES + I \underset{k_{-4}}{\overset{k_4}{\rightleftharpoons}} ESI$	$\mu = \dfrac{\mu_m S}{(K_s + S)\ (1+I/k_I))}$ where $K_I = k_3/k_{-3} + k_4/k_{-4}$	$\mu'_m = \dfrac{\mu_m}{1 + I/K_I}$ $K'_s = K_s$
Substrate	$E + S \underset{k_{-1}}{\overset{k_1}{\rightleftharpoons}} ES \overset{k_2}{\rightarrow} E + P$ $ES + S \underset{k_{-3}}{\overset{k_3}{\rightleftharpoons}} SES$	$\mu = \dfrac{k_2 E_o S}{(K_c + S + S^2/K'_c)}$ $\mu_m = \dfrac{k_2 E_o}{(1 + 2(K_c/K'_c)^{0.5})}$ where K_c & K'_c are the dissociation constants for ES and SES respectively	
Product *	$E + S \underset{k_{-1}}{\overset{k_1}{\rightleftharpoons}} ES \overset{k_2}{\rightarrow} E + P$ $ES + P \underset{k_{-3}}{\overset{k_3}{\rightleftharpoons}} ESP$	$\mu = \dfrac{\mu_m S}{(K_s + S(1+P/k_P))}$ where $K_P = k_3/k_{-3}$	$\mu'_m = \dfrac{\mu_m}{1 + P/K_P}$ $K'_s = \dfrac{K_s}{1+P/K_P}$

where $\mu_m = k_2E_o$ (E_o = initial enzyme concentration), $K_s = (k_{-1} + k_2)/k_1$

* This is the simplest mechanism, other mechanisms could be hypothesized which would lead to alternative rate expressions.

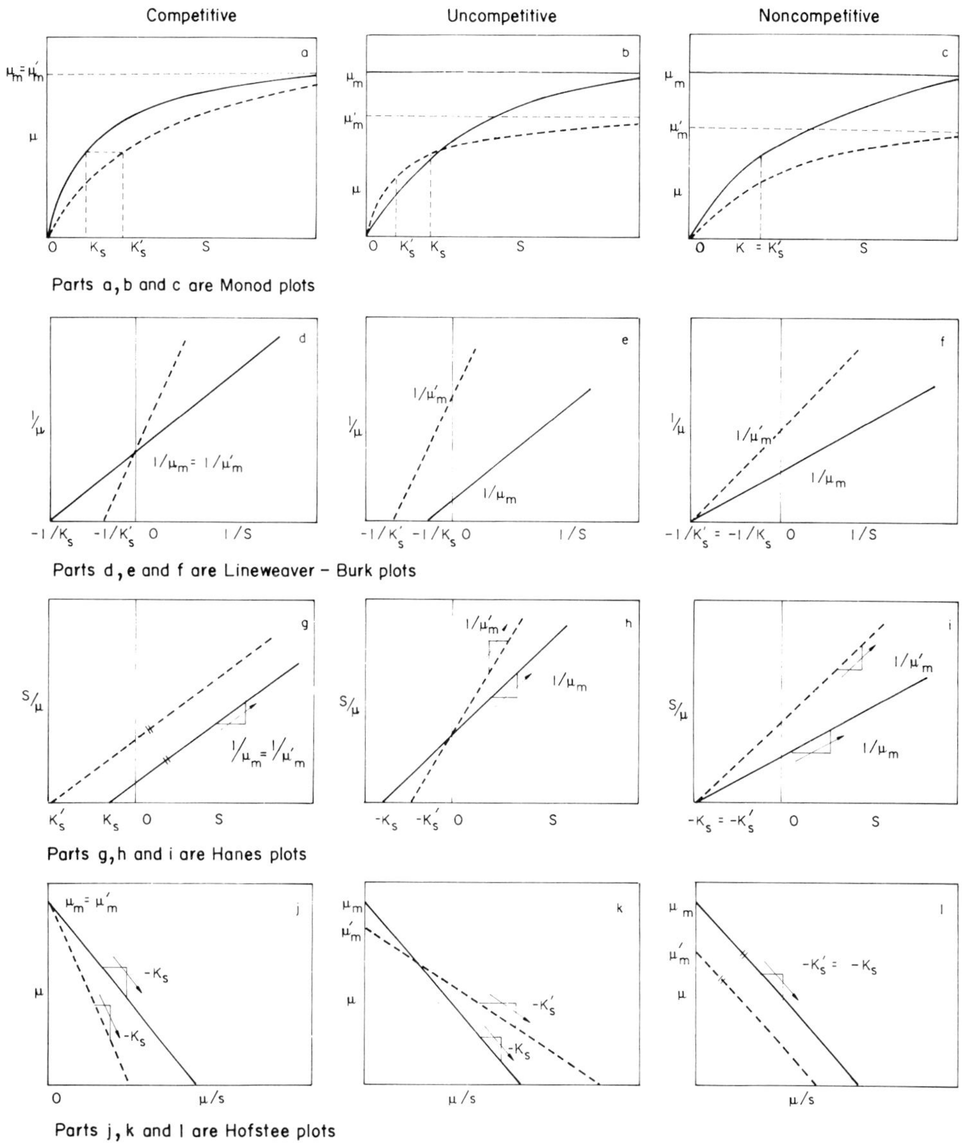

Fig. 13.5: Typical plots for identifying the types of enzyme inhibition. The solid curves represent the uninhibited case, whereas the dashed curves are the inhibited case.

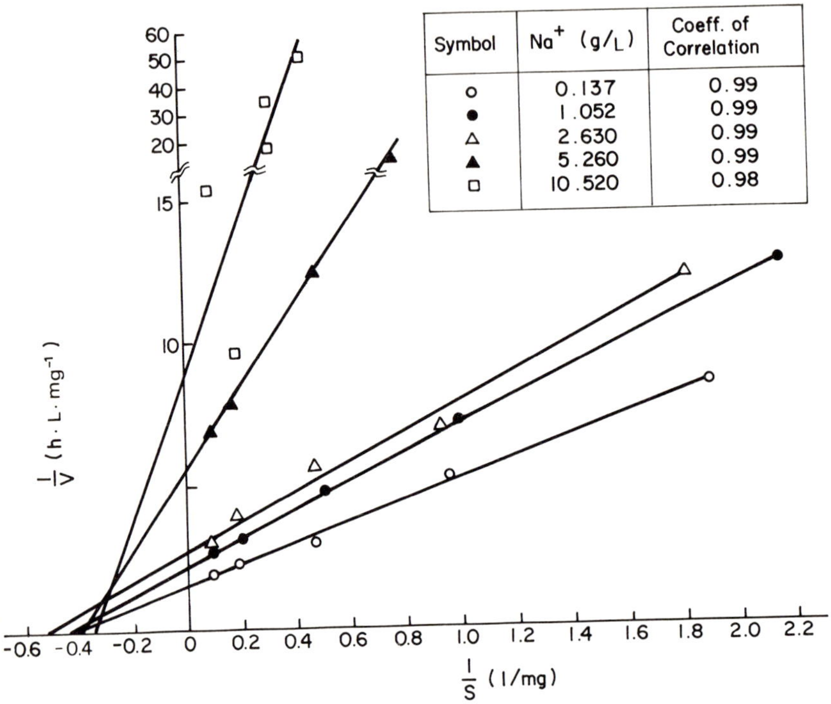

Fig. 13.6: The Lineweaver-Burke plots for identifying the type of inhibition of sodium ion concentration for nitrite oxidizing bacteria (Krittiya, 1984).

13.6 ENVIRONMENTAL FACTORS

Effects of Temperature.

The optimum temperature for growth of nitrifying bacteria appears to be in the range 28-36 °C, and little or no growth occurs below 4 °C (Painter, 1970).

Several workers have described the temperature dependence by an exponential expression.

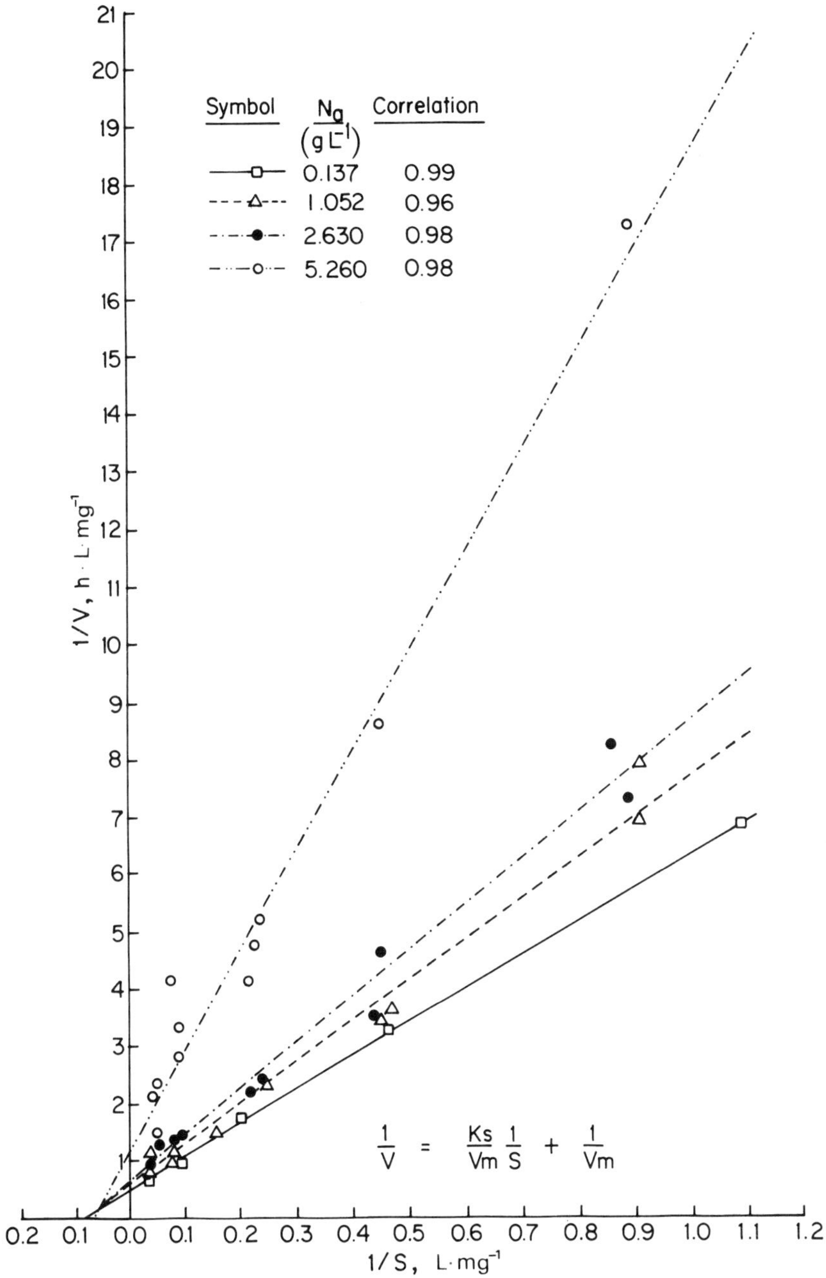

Fig. 13.7: Lineweaver-Burke plots for ammonium-oxidizing bacteria (Visut, 1985)

$$\mu_m = \mu_{m,ref} * O^{(T-T_{ref})}$$

where μ_m and $\mu_{m,ref}$ are the maximum specific constants at temperature T and T_{ref} (0 °C), respectively and O is a constant for a specific temperature range referred to as the "temperature coefficient". The temperature coefficients have been listed in Table 13.4.

TABLE 13.4: Temperature coefficients for nitrifying micro-organisms

Process	Range of temperature	T_{ref}	O	$\mu_{m,ref}$	Reference
Nitrogenous phase in BOD bottle analysis	10 - 22 °C	20 °C	1.097	0.12	Zanoni, 1969
" "	22 - 22 °C	20 °C	0.877	0.12	" "
Nitrification in suspended combined culture	5 - 20 °C	20 °C	1.12	-	Andersen & Poulsen, 1976. Cited in Christensen & Harremoes, 1978
Nitrifying in treatment process	10 - 30 °C	12 °C	1.07	0.5	McCarty, 1975 cited in Smith, 1977
" "	5 - 10 °C	10 °C	1.19	0.25	" "
Nitrosomonas in fill and draw pilot plant activated sludge	8 - 20 °C	-	1.12	0.18	Jenkins, 1969
Nitrosomonas in water from Thames estuary	8 - 30 °C	15 °C	1.099	0.47	Knowles et al., 1965
Nitrobacter in water from Thames estuary	8 - 30 °C	15 °C	1.058	0.79	" "
Nitrosomonas in activated sludge	10 - 25 °C	15 °C	1.123	0.18	Downing & Hopwood cited in Smith, 1977
Nitrosomonas in pure culture	10 - 25 °C	15 °C	1.103	0.47	" "

In addition, the half-saturation constants are also reported to be a function of temperature (Knowles et al., 1965; Charley et al., 1980). The half-saturation constants for *Nitrosomonas* ($K_{S,M}$) and *Nitrobactor* ($K_{S,B}$)

have been analyzed by Knowles et al. (1965) to follow Arrhenius Law as follows:

$$K_{S,M} = 0.405\, e^{0.118\,(T-15)}$$

$$K_{S,B} = 0.405\, e^{0.146\,(T-15)}$$

where T is temperature in °C.

However, since the rates of nitrification usually are found in the zero range, the temperature-dependent saturation constants are rarely reported.

All studies above were concerned with long-term temperature dependence. The effect of rapid changes of temperature on nitrification has also been reported but in fewer cases. For example, in Srna and Baggaley (1975), studies revealed that a 4 °C increase in temperature caused about a 50% increase in the rate of marine nitrification, while a 1 °C drop caused a 30% decrease when compared with calculated values at 21.3 °C.

Effects of pH

The effects of pH on nitrification have been reviewed by Sharma and Ahlert (1977) who showed that the optimal values of pH for the overall nitrification appear to be slightly on the alkaline side.

Boon and Laudelout (1962) have developed a kinetics expression for the effects of pH on the nitrite oxidation by *Nitrobacter winogradskyi*. They suggested that inhibition of high nitrite concentrations results from non-competitive inhibition of nitrous acid, while in the alkaline branch of pH effect curve there is a competitive inhibition of the adsorption of nitrite on the enzyme sites by OH^--ions. The final forms of rate equations in the acid and alkaline sides of the pH effect curve are shown separately in equation (13.1) and (13.2), respectively:

$$\mu = \frac{\mu_m\ S}{(K_s + S)\ (1 + (S[H+] / K_{HNO_2} K_i))} \qquad (13.1)$$

$$\mu = \frac{\mu_m\ S}{K_s\ (1 + (K_b / [H^+])) + S} \qquad (13.2)$$

where

S = nitrite concentration

K_{HNO_2} = equilibrium constant of nitrous acid and nitrite ion dissociation $= 3.98 * 10^{-4}$

K_i = dissociation constant of the enzyme-nitrous acid complex

K_b = basic acid-base dissociation constant of the active enzyme site.

Thus the general rate equation for pH effects was formulated by combining equation (13.1) and (13.2), as in equation (13.3):

$$\mu = \frac{\mu_m \ S}{\{K_s\ (1 + (K_b\ /\ [H+])\ ) + S\}\ \{1 + (S[H+]\ /\ K_{HNO_2}\ K_i\ )\}} \tag{13.3}$$

From their experiments, the values of K_b and K_i were found to be 0.004, and 8.2 μM of NO_2^- concentration, respectively. Suzuki et al. (1974) used the Lineweaver-Burke plot when studying the effect of pH in the range of 6.5-9 on the oxidation of ammonia by *Nitrosomonas europaea.* They found that the value of the Monod saturation ammonia constant decreased when pH increased. When pH is the parameter, the plot shows competitive inhibtion.

Since *Nitrosomonas* and *Nitrobacter* are both sensitive to their own substrates of unionized ammonia and nitrite, and the unionized-ionized nitrogen equilibria depend on pH, Anthonisen et al. (1976) postulated that pH affects nitrification by affecting penetration of the cell membrane. The concentrations of unionized ammonia and unionized nitrite, rather than total ammonia or nitrite ion concentrations, are suggested as inhibiting nitrification. Previous research (Prakasam and Loehr, 1972; Bonn and Laudelout, 1962) also supported this suggestion.

Effects of Heterotrophic Activity

Ammonia removal in the nitrification process is accomplished by autotrophic nitrifying bacteria of a life-mode distinctly different from that of heterotrophic organisms utilizing organic carbon substances. Because the ecosystem in natural water bodies is a mixed culture system, the knowledge of mutual relationships between nitrifying bacteria and heterotrophic bacteria is necessary for the modelling of nitrification. This topic has been studied in the field of wastewater treatment.

The maximum specific growth rate of nitrifying bacteria determined in the treatment process is significantly different from that observed in pure culture. Painter (1977) has listed possible reasons for this difference. The toxic constituents of waste water may inhibit nitrification, but other less likely reasons are the limitation of nutrients or of the physical environment, or genuine differences between isolated stains and those effecting nitrification in the treatment process. Several specific organic compounds which have a direct toxic action on nitrifying bacteria have been investigated in activated sludge (Tomlinson et al., 1966). However, it is unclear whether the products released by heterotrophic activity during the course of biological process treatment have an effect upon nitrification, even if the original waste water lacks known toxic constituents.

Stover and Kincannon (1976) presented experimental results showing the effects of COD/N ratio on nitrification in the activated sludge process applied with non-toxic synthetic media. However, they have not performed the study from the viewpoint of kinetics to identify the mechanism. The competition for nitrogen by heterotrophs, or inhibition, interferes with ammonia removal and reduces the nitrate production under the conditions of high COD/N loading.

In the field of attached growth treatment process, it is expected that nitrification under high organic carbon loading will not occur in the upper part of the trickling filter and on the first disks of a rotating disk unit owing to the limited time available for multiplication of this type of organism in the fast-growing biofilm section (Christensen and Harremoes, 1978).

Tomlinson and Snaddon (1966) described that in the normal development of nitrifying film in the lower half of their percolating filter, the activity of the film remaining adhering to the tube wall of percolating filter was greater than that of the superficial layers of film. The clear reason for this has just been given by recent models involving competition between heterotrophic microorganisms and nitrifying microorganisms (Harremoes, 1982; Kissel, McCarty and Street, 1984; Wanner and Gujer, 1984). All of these recently developed models predicted that the fraction of nitrifiers is greater in the inner layer (near surface of media) than in the outer layer of biofilm.

There is lack of study on the effect of heterotrophic activities on nitrification in natural water bodies, except Torpey´s investigation (1968) on the River Thames. Torpey (1968) demonstrated that nitrification was shifted upstream following the upgrading of sewage treatment plant where more carbonaceous matters were reduced. But it also can be explained by more nitrifiers present in the effluent after improvement of the process.

As a clear explanation of the mechanism of mutual relationships between nitrification and heterotrophic activities has not been found in both fields of wastewater treatment and water quality management, further research is required to develop the mathematical submodel on this matter.

13.7 NITRIFICATION ZONES IN NATURAL WATER BODIES

The appropriate application of growth kinetic submodels to the prediction of nitrification in natural water bodies must be based on the growth mechanisms of nitrifying bacteria, whether they are in the suspended growth state or attached growth state.

Tuffy et al. (1974) described the zones of nitrification in the stream by the interrelation between the bacterial growth mechanisms and characteristics of any regions of the stream. Considering the physical hydraulic characteristics (especially, hydraulic detention time) of stream, nitrification by attached growth is significant in headwaters of a stream where water usually is shallow and swift moving and has a rocky bed.

The water volume of the middle reach and estuary are quite large compared with the surface area of the stream bed and hence the activity of suspended bacteria is more likely to be important. A non-nitrification zone is the region in which the nitrifying bacteria do not have a sufficiently long detention time to be significant in the water phase and the effects of nitrification on dissolved oxygen balance by virtue of attached microorganism is diminished due to the large volume of flow passing over the bed.

Tuffy et al. (1974) proposed the zones of nitrification as follows (refer Fig. 13.8):

1. zone of "surface active" nitrification
2. transitional zone or non-nitrification zone
3. zone of "water phases active" nitrification

where "surface active" means activity of attached nitrifying bacteria on the surface of the stream bed, and "water phases active" means "activity of suspended nitrifying bacteria in the water body".

As regards the "zone of surface activity", Gujer (1976) proposed the model concerning activity of attached nitrifying bacteria in the Glatt River of Switzerland. This river is shallow with a hydraulic radius less than 1 m, and rapid flows with a velocity greater than 0.5 m/sec. Blain (1969), discussing Wezernak and Gannon´s paper (1967), suggested that the nitrification in the Clinton River would be carried out by micro-organisms attached to aquatic plants and on the stream bed. Finstein et al. (1978) used a similar viewpoint to explain the difference of ammonia oxidation occuring in Grand River and in BOD bottle observed by Courchaine (1968).

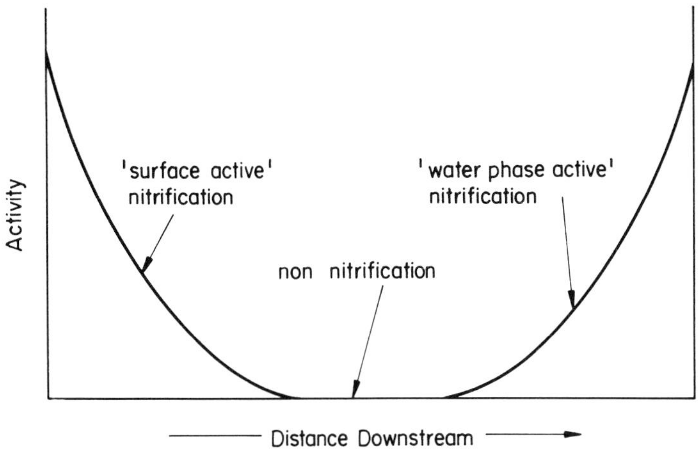

Fig. 13.8: Location of actively nitrifying zones (Tuffy et al., 1974)

Nitrification in rivers classified as "water phase active" reaches have also been investigated recently in such waters as Cherokee Lake (Ruane and Krenkel, 1977), Potomac estuary (Eklins et al., 1979), Trent River (Curtis et al., 1975), James River (Cerco, 1979) etc.

In the case of the Trent River and James River, the investigators reported that nitrifiers were present near the surface layers of stream sediment and suspended solid overlying the sediment, with insignificant amounts of nitrifiers existing in the body of water itself. This densely suspended nitrifer lying over the bed is movable and easily resuspends in the water body. With different physical characteristics between this layer and fixed micro-organisms on the stream bed, the strength of hydraulic characteristics affecting nitrification by each will be different too.

It has been mentioned already in the previous section; section 13.4, that the initial concentration of nitrifying bacteria is important for growth kinetics, as well as specific growth rates. In the particular region which does not have a sufficiently long detention time for nitrifying bacteria growth, however, the nitrification does occur if the nitrifying bacteria are provided continuously into the region. Therefore, it is insufficient if a growth kinetics submodel is applied to a particular region of stream by base only on the environmental characteristics of the region, without considering the supply of nitrifying bacteria from the upper reaches.

The inter-relationship between zones should be taken into account when developing submodels. Garland (1978) showed the significance of the supply of nitrifying bacteria to the lower reaches of the River Trent from the upper reach in which attached growth prevailed.

The "Supply" or "Seeding" of nitrifying bacteria (from the upper reach of stream, from the effluent of a wastewater treatment plant, or, from a tributary) plays an important role in the nitrifying growth kinetics and hence in developing the nitrification submodels.

REFERENCES

Akai, D., Miki, O. and Ohgaki, S. (1983). Nitrification Model with an Inhibitory Effect of Sea Water. Ecological Modelling, Vol. 19, pp. 189-198.

Alleman, J.E. (1984). Elevated Nitrite Occurrence in Biological Wastewater Treatment Systems. Water Science Technology, Vol. 17, pp. 409-419.

Anthonisen, A.C., Loehr, R.C., Prakasan, T.B.S. and Shinath, E.G. (1976). Inhibition of Nitrification by Ammonia and Nitrous Acid, J. Water Pollution control Fed., Vol. 48, pp. 835-852.

Atkinson, B. and Fowler, H.W. (1974). The Significance of Microbial Film in Fermenters, ed. by T.K. Ghose, A. Fiechter and N. Blackebrough, Advance in Biochemical Engineering 3, New York, pp. 246-256.

Bader, F.G. (1978). Analysis of Double-Substrate Limited Growth. Biotechnology and Bioengineering, Vol. 20, pp. 183-202.

Bader, F.G., Meyer, J.S., Fredrickson, A.G. and Tsuchinaya, H.M. (1975). Comments on Microbial Growth Rate. Biotechnology and Bioengineering, Vol. 17, pp. 279-283.

Beckman, W.J., Avendt, R.J., Mulligan, T.J. and Kehrberger, G.J. (1972). Combined Carbon Oxidation Nitrification. J. Water Pollution Control Fed., Vol. 44 , pp. 1916-1931.

Beg, S.A., Siddiqi, R.H. and Ilias, S. (1982). Inhibition of Nitrification of Arsenic, Chromium and Fluoride. J. Water Pollution Control Fed., Vol. 54, pp. 482-488.

Blain, W.H. (1969). Discussion to paper by Wezernak and Gannon, Evaluation of Nitrification in Stream. Journal of the Sanitary Engineering Division ASCE, Vol. 95, pp. 956-958.

Boon, B. and Laudelout, H. (1962). Kinetics of Nitrite Oxidation by *Nitrobacter winogradskyi.* Biochemical Journal, Vol. 85, pp. 440-447.

Bundgay, H.R. (1967). Dynamic Analysis of Microbial Systems. Water Sewage Works, Vol. 114, pp. 190-196.

Cerco, C.F. (1979). Nitrification in the Upper Tidal James River, ed. by B.J. Neilson and L.E. Cronin. Estuaries and Nutrients. Humana Press, New Jersey, pp. 413-424.

Charley, R.C., Hooper, D.G and McLee, A.G. (9180). Nitrification Kinetics in Activated Sludge at Various Temperatures and Dissolved Oxygen Concentrations. Water Research, Vol. 14, pp. 1387-1390.

Christensen, M.H. and Harremoes, P. (1978). Nitrification and Denitrification in Waste Water Treatment, ed. by R. Mitchell. Water Pollution Microbiology, Vol. 2, pp. 391-414.

Chu, G.C.Y., Erickson, L.E. and Fan, L.T. (1973). Dynamic Behavior of a

Complete-Mixing Activated Sludge System. Biotechnology and Bioengineering, Vol. 15, pp. 1101-1121.
Courchaine, R.J. (1968). Significance of Nitrification in Stream Analysis - Effects on the Oxygen Balance. J. Water Pollution Control Fed., Vol. 40, No. 5, pp. 835-847.
Curtis, E.J., Durrant, K. and Harman, M.M. (1975). Nitrification in Rivers in the Trent Basin. Water Research, Vol. 9, pp. 255-268.

Downing, A.L., Painter, H.A. and Knowles, G. (1964). Nitrification in the Activated Sludte Process. Journal of Institute Sewage Purification, pp. 130-153.

Elkins, J.W., Wofsy, S.C., McElroy, M.B. and Kaplan, W.A. (1979). Nitrification and Production of NO in the Potomac; Evidence for variability, ed. by B.J. Neilson and L.E. Cronin. Estuaries and Nutrients, Humana Press, New Jersey, pp. 447-464.
Engel, M.S. and Alexander, M. (1958). Growth and Autotrophic Metabolism of *Nitrosomonas europaea*. Journal of Bacteriology, Vol. 76, pp. 217-222.
Ericsson, B. (1975). Nitrogen Removal in a Pilot Plant. J. Water Pollution Control Fed., Vol. 47, pp. 727-740.

Finstein, M.S., Strom, P.F. and Matulewich, A. (1978). Discussion to paper by Courchaine, Significance of Nitrification in Stream Analysis - Effect on the Oxygen Balance. Journal Water Pollution Control Fed., Vol. 50, No. 8, pp. 2055-2057.

Garland, J.H.N. (1978). Nitrification in the River Trent; In Symposium on the Use of Mathematical Models in Water Pollution Control, University of Newcastle. Ed. bny A. James. Mathematical Models in Water Pollution Control, John Wiley & Sons, pp. 167-191.
Grady, C.P.L. and Lim, H.C. (1980). Biological Wastewater Treatment: Theory and Applications, ed. by P.N. Cheremisinoff; Marcel Dekker Inc., New York.
Gujer, W. (1976), Nitrification in Fliesszewassern-Fall Studie Glatt. Schweizerische Zeitschrift für Hydrologie, 38:2, pp. 171-189.
Gujer, W. (1977). Design of a Nitrificating Activated Sludge Process with the Aid of Dynamic Simulation. Progress in Water Technology, Vol. 9, pp. 337-346.
Gujer, W. and Jenkins, d. (1975). A Nitrification Model for the Contact Stabilization Activated Sludge Process. Water Research, Vol. 9,pp. 561-566.

Hall, E.R. and Murphy, K.L. (1980). Estimation of Nitrifying Biomass and Kinetics in Wastewater. Water Research, Vol. 14, pp. 297-304.
Harremoes, P. (1982). Criteria for Nitrification in Fixed Film Reactors. Water Science and Technology, Vol. 14, pp. 167-187.
Haug, R.T. and McCarty, P.L. (1972). Nitrification with Submerged Filter. J. Water Pollution Control Fed., Vol. 44, No. 11, pp. 2086-2102.
Heukelekian, H. (1942). The Influence of Nitrifying Flora, Oxygen and Ammonia Supply on the Nitrification of Sewage. Sewage Works Journal, vol. 14, No. 5, pp. 964-979.
Hockenbury, M.R. and Grady Jr., C.P.L. (1977). Inhibition of Nitrification Effects of Selected Organic Compounds. J. Water Pollution Control Fed., Vol. 49, pp. 768-777.

Jenkins, S.H. (1969). Nitrification. Water Pollution Control, pp. 610-618.

Kiff, R.J. (1972). The Ecology of Nitrification/Denitrification Systems in

Activated Sludge. Water Pollution Control, pp. 473-484.
Kissel, J.C., McCarty, P.L. and Street, R.L. (1965). Numerical Simulation of Mixed Culture Biofilm. Journal of Environmental Engineering, Vol. 110, No. 2 April, pp. 393-411.
Knowles, G., Downing, A.L. and Barrett, M.J. (1965). Determination of Kinetic Constants for Nitrifying Bacteria in Mixed Culture with the Aid of an Electronic Computer. J. Gen. Microbial, Vol. 38, pp. 263-278.
Krittiya Lertpocasombut (1984). Nitrification Model with Inhibition of Sea Water. Master Thesis, Asian Institute of Technology, Bangkok.

Lamotta, E.J. (1976). Internal Diffusion and Reaction in Biological Films. Environmental Science and Technology, Vol. 10, No. 8, pp. 765-769.

Megee, R.D. III, Drake, J.F, Fredrickson, A.G. and Tsuchiya, H.M. (1972). Studies in Intermicrobial Symbiosis. *Saccharomyces cerevisiae* and *Lactobacillus casei.* Can. J. Microbiology, Vol. 18, pp. 1733-
Müller, J.A., Paquin, P. and Famularo, J. (1980). Nitrification in Relating Biological Contacters. J. Water Pollution Control Fed., vol. 52, No. 4, pp. 688-710.
Murphy, K.L. and Sutton, P.M. (1977). Dynamic Nature of Nitrifying Biological Suspended Growth Systems. Progress Water Technology, Vol. 9(2), pp. 279-290.

Nicholas, D.J.D. and Jones, O.T.G. (1960). Oxidation of Hydroxylamine in Cell-Free Extracts of *Nitrosomonas Europaes.* Nature, Vol. 185, pp. 512-514.

Painter, H.A. (1970). A Review of Literature on Inorganic Nitrogen Metabolism in Microorganisms. Water Research, Vol. 4, pp. 393-450.
Painter, H.A. (1977). Microbial Transfer Transformations of Inorganic Nitrogen. Progress Water Technology, Vol. 8, No. 4/5, pp. 3-29, Pergamon Press, 1977.
Peeters, T.L., Van Goal, A.D. and Laudelout, H. (1969). Kinetics Study of Oxygen Limited Respiration in Nitrifying Bacteria. Bacteriology Proceeding, vol. 141.
Poduska, R.A. and Andrew, J.F. (1975). Dynamics of Nitrification in the Activated Sludge Process. J. Water Pollution Control Fed., Vol. 47, pp. 2599-2619.
Prakasam, T.B.S. and Loehr, R.C. (1972). Microbial Nitrification and Denitrification in Concentration Wastes. Water Research, Vol. 6, pp. 859-869.

Ruane, R.J. and Krenkel, P.A. (1977). Nitrification and Other Factors Affecting Nitrogen in the Holton River. Progress in Water Technology, Vol. 8, pp. 209-224.
Ryder, D.N. and Sinclair, C.G. (1972). Model for the Growth of Aerobic Microorganisms under Oxygen limiting conditions. Biotechnology and Bioengineering, Vol. 14, pp. 787-798.

Saunders, P.T. and Bazin, M.J. (1973). Attachment of Microorganisms in a Packen Column. Metabolite Diffusion through the Microbial Film as a Limiting Factor. J. Applied Chemical Biotechnology, Vol. 23, pp. 847-853.
Sinclair, C.G. and Ryder, D.N. (1975). Models for the Continuous Culture of Microorganisms under Both Oxygen and Carbon Limiting Conditions. Biotechnology and Bioengineering, Vol. XVII, pp. 375-398.
Sharma, B. and Ahlert, R.C. (1977). Nitrification and Nitrogen Removal. Water Research, Vol. 11, pp. 897-925.

Shieh, W.K. and Lamotta, E.J. (1979). The Intrinsic Kinetics of Nitrification in a Continuous Flow Suspended Growth Reactor. Water Research, Vol. 13, pp. 1273-1279.
Smith, A.G. (1977). Nitrification in Activated Sludge Plants Guidelines of Some Operation and Design Aspects. Wastewater Treatment Section, Pollution Control Branch, Ontario, Ministry of the Environment.
Srinath, E.G., Loehr, R.C. and Pakasam, T.B. (1976). Nitrification Organism concentration and Activity. Journal of the Environmental Engineering Division, ASCE, Vol. 102, No. EE2, pp. 449-463.
Srna, R. and Baggaley, A. (1975). Kinetic REsponse of Perturbed Marine Nitrification Systems. J. Water Pollution Control Fed., Vol. 47, pp. 472-486.
Stenstrom, M.K. and Poduska, R.A. (1980). The Effect of Dissolved Oxygen Concentration on Nitrification. Water Research, Vol. 14, pp. 643-649.
Storer, F.F. and Gaudy, A.F. Jr. (1969). Computation Analysis of Transient Response to Quantitatives Stock Loading of Heterogeneous Populations in Continuous Culture. Environmental Science and Technology, Vol. 3. pp. 143-149.
Stover, E.L. and Kincannon, D.F. (1976). Effects of COD: NH-N Ratio on a One Stage Nitrification Activate Sludge System. Water and Sewage Works, September, pp. 120-123.
Stratton, E.E. and McCarty, P.L. (1967). Prediction of Nitrification Effects on the Dissolved Oxygen Balance of Streams. Environmental Science and Technology, Vol. 1, No. 5, pp. 405-410.
Stratton, E.E. and McCarty, P.L. (1969). Discussion to Paper by Werzernak and Gannon, Evaluation of Nitrification in Stream. Journal of the Sanitary Engineering Division, ASCE, Vol. 95, pp. 952-955.
Sutton, P.M., Murphy, K.L., Jnak, B.E. and Monoghan, B.A. (1975). Efficacy of Biological Nitrification. J. Water Pollution Control Fed., Vol. 47, No. 11, pp. 2665-2673.
Suzuki, J., Dular, U. and Kwok, S.C. (1974). Ammonia or ammonium ions as substrate for oxidation by *Nitrosomonas europaea* cells and extracts. Journal of Bacteriology, Vol. 120, pp. 556-558.
Sykes, R.M. (1973). Identification of the Limiting Nutrient and Specific Growth Rate. J. Water Pollution Control Fed., Vol. 45, pp. 359-391.

Tanaka, H. and Dunn, I.J. (1982). Kinetics of Biofilm Nitrification. Biotechnology and Bioengineering, Vol. XXIV, pp. 669-689.
Tomlinson, T.G., Boon, A.G. and Trotman, C.N.A. (1966). Inhibition of Nitrification in the Activated Sludge Process of Sewage Disposal. J. Applied Institute Journal, Vol. 29, pp. 266-291.
Tomlinson, T.G. and Snaddon, D.H.M. (1966). Biological Oxidation of Sewage by Films Microorganisms. Air and Water Pollution Institute Journal, Vol. 10, pp. 865-881.
Torpey, W.N. (1968). Effects of Reducing Pollution of Thames Estuary. Water and Sewage Works, Vol. 115, pp. 295-301.
Tuffy, T.J., Hunter, J.V. and Matulewich, V.A. (1974). Zones of Nitrification. Water Resources Bulletin, Vol. 10, pp. 555-564.

Visut Iam-A-Ram (1985). Effect of Sodium Ion on Nitrification. Master Thesis, Asian Institute of Technology.

Wanner, O. and Gujer, W. (1984). Competition of Biofilm. Water Science and Technology, Vol. 17, pp. 27-44.
Watanabe, Y., Ishiguro, M. and Nisaidome, K. (1980). Nitrification Kinetics in a Rotating Biological Disk Reactor. Progress in Water Technology, Vol. 12, pp. 233-251.
Wezernak, C.T. and Gannon, J.J. (1967). Oxygen-Nitrogen Relationships in

Autotrophic Nitrification. Appl. Microbiology, Vol. 15, pp. 1211-1215.

Wild, H.E., Sawyer, C.N. and McMahan, T.C. (1971). Factors Affecting Nitrification Kinetics. J. Water Pollution Control Fed., Vol. 43, No. 9, pp. 1845-1854.

Williamson, K. and McCarty, P.L. (1976). A Model of Substrate Utilization by Bacteria Films. J. Water Pollution Control Fed., Vol. 48, No. 1, pp. 9-24.

Wong-Chong, G.M. and Loehr, R.C. (1975). The Kinetics of Microbial Nitrification. Water Research, Vol. 9, pp. 1099-1105.

Young, T.B., Bruley, D.F. and Bungay, H.R. (1970). A Dynamic Mathematical Model of the Chemostat. Biotechnology and Bioengineering, Vol. 12, pp. 747-769.

Zanoni, A.E. (1969). Secondary Effluent Deoxygenation at Different Temperatures. J. Water Pollution Control Fed., Vol. 41, No. 4, pp. 640-659.

CHAPTER 14

PREDATOR-PREY INTERACTIONS

by
Tarzan Legovic

14.1 INTRODUCTION

In aquatic as in terrestrial ecosystems the most often found interaction between two biological populations is certainly of the predator-prey type. In such an interaction one population (prey) serves as a food source for the other population (predator). If the two populations do not interact significantly with other populations, then it is said that they form a one predator-one prey system, or just predator-prey system for short. In nature this is rarely the case; more often, one population serves as a food source for more than one predator, and/or the predator serves as a food source for some other, higher predator. Examples are food chains and food webs. In food webs, apart from predator-prey interactions, other interactions between populations such as competitive, mutualistic, commensalistic or amensalistic may be found. Predator-prey interactions have been extensively studied because they are so often found in nature, while the dynamics of prey and predator systems have been investigated also because they form the elementary blocks for building more complicated systems such as food chains or webs.

14.2 PREDATOR-PREY MODELS

The first predator-prey model was formulated by **Lotka** (1924). The same model was independently formulated and extensively studied by **Volterra** (1927). The model involves a single exponentially growing prey population and a predator that feeds on the prey in such a way that the predation is proportional to the density of prey and the density of predators. Furthermore, the growth rate of predators is assumed proportional to the predation rate, and the death rate of the predator population is made proportional to its own density. The equations of the model are:

$$\frac{dN}{dt} = a N - b N P \qquad (14.1)$$

$$\frac{dP}{dt} = c N P - d P \qquad (14.2)$$

where
N = density of prey (number or biomass of prey/volume)
P = density of predators
and a, b, c and d are constants greater than zero.

Equations (14.1) and (14.2) account for the balance of density, and for the numbers of organisms or biomass (since the fixed volume is assumed) of prey and predator populations.

This model marked the beginning of ecological modelling. It has since been generalized and modified for application to specific cases. The original formulation today is mainly of didactical and historical importance. More important is the approach developed by Volterra and Lotka that is used to this day without much modification.
This approach consists of several steps.

First, prey and predator populations are identified. If different prey populations belong to the same trophic level and behave similarly regarding their growth, and if the predator is not able to distinguish different prey populations, one usually groups several prey populations into one variable. The rationale for including different predator species into one variable is: if they feed on the same or similar prey species in the same manner, if their growth rates do not differ and if their mortalities are approximately equal, then they may be grouped into one predator variable. Grouping of either predator or prey populations into one variable must be justified experimentally. One of the weak points of ecosystem models as predictive tools is that grouping of at least some populations is regularly done simply because experimental data on every participating population is not available. Hence, predictions may be at the best, only qualitatively correct. Nevertheless, present ecosystem models will be used as tools to direct research until better experimental evidence is accumulated.

The second step consists of formulating the balance of mass (or numbers or density) of organisms. It also involves assessments of growth, predation and death rates as functions of the environmental parameters and characteristics of the two populations. It is in this step that a model as a complex hypothesis about behavior of predator-prey systems is tentatively formulated.

In the next step the model is compared to independent experimental data. The data usually consist of a time series of measurements of predator-prey dynamics in a well defined region. The comparison is used to validate the model, i.e., to assess the reliability of the model, to accept it in a certain sense or to reject it. Validation of a model is also useful as a guide to further refinement of the model and helps define requirements for data.

The fourth step consists of analysis performed on the model in order to uncover its inner qualitative and quantitative properties. Qualitative properties that are investigated include local and global stability. Quantitative properties are found by simulation of normal and perturbed dynamics, analysis of sensitivity to changes in parameters, and functions or structure of the model.

Finally, advice on managing the predator-prey system is offered; the model is used as a tool in further applications or as a building block for understanding more complicated food webs.

In the Lotka-Volterra model we recognize *four factors* that appear in any predator-prey model: net growth of the prey population, which is simply birth minus natural death rate (first factor on the r.h.s. of (1)); predation rate, which is the first part of predator-prey interaction (second factor in the r.h.s. of (1)); growth rate of the predator population as the second part of the interaction (first factor in the r.h.s. of (2)); and the death rate of predator (second factor in the r.h.s. of (2)). All of the four factors in the model have been generalized in various directions and applied to specific predator-prey systems.

One of the very earliest and the most general model, as far as representation by differential equations is concerned, has been analyzed by Kolmogorov (1936). He studied the following system of differential equations: $dN/dt = G_1(N,P)N$, $dP/dt = G_2(N,P)P$, where G_1 is the net specific rate of change of prey and G_2 is the net specific rate of change of predators. Let us investigate a more specific case:

$$\frac{dN}{dt} = F_1(N)\,N - F_2(N)\,P \tag{14.3}$$

$$\frac{dN}{dt} = F_3(N)\,P \tag{14.4}$$

where $F_1(N)$ is the specific growth rate of prey, $F_2(N)$ the specific predation rate and $F_3(N)$ is the specific rate of change of the predator.

The above equations will represent a predator-prey system if functions F_1, F_2, and F_3, satisfy the following three sets of assumptions:

1. $F_1(0) > 0$, $\partial F_1 / \partial N < 0$, $F_1(\infty) < 0$

The assumptions on F_1 tell us that the specific growth rate of a unit of prey population diminishes from positive to negative values as the prey population increases from small values upward. A decrease in F_1 is interpreted as an increase of intrapopulational competition for limited food resources, so that even without the predator, the prey population cannot grow to infinity (i.e. it is bounded from above). In the absence of predators the equilibrium value (N^+) of prey is found upon solving $F_1(N^+) = 0$.

2. $F_2(0) = 0$, $F_2(N) > 0$ for $N > 0$

These assumptions tell us that if prey is extinct the specific predation rate is zero; otherwise it is positive.

3. $F_3(0) < 0$, $\partial F_3 / \partial N > 0$, $F_3(\infty) > 0$

For some (small) values of the prey density, the net specific rate of change of predators will be negative, because of predator starvation, but as prey increases the specific predation rate increases and, for sufficiently large values of prey population, the specific rate of predation becomes positive.

For most applications in aquatic ecology the model has three equilibrium points: (0,0) - total extinction state, $(N^+,0)$ - predator extinction state and (N^*,P^*) - nonextinction state. The stability of this system is easy to analyze by linearizing the equations around each steady state. The state (0,0) is an unstable saddle point. The state $(N^+,0)$ will be an unstable saddle point if $N^+ > N^*$, i.e. when $F_3(N^+) > 0$. This condition is usually met in predator-prey systems. If $N^* > N^+$ the point is a stable knot.

The stability of the nonextinction state is determined by following two parameters (for a complete treatment see Svirezhev and Logofet, 1983):

$$f = (\partial F_2 / \partial N)^* - N^* (\partial F_1 / \partial N)^* - F_1(N^*)$$

and

$$\pi = 4 P^* F_2(N^*) (\partial F_3 / \partial N)^*$$

If $f < 0$ the state (N^*,P^*) is unstable, while if $f > 0$ it is stable. Furthermore, if $f^2 < \pi$ the state is a focus, and if $f^2 > \pi$ it is a knot.

In case the nonextinction state is an unstable focus or an unstable knot, the system possesses a limit cycle. This means that as $t - \infty$, N(t) and P(t) tend to periodic functions. At the same time $(N^+,0)$ is an unstable saddle point. Most of the models of predator-prey systems found in applications to aquatic ecology fall within the class described above and hence can be readily investigated for stability. The only difficulty may arise from the explicit dependence of any F functions on environmental parameters which change periodically. For that case a theorem has been proved (Butler and Freeman, 1981) which says that the predator-prey system with periodic parameters; described by (14.3) and (14.4), tends into periodic behavior as t - ∞.

One of the implicit premises of the predator-prey models described above is that both prey and predator populations may be considered as separate units in the sense that a number or biomass is a good measure of each population. This premise assumes that the age structure in both populations is stationary, i.e., that the ratio between young and old individuals, which may have different food requirements, is constant in time. A closer look at almost every predator population will reveal, however, that the age distribution is not strictly time stationary. For example, for predator populations in aquatic systems, one would almost certainly uncover seasonal differences in the number of young versus the number of old individuals. Furthermore, some zooplankton populations show several stages, each stage having a distinct growth and death rate, while some stages do not show predator behavior at all. Having this in mind, it is clear that if a stage of a population behaves distinctly, it might be necessary to represent it as a separate population, and a balance equation for it should be set up. When necessary, such modifications require several equations to be set up for each population, but the procedure is a straight extension of the predator-prey model discussed above. In general, such extensions are more difficult to analyze and often are untractable analytically, and one is forced to resort to analysis of particular cases via computer simulations.

14.3 ELEMENTS OF PREDATOR-PREY INTERACTION

Predation rate

Each prey population has developed its own way to hide from predators. On the other side predators have developed their ways of getting prey. Thus we expect that the form of predator-prey interaction may be population or species dependent. This is to say that the form of F_2 and F_3 functions may be dependent on both populations involved. There are furthermore obvious reasons to believe that certain properties of the environment that enable prey to hide, or predators to find prey, may modify F_2 and F_3 functions.

Let us look at some of the applied forms of functions to describe the predation rate proposed so far..

The first and most simple representation for the predation rate is binary collision. The specific rate at which prey is captured by a unit of predator is proportional to the concentration of prey:

$$F_2 = a N \tag{14.10}$$

where a is a constant denoting efficiency of predators in a well-defined environment. As we have already seen, this form has been used in the Lotka-Volterra model (equations 14.1 and 14.2). More recently Canale et al. (1976) used binary collision to represent the rate at which prey are filtered by nonselective filter feeders. These filter feeders are unable to vary their filtering rate, so the specific predation rate is proportional to prey density. This formulation has been applied to a well defined range of prey densities. For example, at very high prey densities the *filtering* mechanism of the predator changes and expression (14.10) may not hold any more.

The binary collision form is, however, inadequate to represent other kinds of predators for two simple reasons. First, given sufficient prey, there is no limit to what one unit of predators can eat in a unit of time. This can easily be fixed in a crude way: one can introduce critical concentration so that $F_2 = a N$ for $N < N_{cr}$ and $F_2 = c$ for $N > N_{cr}$. In applications $N << N_{cr}$ is often the case although situations have been reported where $N \geq N_{cr}$. Second, laboratory experiments have shown that in nature predation rate ceases before concentration of prey decreases to zero. This second drawback is also easy to overcome by defining: $F_2 = a (N-N_0)$, where N_0 is a threshold value and $F_2 = 0$ for $N < N_0$. It is however more likely that F_2 is a smooth function in the entire interval $(0, N_{max})$, where N_{max} $(N_{max} > N_{cr})$ is a theoretical maximum density of prey.

A variety of other forms have been fitted to experimental data. All of them share the property that at high concentrations of prey, the specific predation rate reaches a maximum. The most often applied is the Michaelis-Menten form:

$$F_2 = vN / (k + N) \tag{14.11}$$

where v is maximum specific predation rate which is approached as $N - \infty$, and k is the concentration of prey at which the specific predation rate is half the maximum (v/2); k is known as the half-saturation constant. The form is identical to uptake of a nonaccumulative nutrient into phytoplankton, and it has actually been adopted from phytoplankton research.

As a candidate to describe predation rate, the Michaelis-Menten form fits a number of experimental data.

The second most often used form for predation rate comes from fishery research, and it is known as the Ivlev functional response (Ivlev, 1961)

$$F_2 = P_m [1 - \exp(- k N)] \qquad (14.12)$$

where P_m and k are constants. P_m determines maximum specific predation rate while k determines how fast specific predation is approaching P_m as N increases.

Jost et al. (1973) have studied prey-predator interactions with prey *E. coli* and *Axotebacter vinelahdii* and the predator *Tetrahymena pyriformis.* They have found that on one hand a model using the Michaelis-Menten form underestimates measured minimum prey concentrations, and on the other that it predicts much larger oscillations at low dilution times than was experimantally found. They proposed a formulation which fitted their data better and became known as the multiple saturation form:

$$F_2 = \frac{v_m N^2}{(k_1 + N)(k_2 + N)} \qquad (14.13)$$

where k_1 and k_2 are **two saturation constants**. Notice, however, that to achieve a better fit one more constant was used, so obviously a better fit resulted.

Canale et al. (1976), in their model for Lake Michigan, also distinguished between selective filter feeders and raptors. For selective filter feeders they suggested:

$$F_2 = c_1 \frac{c_2 N + c_3}{N + c_3} N \qquad (14.14)$$

where c_2 is the minimum filtering rate multiplier ($c_2 < 1$) and c_3 is the concentration of prey at which the multiplier is equal to $(c_2 + 1)/2$. When N - 0, the maximum filtering rate c_1 is approached.

For representing the predation rate of raptors Canale et al. used the Michaelis-Menten form.

For a compilation of other functions representing specific predation rate the reader is referred to Table 14.1.

In general, a widespread use of the Michaelis-Menten form to represent predation rate can be explained on the basis of the following: first, it provides a reasonable fit to many experimental populations. Second, it has a common sense feature of monotonically approaching maximum predation rate per unit of predator when prey increases, and it uses only two constants with a clear intuitive meaning. Finally, the constants have been determined for several populations. The availability of these constants is an important feature since most ecosystem models lack adequate experimental data on feeding rates of even the most abundant predators and one is forced to use literature values for "similar" species. On the other hand, in many ecological models, due to lack of data, several "similar" species are lumped together into one compartment, so in practice it is impossible to get experimentally determined parameters, and literature values become even more important. Furthermore, predators may have several fairly well defined age stages so that the parameters for one stage may be different than those for other stages. In such cases intra-populational data are needed and they are even more difficult to find.

All of the above complications are just a tiny part of reality in an aquatic ecosystem. This necessitates a deeper experimental study of the feeding of predators if we are to improve the accuracy of their representation and consequently the predictability of the models.

Table 14.1
Proposed functions for description of dependency of predation rate on prey concentration.

F_2^* (specific predation rate)	Source
$a_1 (1 - \exp(-a_2 N))$	Ivlev (1961)
$b_1 N / (b_2 + N)^2$	Holling (1965) Canale et al. (1976)
$c_1 \; 1 - \exp(-c_2 NP^{1-c_3})$	Watt (1959)
$d_1 N^{d_2}$	Rezenzweig (1971)
$e_1 N^2 / (e_2 + N^2)$	Takahashi (1959)
$g_1 N$	Volterra (1927)
$h_1 \; 1 - \exp(-h_2 N^2 P^{(1-h_3)})$	Watt (1959)
$i_1 (N-N_0) / (i_2 + N)$	Steele (1974)
$j_1 (N-N_0)$	Walsh and Dugdale (1971)
$k_1 \; 1 - \exp(-k_2 (N-N_0))$	O'Brien and Wroblewski (1972) Dugdale (1975)
$e_1 (1 - P/l_2)$	Lotka (1924)
$(n_1 N+n_2) / (N+n_3)$	Canale et al. (1976)
$p_1 N^2 / (p_2 + N)(p_3 + N)$	Jost et al. (1973)
$p_1 (-\exp p_2 N)$	Parker (1973)

* Small letter indicate parameters,
N - prey concentration,
P - predator concentration

Growth rate of predator population

In order to link predation rate to an increase in the biomass of predators, a growth rate for predators needs to be formulated. Usually one calculates it as follows:

$$F_3 = \frac{\text{assimilation}}{\text{efficiency}} * \frac{\text{predation}}{\text{rate}} - \frac{\text{respiration}}{\text{rate}} - \frac{\text{exudation}}{\text{rate}}$$

That is, predation rate is multiplied by assimilation efficiency to get the assimilation rate. The respiration rate and exudation rates have been subtracted from the assimilation rate to yield a new growth rate.

Some other researchers have adopted a modified expression:

$$F_3 = \frac{\text{retention}}{\text{efficiency}} * \frac{\text{assimilation}}{\text{efficiency}} * \frac{\text{predation}}{\text{rate}}$$

In case respiration and exudation rates are proportional to predation rate the two expressions may be made identical. Notice that in both expressions assimilation efficiency is made independent of predation rate. Finally, because of a lack of better experimental data, respiration rate and exudation rate are often taken to be functions of temperature only - and this turns out to be inconsistent, namely, because the growth process involves predation rate as its first step. Ingestion is the second step and it is a function of specific predation rate, temperatuure and perhaps type of prey. Assimilation as a next step may be a function only of ingestion rate, temperature and perhaps the type of prey. Finally, net specific growth rate is only a function of assimilation rate and temperature. The above may be put into a general form for specific growth rate:

$$F_3 = G(A(I(F_2(N,T),T),T)) \tag{14.15}$$

where

I = specific ingestion rate

A = specific assimilation rate

G = specific growth rate

T = temperature

Ingestion

After a prey has been captured a part or all of it may be ingested. In modelling studies so far, ingestion rate has been considered as proportional to predation rate. There have been no attempts to consider that perhaps ingestion rate is a nonlinear function of predation rate, i.e., that when predation rate is small the ingested part is greater and that it decreases as specific predation rate increases. Furthermore, there have been no studies showing that, apart from predation rate, ingestion is also a function of temperature. Lacking such evidence we are forced to adopt the following relationship: $I = a F_2$, where 'a' is a prey and predator specific constant. Given a unit of prey, different predators will be able to ingest different parts of the prey. On the other hand, given a unit of predator, a predator will be able to ingest certain parts of one prey and other parts of other prey.

Assimilation

Not all food that is ingested becomes converted into the biomass of the predator, but only the assimilated part. It is defined as the part that is absorbed across the gut wall of an animal. As reviewed by Steele and Mullin (1977), measurements of assimilation efficiency (assimilation/ingestion) are difficult, and results turned out to be quite variable.

Steele (1974) assumed that assimilation is proportional to ingestion ($A = 0.7\ I$). It is more realistic, however, to take assimilation as a decreasing function of ingestion. For example:

$$A = a_1 I\ (a_2 - I/I_m),\quad 0 < I < I_m \tag{14.17}$$

where a_1, $a_2 > 1$ are constants and I_m is some maximal ingestion limited by physical constraints of the predator. Given that $I = I_m$, the maximum value of the specific assimilation rate is $Amax = a_1, a_2\ I_m$. Steele and Mullin (1977) adopted $a_1 = 0.3$, $a_2 = 3$, so that the assimilation may vary between 60% and 90% of food ingested.

Another approach has been used by Canale et al. (1976). They related assimilation to density of prey and predation rate:

$$A = b_1\ [b_2 / (N + b_2)\ ]\ F_2 \tag{14.18}$$

where b_1 is maximum efficiency and b_2 is half maximum assimilation efficiency. When predation rate is proportional to prey density, such an approach is of course equivalent to the statement that assimilation is regulated according to a Michaelis-Menten function. If, instead, the predation is represented by a Michaelis-Menten function, then as N increases from low values, A increases as well. At the value of $N_c = b_2 b_3$ where b_3 is the half-saturation constant in F_2, the assimilation rate reaches the maximum value. Beyond N_c, as $N - \infty$, the assimilation rate decreases.

Metabolic losses

Part of the assimilated food is lost from the animal through metabolic processes. They involve respiration and excretion. Most of the organic matter is lost through respiration.

Respiration and excretion seem to depend in the same manner on the activity of the predator, so they have been lumped together into a single process.

According to the dependence of metabolic loss, M, it is convenient to split the process into three components (Steele and Mullin, 1977):

M_1 - resting or subsistence metabolism, which may be represented by a constant;

M_2 - respiratory loss by means of foraging, selecting, snatching and devouring the prey, which is a decreasing function of prey density;

M_3 - metabolic loss due to assimilation and metabolic transformation of food.

This last part of the metabolic process is obviously a function of assimilation. Therefore:

$$M = M_1 + M_2(N) + M_3(A) \qquad (14.19)$$

The specific growth rate may finally be represented as:

$$F_3 = A - M \qquad (14.20)$$

Unfortunately, in the literature so far, there do not exist sufficiently precise data to allow one to completely determine $M_2(N)$ and $M_3(A)$ functions.

In models of aquatic ecosystems, loss by metabolic processes have been represented in various ways. Canale et al. (1976) assumed it to be independent of density of prey. Steele (1974), in the model of a phytoplankton-zooplankton system, assumed that the metabolic process is proportional to ingestion. This means that M_1 was assumed to be of minor importance.

14.4 DEPENDENCE OF ACTIVITY OF PREDATORS ON ENVIRONMENTAL FACTORS

Temperature

Temperature is the most important factor that affects the activity of predators in aquatic systems. At low temperatures, predatory acitvity is slower and so are metabolic processes. At high temperatures metabolic losses of some predators increase tremendously. It is even hypothesized that some inefficient predators starve at higher temperatures since they are unable to find sufficient food to meet their metabolic needs (Malej, 1984).

Temperature dependence of predatory activity has been often represented in ecosystem models. Most often a linear, a piece-wise linear, an exponential or a bell-shaped graph showing dependence on temperature is used to represent variations in predation rate and in metabolic losses.

Regularly, temperature dependence (T) has been used as an independent factor multiplying the rate of ingestion. For example, metabolic losses as a function of temperature can be given by:

$$M = \Omega (T) (M_1 + M_2 (N) + M_3 (A)) \tag{14.21}$$

However, there is no law saying that, for example, metabolic losses increase exponentially, $\Omega (T) = a_1 \exp(a_a T)$, with a_1 and a_2 as prespecified constants, which is often used. Rather, forms for Ω (T) have been suggested based on experiments on some predators that may, or may not, hold for other predators.

According to Steele and Moullin (1977) a rather *flexible function* has been developed by Belehradék, and it has been suggested by McLaren (1963) as follows:

$$\Omega (T) = a_1 (T + a_2)^{a_3} \tag{14.22}$$

where a_1, a_2 and a_3 are constants. Sometimes, even a simple exponential function

$$\Omega (T) = a_1 Q_x (T/x - 1) \tag{14.23}$$

may well fit the data where a_1 is the rate measured at some prespecified temperature T^*, and Q_x = (rate of the process at T^*+x) / (rate of the process at T^*). The function is defined on (T^*, T^*+x).

Some functions that have been used in the literature are included in Table 14.2.

Turbidity

Turbidity of waters is caused by a high concentration of inorganic or organic particles (such as phytoplankton or detritus). If high turbidity is caused by higher density of inorganic particles, this may hinder the ability of predators to capture prey. It is well known that sieves of filter feeders may become clogged at high turbidities. On the other hand, it has been shown that algae form aggregates with clay particles (Avnimelech et al., 1983); hence, in turbid waters, filter feeders may feed more heavily on algae that are otherwise too small to be retained on their filters.

In ecological models, turbidity has not been used so far as a factor which may affect predation or grazing rate, although it would not be difficult to incorporate this feature. It is expected that predation rate will be smaller if turbidity is caused by inorganic particles and furthermore that selectivity coefficients may be changed as opposed to values in less turbid waters.

Table 14.2
Source of proposed functions for description of dependency of digestion rate on temperature

Ω (T) *	Source
$K_{opt}\ (T/T_{opt})^n \exp(1-T/T_{opt})^n,\ 0<T<T_{opt}$	Lamanna and Malette (1965)
$K_{opt}\ (1-(T-T_{opt})^m / (T_{max}-T_{opt})^m),\ T_{opt}<T<T_{max}$	
aT	Parker (1973)
$K_{opt} \exp(-a(T_{opt}-T)^2 / (T_{opt}-T_{min})^2),\ T<T_{opt}$	Lehman et al. (1975)
$K_{opt} \exp(-a\ (T-T_{opt})^2 / (T_{max}-T_{opt})^2),\ T>T_{opt}$	
piece-wise linear	Canale et al. (1976)
$K_{opt} \exp(-b\ (T-T_{opt}))$	Jørgensen (1976)
$A_{max} \exp(A_{10}\ T)$	Jørgensen (1976), Kremer and Nixon (1978), Kelly (1975)
$K_1 K_2^{T-20}$	Chen and Orlob (1975)

* Values for most of the parameters may be found in Jørgensen (1979).

14.5 PREDATION ON MORE THAN ONE PREY POPULATION

Assume that a predator population feeds on two types of prey which have densities N_1 and N_2 respectively. Let the specific predation rate, denoted as $s_1^{1,1}$ on the first prey population in a one predator-one prey population experiment follow the Michaelis-Menten expression:

$$\text{specific predation on } N_1 = s_1^{1,1} = \frac{v\ N_1}{k + N_1} \qquad (14.24)$$

Concerning predation of the predator on the second prey population, assume that predation again follows the Michaelis-Menten form with the same v and k constants. The question is: If both populations are offered to the predator, what would be the predation rate on the first prey, second prey, and the total specific predation rate?

Since the **total population** offered to the predators is $N_t = N_1 + N_2$, the total specific predation rate $s_t^{2,1}$ in the one predator-two prey system is:

$$\text{total specific predation rate of P} = s_t^{2,1} = \frac{v\ N_t}{k + N_t} \qquad (14.25)$$

Specific predation rate on i-th prey in one predator-two prey populations system is:

$$\text{specific predation rate on i-th prey} = s_i^{2,1} = \frac{v\ N_i}{k + N_t} \qquad (14.26)$$

When more than two prey populations are present, the expressions (14.25) and (14.26) still hold with the modification that

$$N_t = \sum_{i=1}^{n} N_i$$

It is important to note that expressions (14.25) and (14.26) hold only if v and k are constants having the same value for every prey population, if there are no other predators competing for the same prey, and if there is no preference of predators for a certain prey population.

To generalize expressions (14.25) and (14.26), when $v_i \neq vj$ and $k_i \neq kj$ for $i \neq j$, we need to discuss the underlying model which led to expression (14.24). This model was put forth by Holling (1959) and Rashevsky (1959). The model states that the number of prey eaten, S, in time interval T is the product of prey density, N, the rate of capturing prey while the predator is searching, a, and the time spent in searching. The time interval T is the split into two parts: time spent searching, and the time spent in consuming prey, bS.

Number of prey eaten in the time interval T is then:

$$S = (T - bS)\, aN \qquad (14.27)$$

or

$$\frac{S}{T} = \frac{aN}{1 + abN} \qquad (14.28)$$

By passing to the limit, we get that the rate is given by expression (14.24) where $v = 1/b$ and $k = 1/ab$.

Murdoch (1969) extended the model into the case of one predator and two prey. Assume that rates of capturing each of the two preys a_1 and a_2, and handling times of preys are b_1 and b_2, then:

$$S_1 = (T - b_1S_1 - b_2S_2)\, a_1N_1 \qquad (14.29)$$

$$S_2 = (T - b_iS_1 - b_2S_2)\, a_2N_2 \qquad (14.30)$$

First of all it follows that: $S_1/S_2 = a_1N_1/a_2N_2$. A feature that is not seen in one predator-one prey system is preference. Preference may appear when more than one prey is offered to the predator. In that case the true attack rates are:

$$a_i^t = w_i\, a_i \ , \ i = 1,2 \qquad (14.31)$$

where $0 \leq w_i \leq 1$, $i=1,2$ are preferences. Obviously w_i can not be greater than one because no predator can do better in attacking the prey than in the one predator-one prey case. In case $w_i=1$, $i=1,2$ there is no selection.

When a_i^t is used instead of a_i in (14.29) and (14.30) we get:

$$s_i^{2,1} = \frac{w_ia_iN_i}{1 + w_1a_1b_1N_1 + w_2a_2b_2N_2} \quad i=1,2 \qquad (14.32)$$

In case of one predator feeding on n prey populations (n > 2) specific predation rate on the i-th prey population in the presence of all other prey populations is given by:

$$s_i^{n,1} = \frac{w_iv_iN_i / k_i}{1 + \sum\limits_{l=1}^{n} w_lN_l / k_l} \qquad (14.33)$$

where v_i, k_i are determined from corresponding predator-i-th prey experiments. The expression (14.33) can be further generalized to general food webs with any number of trophic levels and any sharing of prey by predators (Legovic, 1986a). While, some v_i and k_i have been determined from predator-prey experiments, others are exceedingly difficult to measure. It is, therefore, interesting to mention that v_i and w_i/k_i can also be determined from field observations for populations in lower trophic levels. One may consult Vanderploeg and Scavia, 1979.

14.6 EXPERIMENTAL PARAMETERS AND FUNCTION ESTIMATION OF PREDATOR-PREY INTERACTIONS

In recent years research on zooplankton and fish feeding as a function of prey concentration and environmental conditions has been intensified. Whereas a decade ago studies on zooplankton feeding, assimilation and respiration in terms of fluxes were relatively rare, today the research has matured into a quantitative form and the emerging results are readily becoming available for modelling studies.

In what follows we shall mention several studies to give the reader a feeling for the quality of results available today.

In an excellent series of papers Vidal (1980 a, b, c, d) investigated the effects of phytoplankton concentration, temperature and body size on specific growth rate, development and molting rate of *Calanus pacificus* and *Pseudocalanus sp.* and on metabolic rates and net productivity efficiency for *Calanus pacificus.* He reports that the mean dry weight of later stages of the animals increases hyperbolically with food concentration and is inversely related to temperature. The study suggests that the extrapolation of growth rates from on species to another on the basis of similarty in body size is not justified, and furthermore, that growth rate is strongly influenced by temperature. The mean development time decreases hyperbolically with increasing food concentration. The development of early stages is more dependent on temperature than that of later stages (and is delayed at low temperature). Stage duration is not constant but increases logarithmically with increasing body weight. In general, the rate of loss of carbon by molting is low, but it increases hyperbolically with food concentration and linearly with growth rate. As body size increases, growth rate tends to become independent of temperature while development rate remains proportional to temperature in a suboptiomal range. Studies by Vidal show that the relationship between body size and metabolic rate is independent of either temperature or food concentration. The respiration rates were about twice the values reported for unfed, closely related species. Furthermore, he reports on the logarithmic increase in body weight (W), i.e. that the following relationship holds: $R=a\ W^{b}$, where a and b are tempera- ture dependent parameters.

Deason (1980) performed grazing experiments with *Acartia hudsonica* feeding on *Sceletonema costatum.* The ingestion rate was best fitted by the Ivlev equation. The minimum feeding concentration of prey was not significantly different from zero. Maximum filtration rate increased exponentially up to 14-15 °C and then began to decline. Ingestion rate was also an increasing function of temperature.

Downing and Peters (1980) found that cladocera feeding rate may be described by: $I = (1.139 + 1.166L - 0.104N)^4$, where L = body length, N = concentration of algae.

Barthel (1983) found that weight-specific ingestion rate of *Eurytemora affinis* increased with increasing food concentration of algae *Nannochloris sp.* according to the Ivlev curve ($I = I_m \{1 - \exp[-a (N-N_0)]\}$) although the Michaelis-Menten curve with threshold ($I = I_m (N-N_0) / (k_s + N-N_0)$) gave a slightly, but insignificantly, worse fit (N_0 - is the threshold concentration).

In an experiment on feeding juvenile salmon on zooplankton, English (1983) found that net growth rate (G) could be related to zooplankton density by: $G = G_{max} [N / (1+aN)] - b$, where G_{max}, a and b are constants depending on type of prey.

It is quite certain that in the near future there will be a larger number of well documented studies of predation, growth and respiration with regard to their dependencies on environmental factors for the few most abundant species of predators. In turn, this will allow more precise representation of predators in ecosystem models.

14.7 APPLICATIONS: PREDATOR-PREY INTERACTIONS IN ECOSYSTEM MODELS

In models of ecological systems predator-prey interactions have often been included. Reviews of such models and comparison of different concepts they employ may be found in Swartzman and Bentlsy (1979) and Jørgensen (1988). Literature models include from one to ten predators and even more prey populations. Although experimantal data have been included in each of the calibrated models, many parameters may better be described as being within certain (not always narrow) bounds rather than strictly determined.

Published models tend to fit data on predators reasonably well; however, the fit is not unique in the sense that often more than one combination of parameters gives equally reasonable fit. Each set of such parameters may lead to different perturbation dynamics, i.e., different predictions. This, however, is not to say that the models have no predictive value, as would be the case in an interpolation fit. Rather, calibrated and validated models are the most consistent tools available today to judge certain management alternatives. Furthermore, ecosystem models have served yet another very useful purpose, and that is to suggest modification of monitoring studies and needed laboratory experiments in order to explain better the functioning of aquatic ecosystems as a whole.

To some extent there has been an artifical division between scientists investigating properties of predator-prey models and others studying ecosytems models. It has been shown in some examples that qualitative behavior of predators is similar in both types of models (Legovic, 1978). Often research on predator-prey models reveals facts that are later incorporated in larger ecosystem models.

In general predator-prey interactions involving lower trophic levels tend to be better represented. Consequently, the predictive value or reliability of models seems to be higher for lower trophic levels.

14.8 CONCLUSIONS AND RESEARCH NEEDS

Predator-prey interaction is the first interaction between populations that is considered for implementation when an ecological model is developed. As such it has been extensively studied. However since interactions between populations are quite complex neither all aspects nor all effects of environmental varability on this interaction have been studied. As we have said above, the accuracy of representation of this interaction decreases from lower to upper trophic levels. This is because of inaccuracy either of parameters or mathematical formulations, or of grouping significantly different populations into the same variable. All of these reasons can actually be represented by a single one: lack of accurate data. Much of the present research needs include accurate measurements of selectivity constants and accurate synoptic field observations of predatory rates and concentrations of both prey and predator populations.

ACKNOWLEDGEMENT

This research is supported by Comm. for Scientific Research of SR Croatia, Yugoslavia and grants from US NSF BSR-8114823 and BSR-8215587.

REFERENCES

Avnimelech, Y. et al., 1982. Mutual floculation of algae and clay: evidence and implications, Science 216, 63-65.

Barthel, K.G., 1983. Food uptake and growth efficiency of *Euytemora affinis* (Copepoda: Calanoida), Mar. Biol. 74, 269-274.

O´ Brien, J.J. and J.S. Wroblewski, 1972. An ecological model of the lower marine trophic levels on the continental shelf of west Florida, Tec. Rep., Geoph. Fluid Dyn. Inst., Florida State Univ., Tallahassee, 1708.

Butler, G.J. and H.I. Freeman, 1981. Periodic solutions of a predator-prey system with periodic coefficients, Math. Biosci., 55, 23-38.

Canale, R.P. et al., 1976. Plankton based food web model for lake Michigan, In

Modeling Biochemical Porcesses in Aquatic Ecosystems, (Canale, R.P., ed.), Ann Arbor Sci., Ann Arbor, 33-74.
Chen, C.W. and C.T. Orlob, 1975. Ecological Simulation for Aquatic Environments. In System Analysis and Simulation in Ecology (Patten, B.C., ed.), Vol. 3, Academic Press, New York, 475-588.

Deason, E.E., 1980. Grazing of *Acartia hudsonica* (A. clausi) on *Skeletonema costanum* in Narragansett Bay (USA): Influence of food concentration and temperature, Mr. Biol. 60, 101-113.
Downing, J.A. and R.H. Peters, 1980. The effect of body size and food concentration on the in situ filtering rate of *Sida crystallina,* Limnol. Oceanogr., 25, 883-895.
Dugdale, R.C., 1975. Biological Modeling. In Modeling of Marine Systems, (Nihoul, J.C., ed.), Elsevier, Amsterdam, 187-206.

English, K.K., 1983. Predator-prey relationships for juvenile chinook salmon *Oncorhynchus tsawytscha,* feeding on zooplankton in in situ enclosures, Can. J. Fish. Aquat. Sci., 40, 287-297.

Holling, C.S., 1965. The functional response of predators to prey density and its role in mimiery and population regulation, Mem. Eutomol. Soc. Can., 45, 1-60.
Holling, C.S., 1959. The components of predation as revealed by a study of small-mammal predation of the European pine sawfly. Can. Ent. 91, 293-320.

Ivlev, V.S., 1961. Experimental Ecology of the Feeding of Fishes, Yale Univ. Press; New Haven.

Jost, J.L. et al., 1973. Interactions of *Tetrahymena pyriformis, Escherichia coli, Azobaster vinelaudi* and glucose in minimal medium, J. Bacteriol., 113, 834-840.
Jørgensen, S.E., 1976. An eutrophication model for a lake, Ecol. Model., 2, 147-165.
Jørgensen, S.E. (ed.), 1979. Handbook of Environmental Data and Ecological Parameters. ISEM, Copenhagen, 1162.
Jørgensen, S.E., 1988 Fundamentals of Ecological Modelling. Elsevier, Amsterdam.

Kelly, R. The Delaware Estuary, In Ecological Modeling in a Resource Management (Russell, C.S., ed.), Resources for the futrure, Inc., Washington, D.C., 1975, 394 p.
Klomogorov, A.N., 1936. Sulla teoria di Volterra della lotta per l´esistenze, Giorn. Instituto Ital. Attuari., 7, 74-80, (Translated into English in Sculo, F. and J.R. Ziegler, the Golden Age of Theoretical Ecology, 1923-1940, Springer, Berlin, 1978).
Kremer, J.N. and S.W. Nixon, 1978. A Coastal Marine Ecosystem, Springer Verlag, Berlin, 217 p.

Lamanna, C. and M.F. Malette, 1965. Basic Bacteriology, Williams and Wilkins Co., Baltimore.
Lassiter, R.R. and D.K. Kearns, 1974. Phytoplankton population changes and nutrient fluctuations in a simple aquatic ecosystem model, In Modelling the Eutrophications Process (Middle Brooks, E.J. et al., eds.), Ann. Arbor Sci., Michigan.
Legovic, T., 1978. From population to ecosystem models, Period. Biol., 80, 149-161.
Legovic, T., 1986a. Predation in food webs (in prep.)
Legovic, T., 1986b. Determination of parameters in models of food webs.

Lehman, J.T. et al., 1975. The assumptions and rationales of a computer model of phytoplankton population dynamics, Limnol. Oceanogr. 3, 343-364.
Lotka, A.J., 1924. Elements of Mathematical Biology, Reprinted 1956, Dower, New York.

Malej, A., 1984. Metabolism of *Pelagia noctiluca* in the Gulf of Trieste, workshop on Jelly-fish in the Mediterranean sea, Trieste.
McLaren, I.A., 1963. Effects of temperature on growth of zooplankton and the adaptive value of vertical migration. J. Fish. Res. Biol. Can., 20, 685-727.
Murdoch, W.W., 1973. The functional response of predators. J. Appl. Ecol., 10, 335-342.

Parker, R.A., 1973. Some problems associated with computer simulation of an ecological system, In the Mathematical Theory of the Dynamics of Biological Populations, (M.S. Bartlett and R.W. Hiorns, Ed.), 269-287, Academic Press, London.
Pepita, T.S., 1966. On the energy balance of *Calanus helgolandicus* (Claus) in the Black Sea, In Physiology of Marine Animals. Akad. Nauk. S.S.S. 12, Oceanographical Comm., Sci. Pub. House, Moscow, 60-81.

Rashevsky, N., 1959. Some remarks on the mathematical theory of nutrition of fishes. Bull. Math. Biophys. 21, 161-183.
Rosenzweig, M., 1971. Paradox of enrichment: destabilization of exploitation ecosystems in ecological time, Science, 171, 385-387.

Steele, J.H., 1974. The Structure of Marine Ecosystems, Harvard Univ. Press, Cambridge, Mass., 128 p.
Steele, J.H. and M.M. Mullin, 1977. Zooplankton dynamics, In the Sea (E.D. Goldberg et al., eds.), Vol. 6, John Wiley and Sons, New York, 857-890.
Svirezhev, Yu.M. and D.O. Logofet, 1983. Stability of biological communities, Mar. Pub., Moscow.
Swartzman, G.L. and R. Bentley, 1974. A review and comparison of plankton simulation models, ISEM journal, 1, 30-81.

Takahashi, F. Reproduction curve with two equilibrium points: A consideration on the fluctuation of insect populations, Res. Pop. Ecol., 6, 28-36.

Vanderploeg, H.A. and D. Scavia, 1979. Calculation and use of Selectivity coefficients of feeding: zooplankton grazing, Ecol. Model. 7, 135-149.
Vidal, J., 1980a. Physioecology of zooplankton I. Effects of phytoplankton concentration, temperature, and body size on the growth rate of *Calanus pacificus* and *Pseudocalanus sp.* , Mar. Biol., 56, 111-134.
Vidal, J., 1980b. Physioecology of zooplankton II. Effects of phytoplankton concentration, temperature, and body size on the development and molting rates of *Calanus pacificus* and *Pseudocalanus sp.* , Mar. Biol., 56, 135-146.
Vidal, J., 1980c. Physioecology of zooplankton III. Effects of phytoplankton concentration, temperature, and body size on the metabolic rate of *Calanus pacificus* , Mar. Biol., 56, 195-202.
Vidal, J., 1980d. Physioecology of zooplankton IV. Effects of phytoplankton concentration, temperature, and body size on the net production efficiency of *Calanus pacificus* , Mar. Biol., 56, 203-212.
Volterra, V., 1927. Variazione e fluttuazioni del numero d'individui in specie animali conviventi, R. Com. Talassografico Italiano, 142 p.

Walsh, J.J. and R.C. Dugdale, 1971. Simulation model of the nitrogen flow in

the Peruvian upwelling system, Invest. presq., 35, 309-330.
Watt, K.E.F., 1959. A mathematical model for the effect of densities of attached and attaching species on the number of attached, Canad. Eut., 91, 129-144.

CHAPTER 15

PRIMARY PRODUCTIVITY

by
Carol D. Collins and Richard A. Park

15.1 INTRODUCTION

Aquatic primary productivity encompasses the growth (accumulation of biomass) of both phytoplankton and macrophytes - the basis for most aquatic ecosystems. It is of particular concern in water quality management because uncontrolled growth can cause taste and odor problems, unsightly algal scums, oxygen depletion in deeper waters, and aquatic weed beds that impede recreational and navigational uses of water bodies.

The processes that contribute to primary productivity exhibit complex environmental relationships that are often difficult to measure and that do not lend themselves to simple, intuitive analysis. For this reason, simulation models have been used for the past twenty years to provide insights into the mechanisms of primary productivity and to provide predictions of changes in productivity based on changing environmental conditions.

Although measurement of primary productivity involves direct or indirect measurement of photosynthesis or the resulting biomass, in order to model productivity one must usually consider the loss terms as well: respiration, excretion, non-predatory and predatory mortality, and sinking. Equations for all these processes will be given below, based on three models:
MINI.CLEANER (Park et al., 1985), CE-QUAL-R1 (Wlosinski and Collins, in press) and LAKETRACE (Park, in prep.).

Description of Photosynthesis

Photosynthesis is *the fixation* of organic material by plants and *cyanobacteria* (blue-green algae); it is usually represented by the equation

$$6\ CO_2 + 6\ H_2O \xrightarrow{\text{light}} C_6H_{12}O_6 + 6\ O_2$$

The biosynthetic consequences is the light-mediated conversion of CO_2 to organic cell material.

In the first phase of the photosynthetic reaction, solar energy is converted into the chemical energy of ATP. Hydrogen atoms are removed from water molecules and used to reduce $NADP^+$, leaving behind molecular oxygen. The energy-rich products of the first phase are then used as energy sources to bring about the reduction of carbon dioxide to yield glucose in the second phase. The first phase is referred to as the light reaction, and the second phase as the dark reaction of photosynthesis.

The bioenergetics of the photosynthetic reaction seem less significant to the ecological modeller than the quantification of end-products of the reaction (e.g. oxygen and cell mass). Indeed, in an aquatic system, the reductant, water, can be considered as an unlimited resource; and in most systems the carbon source is available as carbon dioxide or bicarbonate. Therefore, in the above equation the light energy needed for the reaction to occur is the primary factor regulating production. Additional factors not included in the equation include macronutrients, and temperature.

15.2 METHODS AND MEASURING

Methods for measuring productivity (see Harris, 1978) impose *temporal and spatial constraints* that must be considered in data interpretation. Under natural conditions phytoplankton respond to many environemtal factors of differing scales.

Circadian rhythms can be observed in productivity rates and cell division. Fluctuations in the external environment impose changes on these physiological processes. External factors affecting productivity include irradiance levels, temperature, and nutrient concentration. Fluctuations in these factors result from cloud cover, vertical mixing processes, settling, and external loadings of nutrients and suspended solids.

Fine-resolution temporal and spatial variations in productivity resulting from the interactions of physiological, environmental and mechanical factors are almost impossible to determine using currently available methods of measurement. Consequently, mathematical models have been structured to be compatible with the methodological constraints.

Productivity Rates

A common method for measuring productivity rates is to measure changes in dissolved oxygen concentrations resulting from photosynthetic activity in bottles containing phytoplankton, macrophytes or macrophyte fragments. This technique has significant temporal limitations. Chemical analysis of oxygen evolved in the bottles often requires long incubation periods (up to 24 hours) for accurate determination. Membrane-covered volumetric (polarographic) sensors for dissolved oxygen determinations have an advantage over chemical techniques in that they provide a continuous record. However, the ability to provide a long-term record may decrease due to accumulation of debris on membrane surfaces that reduces the electronic sensitivity. Furthermore, most studies have utilized high cell densities for this method - densities that may generate artifacts in the data.

A dark bottle (opaque to light) is incubated in parallel with a light (transparent) bottle to determine dark respiration rate. Dark respiration is an energy-producing metabolism of complex carbon compounds which generates CO_2 and consumes O_2.

Assuming that the rate of **dark respiration** is the same in the dark as in the light, the rate of productivity is calculated as the difference between the light and dark bottles.

Carbon uptake, measured as the assimilation of dissolved inorganic carbon into cell matter is another method for measuring photosynthetic rate. The most common technique measures the incorporation of radioactive inorganic ^{14}C into organic material. After incubation with labeled dissolved inorganic carbon, samples are filtered and counted for radioactivity. The filtrate is then acidified and purged to remove unfixed ^{14}C for determination of excretion rates. The incubation time for the ^{14}C technique varies from 2 to 8 hours, although shorter incubation periods have been tested (Lewis and Smith, 1983).

It is a point of controversy as to whether gross or net photosynthetic rate is measured by this technique. Respired CO_2 can be reassimilated during longer incubation periods; if there were no reassimilation of endogenous CO_2 then the ^{14}C technique would measure gross photosynthetic rate. However, it may represent something between net and gross photosynthesis because extracellular material is often labelled with ^{14}C. This release would underestimate uptake rates. **Photorespiration** is a process that diverts light-induced reducing power from the biosynthesis of glucose into the reduction of oxygen. Nearly 50% of the total carbon fixed in the photic zone of a water column may be liberated as the water-soluble organic form (Watt, 1966).

Inaccuracies are magnified by the diurnal pattern of photosynthetic activity. Peak production rates vary and have been observed in the morning (Anderson, 1974; Ganf and Horne, 1975) and in the afternoon (Anderson, 1974). Depression in rates follows the observed maximum. Depending on time of observation and the incubation period, the measured photosynthetic rate may significantly underestimate or overestimate the daily rate.

The mathematical modeller must also consider that *the time scale* that the model uses may be incompatible with the measured data. The most common time scale used in ecosystem models is 24 hours. Photosynthetic rates measured for incubation periods of 3 to 6 hours and extrapolated to daily rates by incorporation of the photoperiod are adequate for models using a 24-hour simulation time step, subject to the qualifications expressed above. If a 1-hour time step is used, careful interpretation of the data is necessary to prevent significant under- or overestimation in the model.

Spatially integrated sampling (e.g. 0 to 10 m) is best used with segmented models such as MINI.CLEANER that can treat the sampling zone as a separate segment. Discrete sampling is best used with models such as CE-QUAL-R1 that use many layers. However, neither sampling method accounts for the effects of vertical mixing on productivity (cf. Lewis et al., 1984). Residence time of algal cells at a particular depth influences the photoadaptation rate and photosynthetic response.

Biomass

The biomass of the primary producers is the variable often of greatest interest to the mathematical modeller. Mathematical models predict dry weight concentration or chlorophyll **a** concentraion. In situ measurements of phytoplankton or macrophyte mass are commonly measured as chlorophyll **a** concentration, cell volume and numbers are poor measures because of intraspecific variability in cell size (Smayda, 1978). Two additional sources of error create significant problems in the comparison of phytoplankton biomass data with predictions in point and one-dimensional models. Advection may cause horizontal patchiness; hundred-fold differences in biomass have been observed on two sides of a small reservoir due to wind-driven currents (D.G. George, personal communication). And chlorophyll **a** varies inversely with light intensity as an exponential function in phytoplankton other than diatoms (Myers, 1946; Nielsen and Jørgensen, 1968). Chlorosis of blue-green algae which often occurs in the top meter of water reduces the chlorophyll **a** to biomass ratio and may result in a significant underestimation of algal biomass.

15.3 MODEL COMPLEXITY

Algal models, like other models, should have the minimum level of complexity consistent with the objectives. The effort expended in programming, calibration, and evaluation increases geometrically with complexity. In the past when programming a model, it was usually easier to simplify code at a later date than to make it more complex. However, modular, structured code that is being produced now by most modelling groups can be modified with relative ease.

Temporal and Spatial Scales

The *temporal* scale of a model can vary by several orders of magnitude. Nutrient uptake and photosynthesis occur at a scale of minutes to hours. Few models try to capture the detailed dynamics of these processes; those that do must represent a host of diurnally varying environmental factors and biotic processes. At the other end of the scale, if seasonal levels of chlorophyll **a** are all that is needed, simple regression models may suffice (Reckhow and Chapra, 1983). Most mechanistic models have daily time steps and are run for the growing season or for a year or more; they are appropriate for representing weekly changes in algal biomass.

Spatial scales also vary depending on the objectives and data. Oftentimes smaller lakes are represented by vertically linked point models for the epilimnion and hypolimnion; LAKETRACE is typical of this type of model. Larger lakes may be represented by both laterally and vertically linked point models. Riverine segments may be linked with reservoir segments, and nearshore eutrophic segments may be linked with offshore mesotrophic segments, to give two examples. In this way, Lakes Erie and Ontario were modelled using 67 segments each (DiToro et al., 1975; Thomann et al., 1976). If high resolution in either the vertical or, less often, the horizontal direction is required, then a one-dimensional model such as CE-QUAL-R1 may be used.

Two-dimensional water quality models (such as CE-QUAL-R2) are used for situations where complex currents present particular problems; however, these usually sacrifice biologic realism for hydrodynamic realism due to computational loads.

State-Variable Complexity

How many algal groups should be modelled? Storage requirements increase approximately linearly with increasing number of state variables.

On mini- and mainframe computers this is not a constraint, but with many models now being implemented on microcomputers, stroage requirements are critical. Furthermore, it is difficult to obtain well-documented parameter values for many groups. Although many models have only one phytoplankton group, these models generally are not able to reproduce both summer and winter biomass levels, or are applicable for only one lake type or season. Most simulations require at least two phytoplankton groups, for example N-fixing blue-greens and other phytoplankton, in order to capture seasonal dynamics.

Scavia (1980) models five groups (N-fixing blue-greens, small diatoms, large diatoms, small others and large others). Many models incorporate four phytoplankton groups, including N-fixing and other blue-greens, large and small diatoms, greens and/or flagellates (for example, Lehman et al., 1975; Canale et al., 1976; Scavia et al., 1976; Park et al., 1985). The use of four or five groups confers generality because one group can replace another as conditions change (through time or from one lake to another). However, adaptive process constructs could achieve the same purpose with appreciable savings in storage requirements and computational load. In the following formulations only temperature response is adaptive, although adaptive light constructs have been proposed (for example, Kremer and Nixon, 1978; Liou and van Eybergen, 1982; Lewis et al., 1984).

An alternative approach is to model algae as a self-adjusting compartment. Straskraba (1979) has done this using an algorithm that changes cell size, optimizing nutrient uptake, self-shading, susceptibility to predation, and sinking. Because optimization routines greatly increase the computational load, at present this approach seems more useful as a theoretical tool, providing insights into ecologic strategies, than as an applied tool.

Process-Level Resolution

Mechanistic models by definition have at least moderately complex process equations with numerous parameters. LAKETRACE is the simplest ecosystem model that we would want to employ for most management problems, yet the phytoplankton equations (for two groups of phytoplankton) require 32 parameter values. A less general ecosystem model would require fewer, and a simple, empirical model would require far fewer parameters. However, a general mechanistic model does not have to be re-calibrated for each site. We have been pleased to find that our models seldom require different parameter values, even when simulating greatly differing lakes and reservoirs. Ironically, simpler mechanistic models, while having fewer parameters, require greater calibration effort because the parameters have to be fine-tuned for specific site conditions. Calibration of a complex model during development may be difficult because of the numerous parameters, but application of the model to other sites should be easier because of the generality of the parameters.

It appears that there is a threshold in model complexity beyond which process resolution and state-variable resolution are represented in enough detail that parameters are of general application. For example, a formulation for temperature limitation that takes into account adaptation to temperature extremes permits the same model with similar or identical parameter values to be used for both temperate and tropical lakes. We refer to such general parameters as "global parameters". The existence of global parameters greatly simplifies the site-specific application of primary productivity models and justifies the additional computational load imposed by more complex formulations.

If a model is intended to be specific for a particular lake or class of lakes, then simplifications may be possible. For example, if a lake is always turbid then photoinhibition can be ignored and a simpler light-limitation equation can be used, or if temperature does not vary appreciably, as in a maritime or tropical climate, then an exponential temperature-limitation equation can be used instead of a more complex saturation equation.

15.4 FORMULATIONS

In the following formulations parameters are indicated by lower case and driving variables and site constants are underlined. These formulations are based on our experience with the **CLEANER** series of models, especially **MINI.CLEANER** (Park et al., 1985); **CE-QUAL-R1**, developed by the U.S. Army Corps of Engineers (Environmental Laboratory, 1982; Wlosinski and Collins, in press); and **LAKETRACE** (Park, in prep.).

Algal Photosynthesis

Photosynthesis is usually represented as a function of multiplicative factors for light, nutrients, and temperature (Park et al., 1985):

$$PPROD = pmax * UI * NUTLIM * TEMP * BIOMASS \qquad (15.1)$$

where

PPROD = phytoplankton photysynthesis (g/m^3/day)
pmax = maximum photosynthetic rate (g/g/day)
UI = light limitation [Eq. 15.8] (unitless)
NUTLIM = nutrient limitation [Eq. 15.4] (unitless)
TEMP = temperature limitation [Eq. 15.2] (unitless)
BIOMASS = phytoplankton biomass (g/m^3)

A detailed explanation of the components of phytoplankton photosynthesis follows.

Temperature Limitation

The ambient temperature exerts a strong effect on the rate at which many biological processes occur. Because a process will proceed at its maximum rate only when the temperature is optimal, thermal effects must be accounted for in the calculation of actual process rates. The reduction factor for non-optimal temperature, TEMP, developed by O'Neill et al. (1972) and modified slightly by Bloomfield et al. (1973) for application to aquatic ecosystems, reflects the response of organisms to variations in temperature (Fig 15.1). This function is:

$$\mathrm{TEMP} = \mathrm{VT}^{(\mathrm{XT} * \exp(\mathrm{XT} * (1.0 - \mathrm{VT})))} \qquad (15.2)$$

where
TEMP = reduction factor for non-optimal temperatures
VT = (tmax - $\underline{T}$) / (tmax - topt)
XT = (WT * (1.0 + SWRT(1.0 + 40.0/WT)))^2 / 400.0
WT = LN (q10 * (tmax - topt)
tmax = maximum temperature at which the process will occur (°C)
topt = optimum temperature (°C)
q10 = rate of change per 10°C temperature rise (unitless)
$\underline{T}$ = ambient temperature (°C)

Most organisms have the capacity to become acclimated to variations in temperature, which effectively alters the optimum and maximum temperatures for each process (Nicol, 1960; Kinne, 1970). A function to represent this acclimation, as a translation of the temperature response curve, was developed by Kitchell et al. (1972). To calculate the magnitude and direction of this acclimation, the difference between the temperature and a specified reference temperature is calculated, giving:

if $\underline{T}$ > tref: OF = ofmax * (1.0 - EXP (-ktemp * ($\underline{T}$ - tref))) if $\underline{T} \leq$ tref: OF = -ofmax (1.0 - EXP (ktemp * ($\underline{T}$ - tref)))

where

$\underline{T}$ = ambient temperature (°C)
tref = reference temperature (°C)
ofmax = maximum temperature acclimation (°C)
ktemp = coefficient for decreasing acclimation as temperature approaches reference temperature (unitless)
OF = temperature translation (°C)

The value of OF approaches that of the maximum possible translation as the difference between the average and reference temperatures increases, reflecting the imperfect acclimation of the organism. The values of OF is then added to the specified TMAX and TOPT parameters, which are then used in the calculation of the correction factor for temperature.

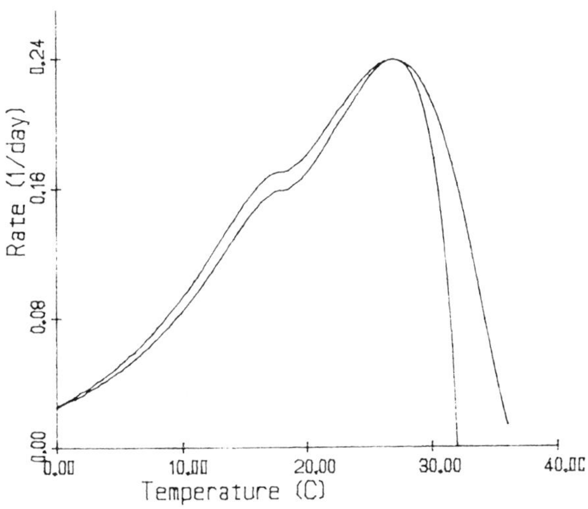

Fig. 15.1: Photosynthesis as a function of temperature, with differing maximum temperatures; from Park et al., 1985.

An alternate formulation (Thornton and Lessem, 1978) that produces a similar response curve (without acclimation) is:

$$R(T) = 0 \qquad \text{if } \underline{T} \leq t_1$$

$$R(T) = K_1 * \text{EXP}(\text{LAMBDA}_1 * (\underline{T} - t_1)) \, (1 + k_1 * (\text{EXP}(\text{LAMBDA}_1 * (\underline{T} - t_1)) - 1)) * k_4 * \text{EXP}(\text{LAMBDA}_2 * (t_4 - \underline{T})) / (1 + k_4 * (\text{EXP}(\text{LAMBDA}_2 * (t_4 - \underline{T})) - 1)) \qquad \text{if } t_1 < \underline{T} < t_4$$

$$R(T) = 0 \qquad \text{if } T \leq t_4$$

where

$$LAMBDA_1 = 1 / (t_2 - t_1) * LN (k_2 * (1 - k_1) / (k_1 * (1 - k_2)))$$
$$LAMBDA_2 = 1 / (t_4 - t_3) * LN (k_3 * (1 - k_4) / (k_4 * (1 - k_3)))$$

The parameters t_1 and t_4 are the temperatures below and above which mortality occurs; t_2 and t_3 define the optimum range of temperature. The parameters k_1 and k_4 are multipliers.

CE-QUAL-R1 (Wlosinski and Collins, in press) assumes values of 0.98 for k_2 and k_3. The maximum reaction rate supplied by the user is multiplied by the reduction factor R(T) to determine k_2 rates corresponding to the local water column temperatures on a layer-by-layer basis.

Nutrient Limitation

Two principal ways of modelling nutrient limitation are used in current models. Growth can be expressed as limited by external nutrient concentrations or by internally stored nutrients. We will consider the simpler of these first.

External Nutrient Limitation

Most extant models use the minimum limiting nutrient construct (cf. Larsen et al., 1973), the main difference being the number and nature of the nutrients considered. For example, CE-QUAL-R1 evaluates the concentration of five nutrients (NH_3, NO_2 & NO_3, Si, PO_4 and CO_2) for limitation of diatoms. In order to determine which material is in least supply, the Monod equation is evaluated for each of the five and the minimum factor is used:

$$NUTLIM = Min (FACTOR) \qquad (15.4)$$
$$FACTOR_i = C_i / (k_i + C_i)$$

where

k_i = half-saturation coefficient for growth factor i

C_i = concentration of potentially growth-limiting factor

Macrophyte Nutrient Limitation

The primary source of both nitrogen and phosphorus for many macrophytes is through root uptake from bottom sediments (Carignan and Kalff, 1980; Best and Mantai, 1978). Other macrophytes may draw nutrients from the water column, depending on nutrient availability (Bole and Allan, 1978).

The submodel for macrophyte photosynthesis used in CE-QUAL-R1 has compartments for sediment and water column nitrogen and phosphorus. Nutrient limitation from these compartements is represented using the Monod function.

Internal Nutrients

By uncoupling algal growth and external nutrient concentrations, one can avoid the persistent time lag between simulated and observed phytoplankton peaks evident in many models. There are several formulations in use at present (for example, Lehman et al., 1975; Bierman, 1976; Jørgensen, 1976; Nyholm, 1976). We prefer an approach based on the work of Rhee (1973, 1974, 1978) and Droop (1974) that is incorporated in MINI.CLEANER (Collins, 1980).

In the following process equations nutrient limitation is assumed to be based on a threshold relationship: the internal nutrient that is most limiting determines the rate of photosynthesis. Algae take up nutrients from the external pools and assimilate the internally stored nutrients for photosynthesis. The specific uptake rate equation for phosphorus and nitrogen was formulated using information published by Rhee (1973, 1974, 1978). Although the kinetic responses to these two nutrients differ somewhat, the equations are identical:

$$V_i = (\text{upmax}/(1.0 + \text{kin}/C_{i4})) * (1.0 + \text{INHIB}/\text{ksurp}) * \text{BIOMASS} * \text{UI} * \text{TEMP} \qquad (15.5)$$

where

V_i = uptake rate of nutrient from external supply (g/g organism/day)
upmax = maximum uptake rate of nutrient (g/g organism/day)
kin = half-saturation constant for nutrient uptake (g/m^3)
C_i = concentration of nutrient
INHIB = concentration of inhibitor (g/m^3)
UI = light-limited growth reduction [Eq. 15.15] (unitless)
TEMP = temperature limitation [Eq. 15.28] (unitless)
ksurp = half-saturation constant for the inhibitor (g/m^3)

The maximum uptake rate is decreased by the inhibitor and the half-saturation constant is unaffected.

Limitation by Internal Nutrients

Nutrient limitation is determined independent of the uptake and assimilation terms. The subsistence quota, which is the internal nutrient per unit algal biomass (Droop, 1974):

$$SQP = INTCONC_i / BIOMASS_i \tag{15.6}$$

is an indication of the nutrient status of the algae. Therefore the nutrient limitation is:

$$UPO4 = SQP / (SQP + pkpo4_i) \tag{15.7}$$

To determine which nutrient is limiting, the atomic ratio of internally-stored nitrogen internally-stored phosphorus is obtained. This is compared with the N:P atomic ratio observed for the group. For example, Rhee (1978) observed an abrupt limitation of growth for *Scenedesmus* spp at an N:P atomic ratio of 30. Since the levels of N:P were varied only by changing the nitrate concentration, the proportional increase below 30 indicates nitrogen limitation and above 30, phosphorus limitation.

Assimilation of Internal Nutrients

The internally stored pool of nutrients is assimilated during photosynthesis according to the Monod function multiplied by the light-limitation factor. The light-limitation factor is used in both the uptake and assimilation equations because numerous studies have shown that light is important in the uptake of phosphorus and nitrogen, providing energy for permeases that are necessary for active transport. There is some evidence that light saturation occurs at a lower intensity than observed for photosynthesis; but, for simplicity, the same parameter values are used in MINI.CLEANER (Collins, 1980).

Light Limitation

A number of formulations for light limitation have been developed over the years. They include hyperbolic saturation equations (Smith, 1936;

Bannister, 1979) and photoinhibition equations (Steele, 1962; Vollenweider, 1965; Steel, 1973). An excellent comparison of formulations is given by Straskraba (1976). The need for a **photoinhibition** term has been demonstrated in laboratory experiments (Liou and van Eybergen, 1982), but may not be significant in eutrophic lakes (Harris and Piccinin, 1977; Jones, 1978; Iwakuma and Yasuno, 1983). Many models use the equation developed by Steele (1962):

UI = EXP(1.0) * (EXP(-(IOSEC / isat * EXP(-EXTCT * Z2))
-EXP(-(IOSEC / isat * EXP(-EXTCT * Z1))) / (EXTCT * (Z2 - Z1))
(15.8)

Z2 = $\underline{Z}$ - $\underline{TOP}$ if pphyz ≤ 0
Z2 = $\underline{TOP}$ + pphyz if pphyz > 0

Z1 = 0 if pphyz ≥ 0
Z1 = Z2 + pphyz if pphyz < 0

where
IOSEC = incident solar radiation for section (ly/day)
isat = saturating light intensity (ly/day)
EXTCT = extinction coefficient for section (l/m)
Z2 = thickness of section (m)
Z1 = distance from top of section for light calculations (m)
$\underline{Z}$ = depth of bottom of section (m)
$\underline{TOP}$ = depth of top of section (m)
pphyz = depth distribution of phytoplankton (> 0 = thickness occupied at top of section, < 0 = thickness at bottom of section, if 0 then evenly distributed)

Non-random distribution of phytoplankton in the water column (and benthic algae on the bottom) is indicated by the variable pphyz. This formulation represents photoinhibition at light intensities above the saturating light intensity; below the saturation level there is also reduction of photosynthesis (Fig. 15.2).

In **CE-QUAL-R1** macrophyte biomass is considered to occupy the bottom of the lake within the euphotic zone, with the biomass distributed vertically according to a species-specific height to weight relationship. The average irradiance reaching a particular model layer determines the photosynthetic response according to the Steele (1962) equation. Light is attenuated for a particular layer depending on the extinction coefficient for water, suspended solids (including inorganic sediment, detritus, phytoplankton, and zooplankton) and overlying macrophytes.

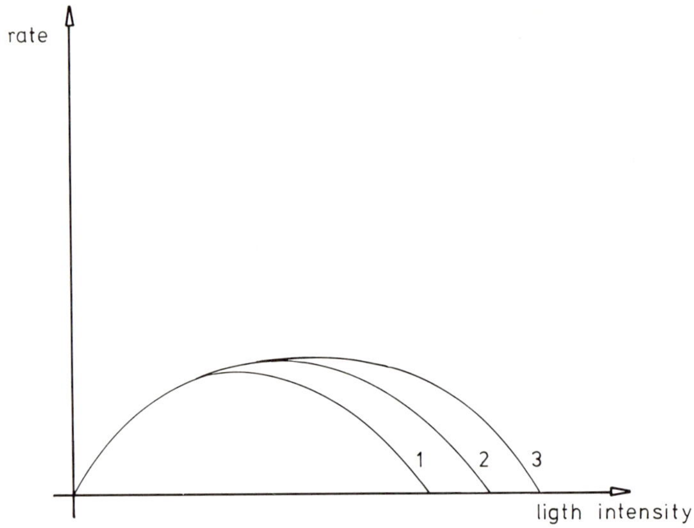

Fig. 15.2: Photosynthesis as a function of light intensity at varying levels of saturating light intensity; from Park et al., 1985. 1 to 3 represent increasing levels of saturating light intensity.

If a time step of less than one day is used, then the photoperiod has to be taken into account. Using an equation developed by Stewart (1975) and modified by Groden (1977) for latitudes between 25 degrees and 60 degrees (Fig. 15.3), the photoperiod is:

$$PPERD = 1.2 + (0.1414 * latud - 2.413) * SIN(360 * TIME/365 + 279) \qquad (15.9)$$

where

PPERD = photoperiod (hours)
TIME = julian date
latud = latitude in degrees (negative if south of equator)

Algal Respiration

Respiration is handled in several different ways in various models; unfortunately, at the present time the process is still not understood well enough for us to choose one formulation over another; this is reflected by the differing treatment in the three models with which we are currently working. In **MINI.CLEANER** respiration is formulated with two terms, a constant proportion (pkresp) of primary production and an endogenous term that increases exponentially with increasing temperature as shown by Riley (1963) and formulated by Groden (1977):

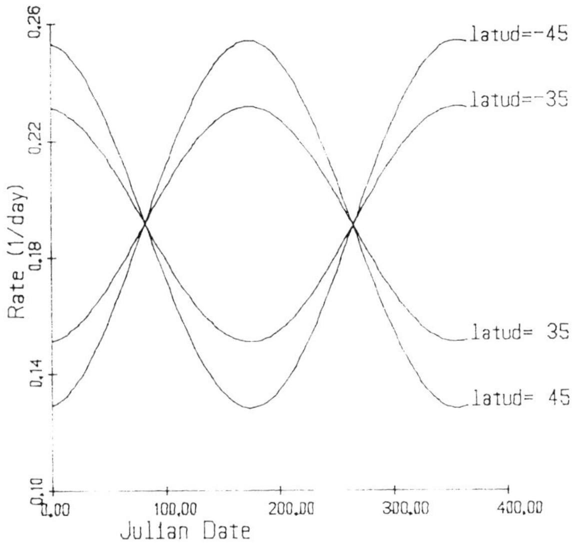

Fig. 15.3: Photoperiod as a function of julian date and latitude; from Park et al., 1985.

RESP = ((kresp * PPROD) + resp0 * EXP(tresp * T)) * BIOMASS (15.10)

where
RESP = respiration ($g/m^3/day$)
kresp = coefficient that relates respiration to photosynthesis (unitless)
PPROD = photosynthesis ($g/m^3/day$)
resp0 = respiration rate at 0°C (g/g/day)
tresp = exponential temperature coefficient (unitless)
T = ambient temperature (°C)
BIOMASS = biomass of phytoplankton (g/m^3)

LAKETRACE assumes respiration is a constant proportion of the maximum photosynthetic rate, reduced by suboptimal temperature:

RESPIRATION = kresp * pmax * TEMP * BIOMASS (15.11)

where
kresp = proportionality constant for respiration (unitless)
pmax = maximum photosynthetic rate (g/g/day)
TEMP = nonlinear temperature response factor which is purely exponen-

tial only at low temperatures and which decreases at temperatures beyond an optimal

CE-QUAL-R1 uses a formulation for dark respiration of algae and macrophytes that is similar to LAKETRACE except that the temperature response is exponential (as in MINI.CLEANER):

RESP = presp * RTMLT1 * BIOMASS (15.12)

where
presp = maximum dark respiration rate (g/g/day)
RTMLT1 = rising limb of the CEQUAL temperature function (unitless)
BIOMASS= biomaaa od algae (g/m^3) or macrophyte (g/m^2)

Algal Excretion

The excretion of photosynthate and photorespiratory compounds occurs under a variety of environmental conditions. In general, any environmental condition that inhibits cell multiplication, but still permits photoassimilation, results in release of organic compounds. This is especially true for inhibitory low or high levels of light (Fogg, Nalewajko and Watt, 1965; Watt, 1966; Nalewajko, 1966). Therefore, the equation for algal excretion formulated by Desormeau (1978) is:

EXCR = PPROD * (1.0 - UI) * pkexcr (15.13)

where
EXCR = excretion ($g/m^3/day$)
PPROD = photosynthetic rate ($g/m^3/day$)
UI = light limitation [Eq. 15.15] (unitless)
pkexcr = coefficient of proportionality between excretion and photosynthesis (unitless)

With additional data this equation could be expanded to include the effects of low dissolved oxygen, carbon dioxide, and temperature.

In modelling macrophytes the combination of nutrient uptake from the sediments and excretion into the water column represents "nutrient pumping" (Wetzel and Manny, 1975).

Sinking

In its simplest formulation, sinking can be represented as a constant, linear function with provision for the thickness of the layer (Wlosinski and Collins, in press):

$$\text{Settling} = \text{FRAC} * \text{set1} * \text{VOL}((I+1) * \text{ALGAE}(I+1) / \text{SDZ}(I+1)) - \text{set 1} * \text{DVOL} * \text{ALGAE} / \text{SDZ} \quad (15.14)$$

There is a transfer term due to settling of algae into lower layers based on a settling rate coefficient defined for each algal compartment.

However, *sinking rate can increase* by an order of magnitude as algae are stressed (Smayda, 1970, 1974). This can be represented by incorporating an exponential factor based on photosynthetic limitation (Park, in prep; cf. Scavia et al., 1976);

$$\text{SINKING} = \text{ksed} / \text{DEPTH} * \text{BIOMASS} * \text{SEDACCEL} \quad (15.15)$$

ksed = intrinsic settling rate

DEPTH = thickness of layer

and

$$\text{SEDACCEL} = e^{(\text{esed} * (1.0 - \text{PPRODLIMIT}))}$$

where

esed = exponential settling coefficient

PPRODLIMIT = photosynthesis reduction factor based on light, temperature and nutrient limitation

Non-Predatory Mortality of Algae

Under nutrient- and possibly light-limiting conditions phytoplankton exhibit a pattern of decreasing viability, accompanied by an increase in colonization by decomposers, and leading to the eventual loss of cell contents (Barica, 1975; Jones, 1976). Abnormally high temperature also can lead to cell lysis. Mortality is difficult to measure in the field, but it is generally regarded as being less than 10% per day. These effects are represented as (Park et al., 1985):

$$\text{MORT} = (\text{kmort} * \text{EXP}(\underline{T} - \text{tcrit}) + (1.0 - \text{EXP}(-\text{emort} * (1.0 - (\text{NUTLIM} * \text{UI}))))) \quad (15.16)$$

where

kmort = intrinsic mortality rate (g/g/day)

tcrit = critically high temperature above which exponential mortality occurs (°C)

emort = maximum mortality rate (g/g/day)

NUTLIM = nutrient limitation [Eq. 15.4] (unitless)
UI = light limitation [Eq. 15.8] (unitless)

The first term accounts for mortality at high temperatures (Scavia and Park, 1977) as shown in Fig. 15.4. Low temperatures are not assumed to be a factor in phytoplankton mortality. The second term accounts for the effects of nutrient and light limitation, similar to the way that nutrient, light and temperature limitation effects on sinking are modelled (Eq. 15.15). The complex formulation is necessary so that emort will have physical meaning.

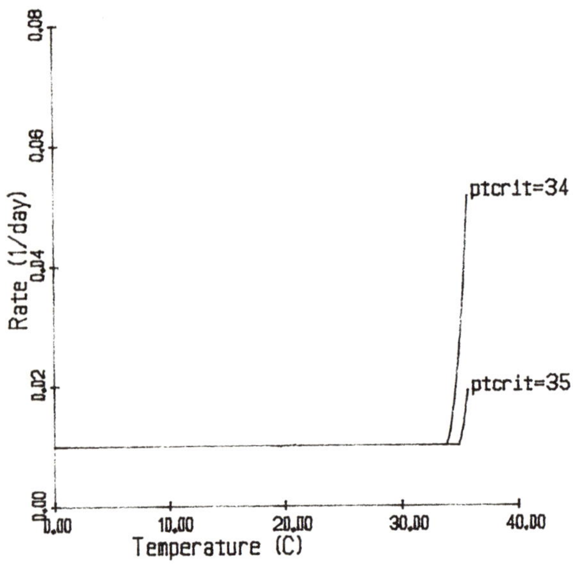

Fig. 15.4: Phytoplankton mortality as a function of critically high temperature; from Park et al., 1985.

A much simpler formulation with a constant intrinsic mortality rate being increased exponentially by high temperatures is (Wlosinski and Collins, in press):

$$D = c_t * e^{(I - c_p)} \qquad (15.17)$$

where D = fraction of algae that dies
c_t = maximum mortality
I = temperature of layer
c_p = temperature at which entire population dies

15.5 MODEL APPLICATION AND EVALUATION

Primary productivity models have been built and used for a variety of reasons. Oftentimes application is primarily for scientific research, with the model being either a major result of the project or a rigorous means of synthesizing data from an intensive field study. In these cases the choice of the model is dictated by the scientific objectives of the study. For example, a study of diurnal patterns of productivity would require a model with very detailed process equations; a study of seasonal succession in a particular category of lake would require a model with a lesser degree of process resolution but with several state variables for algae.

Many modellers have been concerned with pre-impoundment studies - forecasting possible algal problems in reservoirs that have not yet been constructed. Traditionally, the concern has been with maximum algal levels, and empirical models have been sufficient. In recent years regulatory pressures in North America and Europe have required far more detailed predictions for time-varying algal blooms and dissolved oxygen demand in pre-impoundment studies. As a result, more complex models have been necessary. **CE.QUAL-R1** was developed largely because of this need for detailed predictions.

Detailed models are being used with increasing frequency in guiding management decisions in the mitigation of existing algal and associated detritus and dissolved oxygen problems. Waste-load allocations to prevent the occurrence of nuisance algae require models capable of simulating seasonal succession, with particular attentions to the maximum levels of nuisance algae and consequent dissolved oxygen levels at depth under various loadings of nutrients. **LAKETRACE** was developed with this particular requirement in mind. Design of reaeration and selective withdrawal systems usually require one- and two-dimensional models. **CE-QUAL-R2** is presently being used for this type of problems by the U.S. Army Corps of Engineers.

MINI.CLEANER

This model is one of the latest of a series that dates back to the **CLEAN** model (Park et al., 1974). It is a one-dimensional water quality model that simulates 31 biological and chemical constituents through time. As many as 5 vertical and horizontal segments can be represented. The model was calibrated for Ovre Heimdalsvatn, a subalpine lake located in central Norway (Collins, 1980). Calibration data are from Larsson et al. (1978) and Brettum (1972).

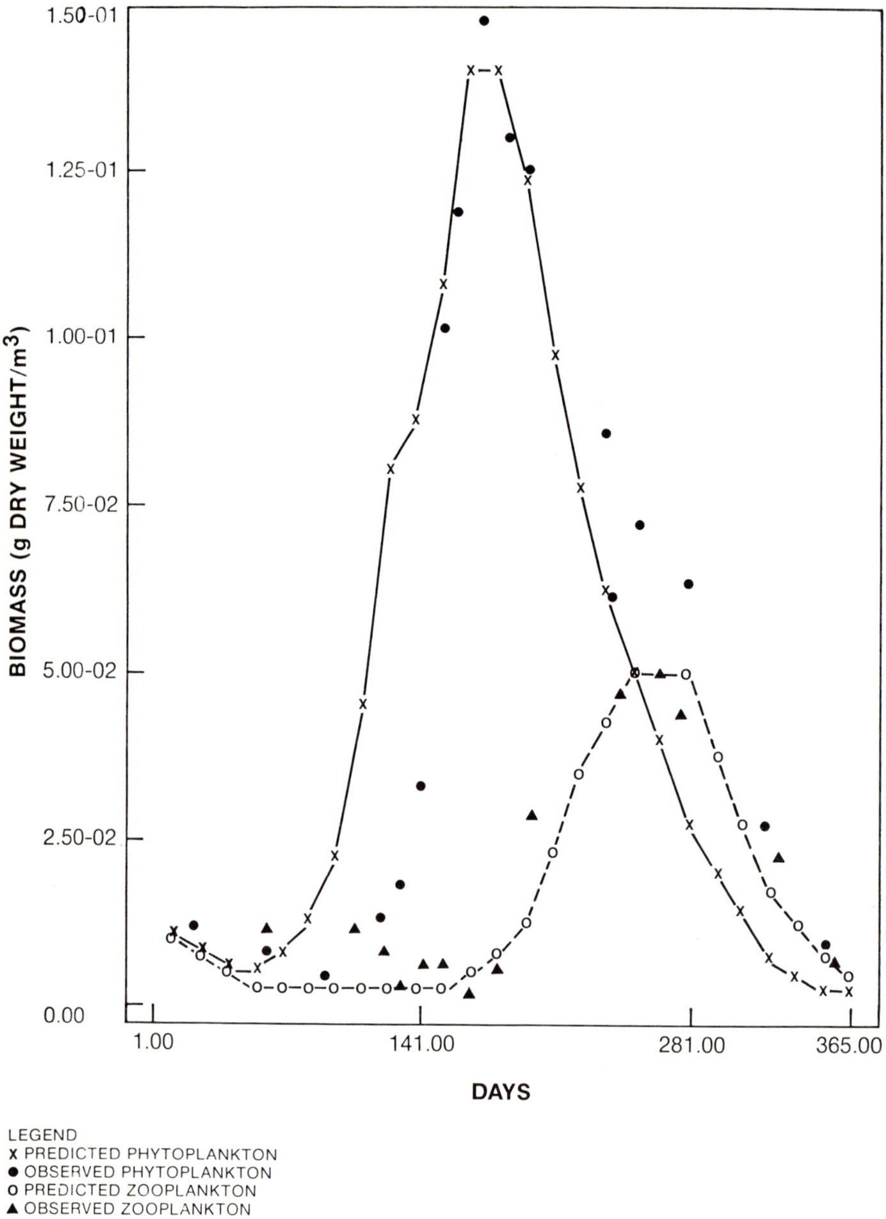

Fig. 15.5: Comparison of predicted and observed values for phytoplankton and zooplankton biomass in Ovre Heimdalsvatn, Norway; after Collins, 1980.

The predicted phytoplankton and zooplankton biomass of Ovre Heimdalsvatn for 1972 are plotted against the observed biomass values during this time period (Fig. 15.5). During the periods of ice cover growth is light-limited. During the summer months photoinhibition appears to be one of the factors that causes the decline of the phytoplankton bloom. Prior to the sharp rise in biomass there is an increase in nutrient limitation; however, during the bloom, when cells are most actively dividing, internal nutrients appear to limit growth most severely. The Monod-type growth equation relating growth directly to external nutrient levels was tested, and simulations erroneously predicted phytoplankton growth in phase with nutrient loadings.

The model was validated using data from Vorderer Finstertaler See, a high-mountain lake situated in the central Alps in Austria. The initial conditions and driving variables were specified according to data from 1970 (G. Bretschko, personal communication). Parameter values used in the calibration for Ovre Heimdalsvatn were not changed. Results are given in Figure 15.6, where predicted and observed biomass values are plotted for both phytoplankton and zooplankton. The total seasonal biomass of phytoplankton was higher in 1969 and lower in 1971, so the initial and final biomass values for 1970 are reasonable. There is no significant difference in the predicted and observed mean phytoplankton biomass in the validation, nor is there a significant difference in the weighted time of predicted and observed peak phytoplankton biomass. One can never truly validate a model, but this seems to be as rigorous an evaluation as is possible (cf. Cale and Shugart, 1981).

CE-QUAL-R1

This model is a one-dimensional (horizontally-averaged) water quality model that computes the vertical distribution of temperature and 37 biological and chemical constituents in a reservoir through time. In the model a reservoir is conceptualized as a vertical series of horizontal layers where thermal energy and mass are uniformly distributed in each layer. Layer thickness is variable and is dependent on the balance of inflowing and outflowing waters.

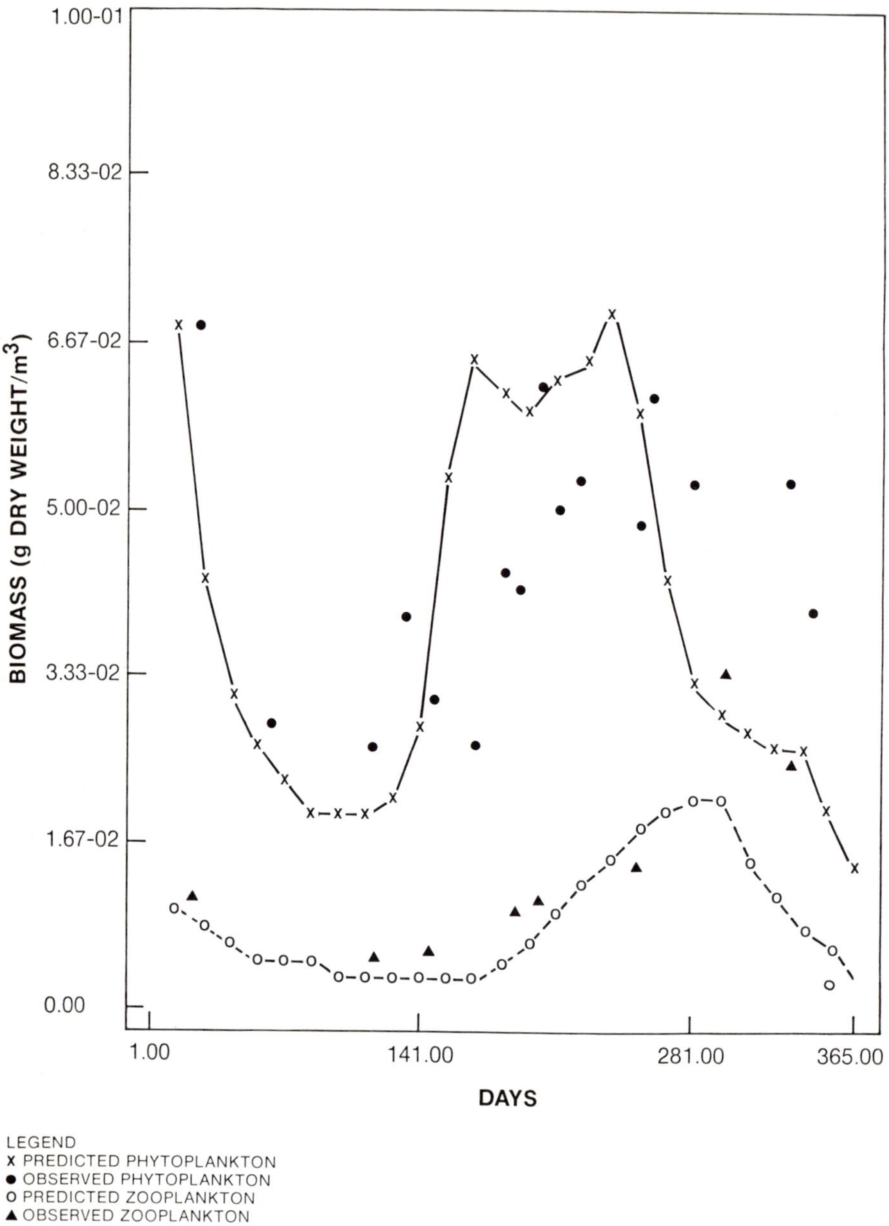

Fig. 15.6: Comparison of predicted and observed values for phytoplankton and zooplankton biomass in Vorderer Finstertaler See, Austria; after Collins, 1980.

Model predictions are evaluated both graphcally and statistically. Although a number of statistical tests are available (see Wlosinski, 1984), most of the comparisons are made using the Reliability Index (RI) of Leggett and Williams (1981). As predicted versus measured values diverge, the RI becomes larger. The value of the RI is 1.0 in the case of perfect prediction. If all comparisons are within a factor of 2 of each other, the RI is 2.0. Because the RI does not depend on whether the observed or predicted value is greater, and because it is scale variant, it appears to be the best statistic for aggregating and comparing results of different variables. Statistics are calculated for vertical profiles for each variable on those dates when data were collected, and for each variable summed over depths and dates.

As part of the model evaluation, **CE-QUAL-R1** was applied to DeGray Lake, Arkansas. Data collected in 1979 were used for calibration and data for 1980 were used for verification. Both qualitative and quantitative tests were used for comparing model predictions with measured values (Wlosinski and Collins, in press).

Results from the original model after final calibration indicated the need for a number of improvements to the model. These changes included the replacement of the Monod equation with Steele´s (1962) equation to represent photosynthesis as a function of irradiance. Loss terms for photorespiration and non-predatory mortality were also added to the phytoplankton compartment. The improved model had an overall average RI of 2.84 for 1979, and the verification simulation (1980) had an RI of 2.59. The phytoplankton compartment had an RI of 2.18 for the verification simulation.

The model was then evaluated for Eau Galle Reservoir, Wisconsin, which is physically, chemically and biologically different from DeGray. DeGray Lake is 32 km long and dendritic, with a normal depth of 57 meters, an area of 5.3×10^7 m^2 and a volume of 7.9×10^8 m^3. Eau Galle is 1 km long and circular, with a normal depth of 10 m, an area of 4.5×10^5 m^2 and a volume of 1.6×10^6 m^3. DeGray is monomictic, while Eau Galle is dimictic. Eau Galle is much more productive than DeGray.

Differences in the two sites necessitated the inclusion of specific process equations to represent the contrasting characteristics of the two reservoirs. The changes made to the model for application to Eau Galle included the additions of a macrophyte submodel, a third algal compartment and a dissolved silica compartment.

Final calibration results for Eau Galle had an average RI value of 2.57. Final verification produced an average RI of 2.62. The seasonal succession of algae predicted by the model is shown in Fig. 15.7. Measured data (solid lines) were obtained from Barko et al. (in press). Their results showed that the epilimnetic algal concentration in 1981 was composed primarily of diatoms, bluegreens and dinoflagellates. The success in predicting the total phytoplankton biomass is attributed to using 3 compartments to represent the dominant algae and their seasonal succession.

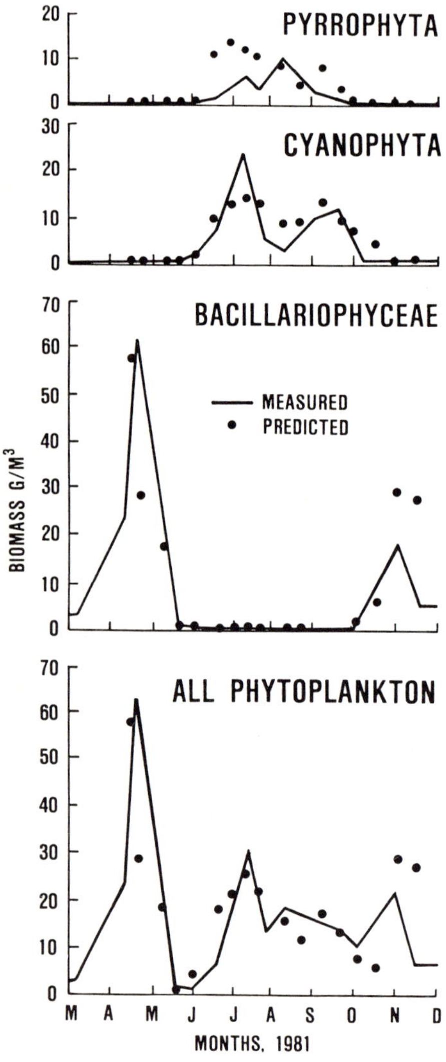

Fig. 15.7: Measured (circles) and predicted (solid line) algal biomass values in the epilimnion for Eau Galle Reservoir, 1981. Station 20.

Luxury uptake is not modelled in **CE-QUAL-R1**, and to do so would make the model more complex than is needed for the purpose of the Army Corps of Engineers. To overcome this problem, the initial conditions were increased for orthophosphate to a value between that observed for orthophosphates and that observed for total phosphorus. For the 1982 verification simulations the measured values for orthophosphate were used.

Eau Galle Reservoir had a significant macrophyte population dominated by *Ceratophyllum demersum.* Macrophyte biomass and associated epiphytes occupied about 17% of the surface area of Eau Galle. A macrophyte submodel developed by Collins et al. (in press) was incorporated in **CE-QUAL-R1** and evaluated for Eau Galle (Wlosinski and Collins, 1985). Macrophyte biomass measurements were taken on an aerial basis in 1981. Macrophyte biomass is represented in the model as g/m^2; therefore, a separate integration routine (the same one used for the sediment and fish compartments) was used to compute changes in mass on a m^2 basis. A comparison of predicted versus measured macrophyte biomass is presented in Fig. 15.8.

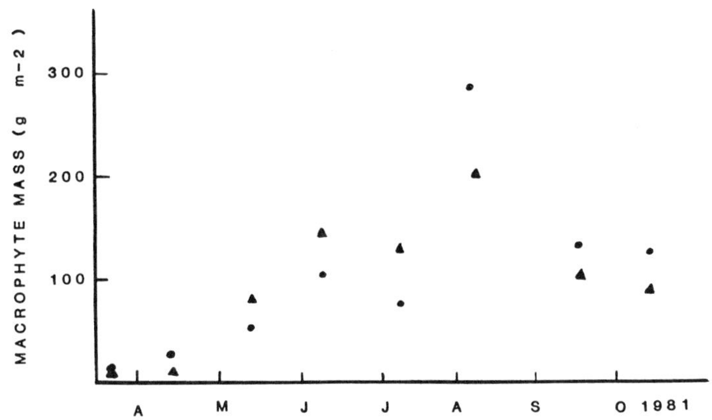

Fig. 15.8: Measured (circles) and predicted (triangles) values for macrophyte biomass for Eau Galle Reservoir in 1981.

The user is able to choose a time step in **CE-QUAL-R1** ranging from 1 to 24 hours. Shorter time steps show some of the dynamics occuring during a 1-day period, but are expensive. All the evaluative simulations reported here used a 24-hour time step. To test the model using a shorter time step, a number of simulations were performed with a 3-hour time step.

Initially the 1981 calibration data from Eau Galle were used without any changes to coefficients. Driving variables were not changed; the model simply interpolated the 24-hour data to supply values for 3-hour periods. The results were unsatisfactory, indicating that the same set of coefficients cannot be used with time steps less than 24 hours.

Eight coefficients had to be changed to obtain resonable results with a 3-hour time step. The coefficient changed were the mixing coefficient for wind, the light saturation and maximum gross production coefficients for the algal compartments, and the ingestion coefficient for zooplankton. Although the predictions for macrophytes were not evaluated, it is likely that the light saturation and maximum gross production coefficients should be changed for macrophytes as well for simulations with time steps under 24-hours. The RI for the final 3-hour simulation was 2.69.

Predictions of Flux

Scavia (1980), Collins and Wlosinski (1984) and Wlosinski (in press) have shown the need to compare predicted and measured flux values in addition to comparing mass or concentration values. Because concentrations are the result of both positive and negative fluxes that can cancel out, reasonable concentrations may be predicted for the wrong reasons. As part of the Eau Galle study, a number of process rates were measured in situ and can be compared to model predictions. These included algal productivity, sediment oxygen demand, and algal sedimentation rates.

Productivity rates vary extensively over a 24-hour period and from day to day and are dependent on local conditions such as light, temperature and the concentrations of nutrients. CE-QUAL-R1 simulates diurnal variations in temperature and light, but process mechanisms and rates are based on average daily values in a simulation with a 24-hour time step. Studies reveal that photosynthetic capacity, cell division, nutrient uptake, respiration and grazing vary according to algal circadian rhythms (Prezelin et al., 1977; Chrisholm, 1981). Nevertheless, comparisons should show rates that are similar.

Productions rates were measured from noon until 2 p.m. at discrete intervals between the surface and 3 m on a biweekly sampling schedule. The measured values of mg $O_2/m^2/hr$ were converted to g $O_2/m^2/day$ for comparison with predicted values, assuming a 10-hour day for photosynthesis. The comparison (Fig. 15.9) used predictions from approximately 2 m depth. Considering the variability in measured data at different stations and the fact that the measured data are for a much shorter period than the 24-hour time step used, predictions are reasonable during the spring diatom bloom. Predictions at 2 meters undersectimate productivity during summer because blue-green algae, the dominant group in summer, are concentrated near the surface.

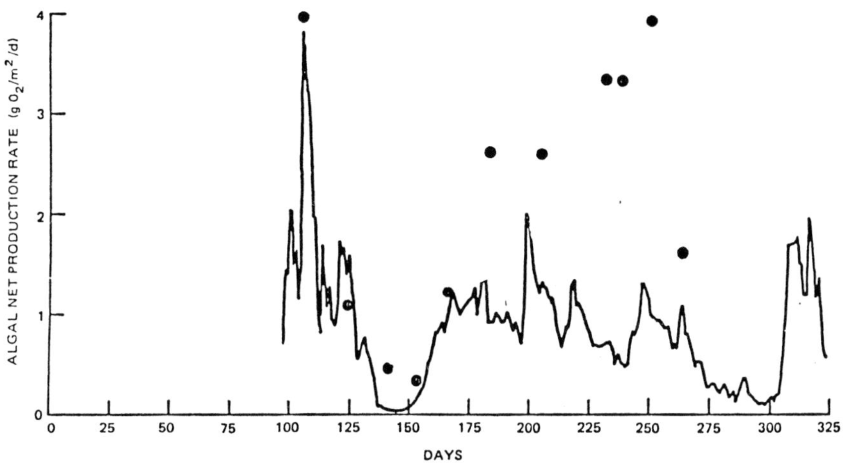

Fig. 15.9: Algal net production rates for Eau Galle Reservoir, 1981, station 20. Circles represent observed rates from 0 to 3 meters. The solid line represents predicted algal net production rate at approximately 2 meters.

Sedimentation rates af algae were also available. Sediment traps were deployed at 4 and 8.5 m below the water surface and were retrieved after approximately 2 weeks. Direct comparison of predicted and measured values was not possible because of the variable-layer scheme employed in the model. The model predicts values on a per-layer basis; because the thickness of each layer varies over time, a given depth contour may be located in different layers during different time steps. For comparative purposes the predictions from layer 4 and layer 8 were used. In general, the predictions were more uniform than the measured data; and, with one exception, the model predictions were lower than the measured values. The discrepancy may be due in part to the slower decomposition of chlorophyll **a** (the basis for the measurements) in comparison with the decomposition of biomass as represented in the model.

15.6 PARAMETER ESTIMATION AND DATA REQUIREMENTS

Parameter values are given in Table 15.1 for CE-QUAL-R1. Although some of the values are the same as found in other models, one should be cautious in transporting parameter values from one model to another. Every model is a unique simplification of the real world, and specific parameter values are appropriate only in the context of the overall formulation. When it is developed, a model is calibrated so that parameters compensate for the structure and unique formulations of that particular model.

TABLE 15.1
Representive parameter values from CE-QUAL-R1

Parameter	Units	Description	Blue-gr.	Dinofl.	Diatoms
PMAX	l/day	max. production rate	0.99	1.4	1.6
SETL	m/day	settling rate	0.05	0.10	0.12
KPO_4	mg/l	phosphorus half-sat.	0.02	0.02	0.004
KN	mg/l	nitrogen half-sat.	0.06	0.09	0.07
KCO_2	mg/l	carbon half-sat.	0.10	0.10	0.08
KSi	mg/l	silica half-sat.			0.05
ISAT	$kcal/m^2/hr$	light saturation	85	115	45
PKEXCR	l/day	max. excretion rate	0.05	0.04	0.04
C_t	l/day	max. mortality rate	0.07	0.06	0.01
KRESP	l/day	max. respiration rate	0.14	0.17	0.14
T_1	°C	low temp. threshold	7	12	0
T_2	°C	low temp. optimum	15	19	8
T_3	°C	high temp. optimum	28	25	12
T_4	°C	high temp. threshold	35	35	17

15.7 CONCLUSIONS AND RESEARCH NEEDS

Future models are likely to be less complex in terms of number of state variables, but more complex in terms of process interactions. Self-adjusting compartments are a logical extension of this trend. The influence of external driving variables on photosynthesis will undoubtedly be better understood and better represented in future models. Efforts in reducing complexity as an artifact of model structure will require the incorporation of mechanistic formulations for physical phenomena. For example, more realistic formulations for the effects of vertical mixing on photosynthesis need to be investigated; this will entail improved methods for measuring photosynthesis during mixing events in order to distinguish between instantaneous and adaptive responses.

REFERENCES

Anderson, R.S., 1974. Diurnal primary production patterns in seven lakes and pond in Alberta (Canada). Oecologia 14: 1-17.

Bannister, T.T., 1979. Quantitative description of steady-state, nutrient-saturated algal growth, including adaptation. Limnol. Oceanogr., 24: 76-96.

Barica, J., 1975. Collapses of algal blooms in prairie pothole lakes: their mechanism and ecological impact. Verh. Int. Verein. Limnol., 19: 606-615.

Barko, J.W., D.J. Bates, G.J. Filbin, M.S. Hennington and D.G. McFarland, in press. Seasonal growth and community composition of phytoplankton in a eutrophic Wisconsin impoundment. J. Freshwater Ecology.

Best, M.D. and K.E. Mantai, 1978. Growth of *Myriophyllum* : sediment or lake water as the source of nitrogen and phosphorus? Ecology, 59: 75-80.

Bierman, V.J., 1976. Mathematical model of the selective enrichment of blue-green algae by nutrient enrichment. **In:** R.P. Canale (ed.), Modelling of Biochemical Processes in Aquatic Ecosystems. An Arbor Science Pub., Ann Arbor, MI.

Bloomfield, J.A., R.A. Park, D. Scavia and C.S. Zahorcak, 1973. Aquatic modelling in the Eastern Deciduous Forest Biome, U.S. International Biological Program. **In:** E.J. Middlebrooks, D.H. Falkenborg and T.E. Maloney (eds.), Modelling the Eutrophication Process. Utah State University, Logan, Utah, pp. 139-158.

Bole, J.B. and J.R. Allan, 1978. Uptake of phosphorus from sediment by plants, *Myriophyllum spicatum* and *Hydrilla verticillata.* Water Research, 12: 353-358.

Brettum, P, 1972. The phytoplankton of lake Ovre Heimdalsvatn, central south Norway in 1969-70. Norwegian Jour. Botany, 19: 1-30.

Cale, Jr., W.G. and H.H. Shugart, 1983. Ecological reality and model validation. **In:** G.E. Lasker (ed.) Proc. International Congress Applied Systems Research and Cybernetica, IV, pp. 1911-1918.

Canale, R.P., L.M. DePalma and A.H. Vogel, 1976. A planktonbased food web model for Lake Michigan. **In:** R.P. Canale (ed.), Modelling of Biochemical Processes in Aquatic Ecosystems. Ann Arbor Science Pub., Ann Arbor, MI.

Carignan, R. and J. Kalff, 1980. Phosphorus sources for aquatic weeds: water or sediments? Science, 207: 987-989.

Chrisholm, S.W., 1981. Temporal patterns of cell division in unicellular algae. **In:** T. Platt (ed.), Physiological Basis of Phytoplankton Ecology. Can. Bull. Fish. Aq. Sci., 210: 150-181.

Chapra, S.C. and K.H. Reckhow, 1983. Engineering Approaches for Lake Management, Volume 2: Mechanistic Modelling. Butterworth Pub., Boston, 492 pp.

Collins, C.D., 1980. Formulation and validation of a mathematical model of phytoplankton growth. Ecology, 6(3): 639-649.

Collins, C.D., R.A. Park and C.W. Boylen, In: press. A mathematical model of submersed aquatic plants. Miscell. Paper A-85, U.S. Army Engineers Waterways Experiment Station, Vicksburg, Miss.

Collins, C.D. and J.H. Wlosinski, 1984. Verification of the reservoir water quality model CE-QUAL-R1, using daily flux rates. Lake and Reservoir Management. EPA 440/5/84-001.

Desormeau, C.J., 1978. Mathematical Modeling of Phytoplankton Kinetics with Application to Two Alpine Lakes. Rpt. No. 4, Center for Ecological

Modeling, Rensselaer Polytechnic Institute, Troy, NY, 21 pp.
DiToro, D.M., D.J. O'Connor, R.V. Thomann and J.L. Mancini, 1975. Phytoplankton-zooplankton nutrient interaction model for western Lake Erie. **In:** B.C. Patton (ed.), Systems Analysis and Simulation in Ecology, Vol. III. Academic Press, New York, NY.
Droop, M.R., 1974. The nutrient status of algal cells in continuous culture. Jour. Mar. Biol. Assoc. U.K., 54: 825-855.

Environmental Laboratory, 1982. CE-QUAL-R1: a numerical one-dimensional model of reservoir water quality; a user's manual. Instruction Report E-82-1, U.S. Army Engineer Waterways Experiment Station, Vicksburg, Miss.

Fogg, G.E., C. Nalewajko and W.D. Watt, 1965. Extracellular products of phytoplankton photosynthesis. Proc. Royal Soc. B., 162: 517-534.

Ganf, G.G. and A.J. Horne, 1975. Diurnal stratification, photosynthesis and nitrogen fixation in a shallow equatorial lake (Lake George - Uganda). Freshwater Biol. 5: 13-39.
Groden, T.W., 1977. Modeling Temperature and Light Adaptation of Phytoplankton. Rpt. No. 2, Center for Ecological Modeling, Rensselaer Polytechnic Institute, Troy, NY, 17 pp.

Harris, G.W., 1978. Photosynthesis, productivity and growth. The physiological ecology of phytoplankton. Ergebnisse Limnol. 10: 1-171.
Harris, G.P. and B.B. Piccinin, 1977. Photosynthesis by natural phytoplankton populations. Arch. Hydrobiol. 86: 405-457.

Iwakuma, T. and M. Yasuno, 1983. A comparison of several mathematical equations describing photosynthesis-light curve for natural phytoplankton populations. Arch. Hydrobiol. 97: 208-226.

Jones, J.G., 1976. The microbiology and decomposition of seston in open water and experimental enclosures in a productive lake. Jour. Ecology, 64: 241-278.
Jones, R.I., 1978. Adaptation to fluctuating irradiance by natural phytoplankton communities. Limnol. Oceanogr. 23: 920-926.
Jørgensen, S.E., 1976. A eutrophication model for a lake. Ecol. Modelling, 2: 147-165.

Kinne, O., 1970. Temperature. **In:** O. Kinne (ed.), Marine Ecology. Wiley Interscience, New York, NY, Vol. 1, Part 1, pp. 407-514.
Kitchell, J.F., J.F. Koonce, R.V. O'Neill, H.H. Shugart, Jr., J.J. Magnuson and R.S. Booth, 1972. Implementation of a Predator-prey Biomass Model for Fishes. Eastern Deciduous Forest Biome Memo Rept. No. 72-118.
Kremer, J.N. and S.W. Nixon, 1978. A Coastal Marine Ecosystem. Springer-Verlag, New York, NY, 217 pp.

Larsen, D.P., H.T. Mercier and K.W. Malueg, 1973. Modelling algal growth dynamics in Shagawa Lake, Minnesota, with comments concerning projected restoration of the lake. **In:** E.J. Middlebrooks, D.H. Falkenborg and T.E. Maloney (eds), Modelling the Eutrophication Process. Utah State University, Logan, Utah, pp. 15-31.
Larsson, P., J.E. Brittain, L. Lein, A. Lillehammer and K. Tangen, 1978. The lake ecosystem of Ovre Heimdalsvatn. Holarctic Ecology 1: 304-320.
Leggett, R.W. and L.R. Williams, 1981. A reliability index for models. Ecol. Modelling 13: 303-312.
Lehman, T.D., D.B. Botkin and G.E. Likens, 1975. The assumptions and

rationales of a computer model of phytoplankton population dynamics. Limnol. Oceanogr. 20: 343-364.
Lewis, M.R., J.J. Cullen and T. Platt, 1984. Relationships between vertical mixing and photoadaptation of phytoplankton: similarity criteria. Mar. Ecol. Progr. Ser. 15: 141-149.
Lewis, M.R. and J.C. Smith, 1983. A small volume, short incubation-time method for measurement of photosynthesis as a function of incident irradiance. Mar. Ecol. Prog. Ser. 13: 99-102.
Liou, J.K. and C. van Eybergen, 1982. Light adaptation and inhibition: processes important in modelling the growth of algae in drinking water basins. Water Research 16: 765-774.

Myers, J., 1946. Culture conditions and the development of the photysynthetic mechanism - IV. Influence of light intensity of photosynthetic characteristics of *Chlorella.* Jour. Gen. Physiol. 29: 429-440.

Nalewajko, C., 1966. Photosynthesis and excretion in various planktonic algae. Limnol. Oceangr., 11: 1-10.
Nicol, J.A.C., 1960. The Biology of Marine Animals. Interscience Publishers, Inc., New York, NY, 707 pp.
Nielsen, E.S. and E.G. Jørgensen, 1968. The adaptation of plankton algae I. General part. Physiologia Plantarum 21: 401-413.
Nyholm, N., 1976. A mathematical model for microbial growth under limitation by conservative substrates. Biotech. & Bioeng. 28: 1043-1056.

O'Neill, R.V., R.A. Goldstein, H.H. Shugart, Jr. and J.B. Mankin, 1972. Terrestrial Ecosystem Model. Eastern Deciduous Forest Biome Memo Rept. No. 72-19, 39 pp.

Park, R.A., R.V. O'Neill, J.S. Bloomfield, H.H. Shugart, Jr., R.S. Booth, R.A. Goldstein, J.B. Mankin, J.F. Koonce, D. Scavia, M.S. Adams, L.S. Clesceri, R.M. Colon, E.H. Dettmann, J. Hoopes, D.D. Huff, S. Katz, J.F. Kitchell, R.C. Kohberger, E.J. LaRow, D.C. McNaught, J. Peterson, J. Titus, P.R. Weiler, J.W. Wilkinson and C.S. Zahorcak, 1974. A generalized model for simulating lake ecosystems. Simulation, 23(2): 33- 50.
Park, R.A., B. Indyke, B.B. MacLeod, C.D. Collins and J.R. Abanese, 1985. Documentation of the Aquatic Ecosystem Model MINI.CLEANER. U.S. Environ. Protection Agency, Athens, Ga.
Prezelin, B.B., B.W. Meeson and B.M. Sweeney, 1977. Characterization of photosynthetic rhytms in marine dinoflagellates. I. Pigmentation, photosynthetic capacity and respiration. Plant Physiology 60: 384-387.

Reckhow, K.R. and S.C. Chapra, 1983. Engineering Approaches for Lake Management Volume 1: Data Analysis and Empirical Modeling. Butterworth Pub., Boston, Mass. 340 pp.
Rhee, G-Yull, 1973. A continuous culture study of phosphate uptake, growth rate and polyphosphate in *scenedesmus spp.* J. Phycol., 9: 495-506.
Rhee, G-Yull, 1974. Phosphate uptake under nitrate limitation by *Scenedesmus spp.* and its ecological implications. J. Phycol., 10: 470-475.
Rhee, G-Yull, 1978. Effects of N:P atomic ratios and nitrate limitation on algal growth, cell composition, and nitrate uptake. Limnol. Oceanogr., 23: 10-25.
Riley, G.A., 1963. Theory of food-chain relations in the ocean. The Sea, vol. 2.

Scavia, D., B.J. Eadie and A. Robertson, 1976. An Ecological Model of Lake Ontario: Model Formulation, Calibration and Preliminary Evaluation.

NOAA Tech. Rept. ERL 371GLERL 12, Ann Arbor, Mich., 63 pp.

Scavia, D., 1980. The need for innovative verification of eutrophication models. Workshop on Verification of Water Quality Models. EPA 600/9-80-016.

Scavia, D. and R.A. Park, 1976. Documentation of selected constructs and parameter values in the aquatic model CLEANER. Ecol. Mod., 2(1): 33-58.

Smayda, T.J., 1970. The suspension and sinking of phytoplankton in the sea. Oceanogr. Mar. Biol. Ann. Rev., 8: 353-414.

Smayda, T.J., 1974. Some experiments on the sinking characteristics of two freshwater diatoms. Limnol. Oceanogr., 19: 628-635.

Smayda, T.J., 1978. Form phytoplankters to biomass. **In:** A. Sournia (ed.), Phytoplankton Manual. Monographs on Oceanographic Methodology Vol. 6. UNESCO, U.K.

Smith, E.L., 1936. Photosynthesis in relation to light and carbon dioxide. Proc. Natl. Acad. Sci. Wash., 22: 504-511.

Steel, J.A., 1973. Reservoir algal productivity. **In:** A. James (ed.), Proc. of Symposium on the Use of Mathematical Models in Water Pollution Control. Newcastle-upon-Tyne, England, pp. 1-19.

Steele, J.H., 1962. Environmental control of photosynthesis in the sea. Limnol. Oceanogr., 7: 137-150.

Stewart, D.C., 1975. Mathematical Modelling of the Ecosystem of Lough Neagh. Ph. D. dissertation, Queens University, Belfast.

Straskraba, M., 1976. Development of an analytical phytoplankton model with parameters empirically related to dominant controlling variables. **In:** R. Glaser, K. Unger and M. Koch (eds.), Umweltbiophysik. Academie Verlag, Berlin, pp. 33-65.

Straskraba, M., 1979. Natural control mechanisms in models of aquatic ecosystems. Ecol. Modelling, 6: 305-321.

Thomann, R.V., R.P. Winfield and D.M. DiToro, 1976. Mathematical Modelling of Phytoplankton in Lake Ontario. 2. Simulations Using Lake 1 Model. U.S. Environ. Protection Agency Rept. EPA-600/3-76-065, Duluth, Minn.

Thorton, K.W. and A.S. Lessem, 1978. A temperature algorithm for modifying biological rates. Trans. Am. Fish. Soc., 107: 284-287.

Vollenweider, R.A., 1965. Calculation models of photosynthesis-depth curves and some implications regarding day rate estimates in primary production measurements. **In:** C.R. Goldman (ed.), Primary Productivity in Aquatic Environments, Univ. California Press, Berkely, Cal. pp. 425-457.

Watt, W.D., 1966. Release of dissolved organic material from the cells of phytoplankton populations. Proc. R. Soc. B., 164: 521-551.

Wetzel, R.G. and B.A. Manny, 1975. Secretion of dissolved organic carbon and nitrogen by aquatic macrophytes. Verh. Internat. Verein. Limnol. 18: 162-170.

Wlosinski, J.H., <u>in press</u>. Flux use for calibrating and verifying models. ASCE Jour. Environ. Eng.

Wlosinski, J.H. and C.D. Collins, <u>in press</u>. Evaluation of a water quality model (CE-QUAL-R1) using data from a small Wisconsin reservoir. Ecol. Modelling.

Wlosinski, J., 1984. Evaluation techniques for CE-QUAL-R1: A one-dimensional reservoir water quality model. Miscellaneous paper E-84-1, U.S. Army Engineers, Waterways Experiment Station, Vicksburg, Mississippi.

CHAPTER 16

FISH GROWTH

by
Jon From and Gorm Rasmussen

16.1 INTRODUCTION

The growth of a fish is considered as an interaction between the specimen and the environment. The **body size** is a concept of the model because no realistic growth model of any application can ignore the influence of body size upon the growth processes. A growth model can be of the metabolic type stressing the fate of food items. Increase of body weight by intake of water or rearranging of body constituents is not here considered as growth.

Growth has many aspects. Growth phenomena can take place at the cellular and at the organ level. Growth in relation to e.g. age, temperature, ration, and body size can be described by entirely empirical mathematical equations and their importance as analytical models of growth are determined by the information contained in the parameters.

Earlier growth models have been more or less **empirical equations** fitting a course of growth in relation to time or age, e.g. the **logistic**-, the **Gompertz-**, the **Johnson**-, and the **Richard growth curve**. These models are all discussed by Ricker (1979). The purpose was to get the best fit without considering the meaning of the parameters. Besides that, the collected data have been more or less biased from the sampling procedures, due to selective mortality of individuals within a year-class/population, and gear selectivity. In this way the computed course of the population growth curve is less than the "true average growth rate of the fish themselves", (Ricker, 1979). It was also generally observed that the growth curve in temperate climates varied seasonally with changes in temperature and food availability with a generally **sigmoid course of growth** when the fish approached what was called the **asymptotic body size**. Changes in the environment to more favorable conditions increases the growth of fish to a new and higher asymptotic body size. These distinctive patterns of growth in the life of a fish were called **growth stanzas** which was separated by physiological and ecological thresholds, Parker & Larkin (1959).

A growth model ought to consider all the factors that might influence growth. These factors are **1. intrinsic**: fish species and race, fish size, swimming activity, maturity, age, and **2. extrinsic**: which can be subdivided in **a. abiotic** and **b. biotic**.

Abiotic: photoperiod, temperature, oxygen content of the water, pH, carbon dioxide, various toxic substances such as ammonia, nitrite, heavy metals etc., salinity, light intensity, and care. Concerning growth in hard and soft streams it has been found (Edwards, Densem & Russel,1979) that in particular the high growth rates of trout in chalk streams may be related almost entirely to the thermal properties of such waters and not to direct effects of calcium.

Biotic: diets, ration, feeding frequency, care, diseases, and social hierarchy.

To incorporate all these factors in a growth model will demand an enormous amount of experimentation. Therefore it is necessary to reject many of the factors. As a beginning we must concentrate on the factors that are most important.

Stauffer (1973) concludes that: "any attempt at modelling growth must include the three factors: **ration**, **fish size** and **temperature**, as variables that have the most influence on the growth for a given species and diet."

16.2 THE METABOLIC GROWTH MODEL

The basis for animal life and by this growth is a food consumption. Hence a growth model will partly be a description of the fates of the food consumed. They can be represented as in Fig. 16.1.

Some of the <u>food consumed</u> are after digestion, <u>assimilated</u> through the intestinal wall. The rest of the food is passed out as <u>faecal loss</u>. Following ingestion of a meal, the rate of metabolism, expressed in units of **heat production**, increases. This increase is generally known as "**specific dynamic action**". Energy requirements for absorption, digestion, transportation, and deposition of food materials are distinct from those for specific dynamic action but experimentally difficult to separate. Where the distinction is not made the term "apparent specific dynamic action" is appropiate (Beamish, Niimi & Lett,1975). Not all of the assimilated materials can be used for the physiological work or growth, because some nitrogenous materials are not metabolizable but are excreted through the gills or kidneys as <u>excretional loss</u>. The remaining **metabolizable materials** are either used for **basal metabolism** and **activity** or appear as **growth.**

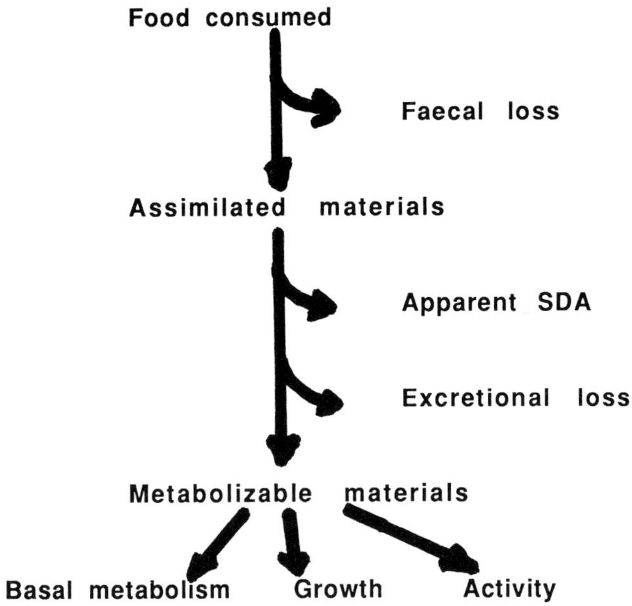

Fig. 16.1: Redrawn from Beamish, Niimi & Lett (1975).

Or in another way (from Davis & Warren, 1971):

$$C = F + U + B + R$$

C = energy value of **food consumed**
F = energy value of **faeces**
U = energy value of materials excreted in the **urine** or through the gills or skin
B = total change in energy value of materials of body (**growth**)
R = total energy of matabolism; this can be subdivided as follows:
R = Rs + Rd + Ra
Rs = energy equivalent to that released in the course of metabolism by an unfed and resting fish (**standard metabolism**)
Rd = additional energy released in the course of digestion, assimilation, and storage of materials consumed (including **specific dynamic action** or **SDA**)
Ra = additional energy released in the course of **swimming** and other activity

Some of the terms can be subdivided as follows:

F (total amount of **faeces**) consists of a mixture of non-assimilated food ("**true faeces**"), plus different non-reabsorbed residues ("**metabolic residues**") of body origin, from the intestine (mucosal cells, digestive enzymes, other secretions and microflora). Further, the "true faeces" consist of settable, suspended and dissolved faeces.

Rs consists both of an *oxygen consumption* and an excretional loss called *endogenous excretion,* U1. The excretional loss caused by feed intake is called *exogenous excretion,* U2. In this way, total excretion = U1 + U2. See, e.g. Brett & Groves (1979).

Some authors have failed to recognize U1 and hence they have claimed that the basal or *resting metabolism* can be measured solely as oxygen consumption. E.g. Warren & Davis (1967), Warren & Doudoroff (1971), Beamish, Niimi & Lett (1975).

Growth of a specimen can be considered as the difference between what enters the body and what leaves it: Growth = assimilated part of the food minus the part of food assimilated which gives energy to the different functions of the organism, so that:

$$\text{Growth} = \text{In} - \text{Out}$$

This reflection of growth may be developed and formalized in many ways.

From & Rasmussen (1984) have worked along the lines laid down by Ursin (1967) who elaborated the principles worked out by Pütter (1920). Pütter has:

$$\text{growth} = kl^2 - k'l^3$$

where k and k′ are constants and l is length.

Assuming isometrical growth Pütter says that the dimension of l is $G^{1/3}$ where G is weight. This gives:

$$\text{growth} = kaG^{2/3} - k'aG$$

where a is a constant.

In the following **Ursin′s growth model** will be described. It is mathematically consistent and applicable to parameter estimations based on relatively simple experimental designs.

Let **t** be the time variable, w(t) the weight of a fish at age t and Δt the length of the time period. And let Δt be considered as a relatively small period of time, say one day.

Formally the statement above can be written:

$$w(t + \Delta t) = w(t) + IN - OUT \qquad (16.1)$$

where the terms are measured in the same unit (e.g. energy or nitrogen). Rearranging (16.1) we have:

$$w = IN\,\Delta t - OUT\,\Delta t$$

Assuming as an approximation that a growth curve is a smoothed continuous curve, the difference quotient $\Delta w/\Delta t$ can be replaced by the differrential quotient dw/dt.

The basic equation is:

$$dw/dt = IN - OUT, \text{ or}$$

$$dw/dt = H(dR/dt) - K(w(t), H(dR/dt)), \qquad (16.2)$$

where

IN = H(dR/dt)

OUT = K(w(t), H(dR/dt))

Thus

dw/dt = weight change per unit time

w(t) = weight of fish at time t, a variable

dR/dt = weight of food consumed per unit time, **feeding rate**

H(dR/dt) = the **anabolic term** ("the build up term")

K(w(t), H(dR/dt)) = the **catabolic term** ("the break down term")

The anabolic term expressed that the quantity absorbed is a function of the quantity eaten. The catabolic term is described by two terms representing I. the catabolism of a starving fish, and II. the catabolism resulting from feeding and its subsequent processes.

The anabolic term

(I) Feeding

The functional coherence is assumed to be valid

$$dR / dt = fh(T)w(t)^m \qquad (16.3)$$

where

h(T) = **coefficient of anabolism**, temperature dependent
T = **temperature**, a variable
m = **exponent of anabolism**, a real number
f = **feeding level**, a variable
t = **time**

The **feeding level** is defined as the fraction eaten of the maximum quantity which could be eaten ($0 \leq f \leq 1$). The feeding level for a starving fish is 0, and for a fish eating the maximum ration $f = 1$.

(II) Assimilation
Efficiencies of the **absorption** of the nutrients in the diet are a fundamental part of dietary formulations (Fänge & Grove, 1979) but from a general point of view **energy** and/or **nitrogen assimilation** has gained wide application (Brett & Groves, 1979). Assimilation, ß, can be taken as the fraction of the food which is assimilated. It is generally realized that the efficiency of assimilation must be a function of food composition (both quantitatively and qualitatively), **feeding level**, **temperature** and **fish size**. At most instances the food is considered to be approximately the same. That means:

$$\beta = B(f,T,w) \qquad (16.4)$$

(III) Thus the **anabolic term** (IN) becomes:

$$H(dR/dt) = \beta f h(T) w(t)^m \qquad (16.5)$$

The catabolic term

This consists of the following two terms: (I) and (II):

(I) Starving catabolism
The catabolism of a starving fish ($f = 0$):

$$(dw/dt) = k(T) w(t)^n \qquad (16.6)$$

where
k(T) = **coefficient of catabolism**, temperature dependent
n = **exponent of catabolism**, a real number

(II) Feeding catabolism
The feeding catabolism assumed to be:

$$A\beta dR/dt \quad \text{where} \quad A = a + U2 \qquad (16.7)$$

where A represents the fraction of the assimilated food producing energy for the catabolic processes resulting from feeding. The value of A depends on the food type and **feeding level**, and to a lesser degree on **temperature** and **fish size**. Apart from the additional energy required for eating a corresponds to what Beamish, Niimi & Lett (1975) call the **"apparent specific dynamic action"**, measured as oxygen consumption see paragraph "Oxygen consumption, a + k´(T)". And U2 is the energetics loss from the exogenous nitrogen excretion, see paragraph "Nitrogen excretion, U".

In energy terms: A = a + U2, where U2 is recalculated from exogenous excretion to energy.

In nitrogen terms: A = U2.

(III) Total catabolism

(I) + (II) give the catabolic term (OUT):

$$K(w(t),H(dR/dt)) = k(T)w(t)^n + A\beta dR/dt \qquad (16.8)$$

Nitrogen excretion, U

The nitrogen excretion, U, is the sum of the **endogenous excretion**, U1, from starving fish, and the **exogenous excretion**, U2, from feeding fish, i.e. U = U1 + U2, e.g. Brett & Groves (1979).

The nitrogen excretion U2 is analogous to the increase in oxygen consumption after feeding.

In fish U will consist of NH_3, urea and negligible amounts of uric acid, amino acids, and other nitrogen containing compounds (Forster & Goldstein, 1969; Goldsteing & Forster, 1970; Fische, 1977). The composition of the nitrogen excretion differs in salt- and fresh-water fish, but there are also differences from species to species within the two ecological groups (Brett & Grove, 1979).

Oxygen consumption, a + k´(T)

The oxygen consumption of a fish is considered as the sum of: (1) the oxygen consumption k´(T) of a starving fish (f = 0), and, (2) the oxygen consumption of a feeding or fed fish (f > 0).

The oxygen consumption of a feeding or fed fish depends on **feeding level, temperature** and **fish weight**, in such a way that a – 0 for f – 0, and a is maximum for f = 1.

The unit used in a growth equation

The only **units** in which all the quantities can be measured are **energy** and **nitrogen**. If units of g body wet weight is used the "in" and "out" cannot be split up in faeces, excretory products, A, and so on. Further, it wet weight alone is used it must be assumed with Ursin (1967), that the food has the same chemical constitution as the fish. If a model shall be used in connection with **fish farming**, where pelleted food is used this assumption is clearly not permissible.

Therefore we can write the following **balanced equations**:

Nitrogen

dw/dt = consumed - faecal - excreted (exogenous + endogenous)

Energy

dw/dt = consumed - faecal - excreted nitrogen (exogenous + endogenous recalculated to energy) - feeding respiration (recalculated to energy) - starving respiration (recalculated to energy)

Energy is most often measured in **calories** or **Joules**, but in experiments most often energy of metabolism will be measured as an **oxygen consumption**. When the oxygen consumption of a living animal is to be converted to calories it is not only necessary to consider the consumption of the food but also of the excretory product. Krokhin (1959) uses a coefficient of 3.38 cal/mg oxygen, Davis & Warren (1971) use 3.42 cal/mg oxygen. But Elliott & Davison (1975) say that 3.42 cal/mg oxygen may be applicable to some herbivorous fish but 3.24 cal/mg oxygen is more appropriate for a carnivorous fish that utilizes ammonia as its chief excretory product. For a proteinaceous diet Brafield & Solomon (1972) find a value of 3.20 cal/mg oxygen. All these considerations and inaccuracies are avoided if **mg oxygen** is used as energy unit. But here the problem arises that the inorganic material (NH_3-N) in the **excretory product**, cannot be measured in this unit. **Ammonia** can be converted to energy by using a value of 5.94 cal/mg, Elliot & Davision (1975). This figure can then be converted to mg oxygen by using the factor 1/3.4 mg oxygen/cal = 0.3 mg oxygen/cal

i.e. X mg NH_3-N = X * 5.94 * 0.3 mg oxygen = X * 1.8 mg oxygen.

The inaccuracy by using these two conversion factors is not important as the fraction of ammonia usually constitutes less than 5% of the ingested food measured as energy. See Table 16.1.

The model

Inserting (16.3) in (16.8) and then (16.5) and (16.8) into (16.2) gives the **growth equation**

$$dw/dt = (1 - A)\, ßfh(T)w(t)^m - k(T)w(t)^n \qquad (16.9)$$

This **growth model** can describe the course of a growth curve according to varying factors, i.e. **temperature, ration size and fish size**. Further, it may predict **maintenance ration** as a function of temperature and fish size; amount of **faeces** and **ammonia** led into recipients from aquaculture systems, and so on.

16.3 PARAMETER ESTIMATION

The reduced metabolic growth model

The growth model to be described implies that **temperature**, T and **feeding level**, f are approximately constant in time. This reduces (16.9) to the form:

$$dw/dt = Hw(t)^m - kw(t)^n \qquad (16.10)$$

where
$H = (1 - A)\, ßfh(T)$ and
$k = k(T)$

Only if $m = n$ or $m = 1$ or $n = 1$ will there exist an analytical solution of (16.10). The shape of the growth curve depends on m and n.

(i) $m < n = 1$

The **von Bertalanffy equation** (Pütter, 1920; von Bertalanffy, 1957) states that for fish $m = 2/3$ and $n = 1.0$.

Putting $dw/dt = 0$ gives:
$w = W_{oo} = (H/k)^{1/(n-m)}$

W_{oo} is an **asymptote** for $n > m$ and the curve has an inflection at

$$w = W_I = W_{oo}\,(m/n)^{1/(n-m)} = 8/27W_{oo}$$

The value 2/3 is based on the assumption that the quantity of food absorbed is proportional to the absorbing surface which in the case of isometrical growth is proportional to $w^{2/3}$. The value 1 for n stems from the assumption that catabolic processes take place all over the body with no respiratory limitations and therefore are proportional to w. With these values for m and n the equation has been widely used (Pütter, 1920; Bertalanffy, 1957 and Beverton & Holt, 1957). However, there is extensive evidence (Parker & Larkin, 1959; Hemmingsen, 1969; Winberg, 1960, 1961; Sperber, From & Sparre, 1977; Rasmussen & Therkildsen, 1979 and From and Rasmussen, 1984) that $n < 1$ ($0.73 \leq n \leq 0.81$) and so fasting catabolism is not proportional to weight. The size of m has not often been estimated. For **salmonids** m is found to vary from 0.68 to 0.84, Elliott (1976), Sperber, From & Sparre (1977), and From & Rasmussen (1984). For **european eel, *Anguilla anguilla***, m could be approximated to the value 0.82, Rasmussen & Therkildsen (1979).

However, the growth equation

$$dw/dt = Hw^{2/3} - kw \qquad (16.11)$$

has been used very successfully for commercially exploited fish populations (Beverton & Holt, 1957).

Integration of (16.11) gives:

$$w(t) = W_{oo} \, (1 - \exp(-(t - t_o) \, K)) \qquad (16.12)$$

where

$K = k/3$

t_o = the **hypothetical age** at which the fish is of length zero

Growth in weight is transformed into growth in length on the assumption that

$$w(t) = ql(t)^3$$

where

l(t) is body length and q is a constant (the **condition factor**).

Equation (16.13) gives the growth equation in length:

$$l(t) = l_{oo} \, (1 - \exp(-(t - t_o) \, k))^3 \qquad (16.13)$$

(II) m = n

Putting m = n reduces equation (16.10) to

$$dw/dt = aw^n \qquad (16.14)$$

where $a = H - k$

(16.14) might easily be integrated and gives:

$$w(t) = (w(0)^{(1-n)} + at)^{1/(1-n)} \qquad (16.15)$$

Equation (16.15) has no asymptote and has been used by Sparre (1979) and Rasmussen (1983) for **european eel.**

(III) m = n = 1

For m = n = 1 equation (16.10) reduces to:

dw/dt = aw, which integrated gives

$$w_t = w_0 \exp(at) \qquad (16.16)$$

Equation (16.16) has been used widely, e.g. Ricker (1975) and Chapman (1978), where a = G (the **instantaneous growth rate**).

(IV) Spawning

Equation (16.10) does not include spawning loss. The growth curve is thus considered as an approximation to the actual weights at the different times.

Growth modified by spawning is achieved by adding the loss of **gonadal products** as an additional OUT term in the growth equation:

$$dw/dt = Hw^m - kw^n - OUT_{gonadal} \qquad (16.17)$$

Spawning by a fish population is normally distributed about a certain day so that the average spawning rate of the population gradually increases to a maximum after which it begins to decrease until it again reaches zero at the end of the spawning season. At the present time no spawning model can be formalized with parameters from the literature. Therefore spawning is considered to be momentaneous. Body weight is reduced by a fraction Π

simulating the loss of gonadal products. This has been used by ,e.g. Stewart, Weininger, Rottiers & Edsall (1983).

Thus equation (16.17) is integrated until the age of first spawning to produce a continuous growth of the individual fish but at time ts the body weight suddenly drops due to spawning, i.e.

$$w(ts+) \quad = \quad w(ts-) * (1 - \Pi)$$

where w(ts-) is the body weight immediately before spawning, Π is the gonadal products as a fraction of the body weight and w(ts+) is the body weight immediately after spawning.

The extended metabolic growth model

Describing growth during a shorter interval than a year will normally imply that temperature T and feeding level f are not constant. It is necessary to introduce additional parameters for h(T), ß, A and k(T).

Only controlled experiments (e.g. in aquaria or ponds) where f and T are known can be used to estimate the parameters.

At the present time a very limited number of controlled experiments, where nearly all the parameters in relation to fish **weight, temperature** and **feeding level** are estimated, have been carried out, e.g. Ellitoo (1976) and From & Rasmussen (1984).

(I) The anabolic term

Feeding

Ursin (1979) describes the **coefficient of anabolism** as

$$h(T) \quad = \quad \frac{1}{h1 * \exp(h2*T) \; + \; h3 * \exp(h4*T)} \qquad (16.18)$$

and says that the expression "is derived from the **Michaelis-Menten expression** for the rate of enzymatic processes and the **Arrhenius equation** for the temperature dependence of simple chemical processes".

According to this expression the **feeding rate** increases with increasing temperature up to a maximum point beyond which it decreases.

Equation (16.18) could be substituted by the **hyperbolic caternary curve** of Janish (1927). But the **catenary curve** should be used with reservation, because up till now it has not been possible to produce a symmetrical curve based on experiments relating feeding rate and temperature.

Most often, however, only the ascending part of the curve is used, i.e. at temperatures below the **maximum feeding rate**, so that an approximation of h(T) can be described as:

$$h(T) = h1 * \exp(h2*T) \quad (16.18a)$$

From an "estimating point of view", (16.18) can be substituted by a purely empirical equation, e.g. a second - or, better, a third order polynomial, where it is simple to estimate the parameters. And in practice it gives the same relationship between observed and calculated observations:

$$h(T) = h1 + h2*T + h3*T^2 + h4*T^3 \quad (16.18b)$$

Assimilation

The total amount of **faeces** = (1 - ß) ΔR. At a given **fish weight** and **temperature** the amount of faeces is expected to be maximum for f = 1 and 0 for f = 0.

If ß is calculated as 1 - (total-faeces/food) and expressed as a function of f, temperature, and weight, the size of ß will be underestimated, depending on the amount of the **metabolic residues** in the faeces. Mainly, the settable faeces will originate from non-assimilated food.

The amount of settable faeces can therefore be described as:

$$\text{Settable faeces} = b1 * f^{b2} * \exp(b3*T)\ w^{b4} \quad (16.19)$$

(16.19) expresses that settable faeces only will occur for f > 0. As the suspended and dissolved faeces from a feeding or fed fish cannot be separated analytically into "**true faeces**" and "**metabolic residues**" this fraction of faeces for $0 \leq f \leq 1$ can best be described as:

$$(\text{Suspended} + \text{dissolved})\ \text{faeces} = b1 * \exp(b2*f)\ \exp(b3*T)w^{b4} \quad (16.20)$$

Equation (16.20) expresses that this fraction of the total amount of faeces has contributions from both **starving** and **feeding fish**.

In this way the amount of suspended + dissolved faeces is:

$$(\text{Suspended} + \text{dissolved})\ \text{faeces, f= 0: } b1 * (b3*T)w^{b4} \quad (16.20a)$$

so that

$$(\text{Suspended} + \text{dissolved})\ \text{faeces, f > 0: } b1 * (\exp(b2*f)-1)\ \exp(b3*T)w^{b4} \quad (16.20b)$$

Subsequently for f > 0 the total amount of faeces originating from feeding is:

Total faeces = (16.19) + (16.20b)

$ß = 1 - \text{total faeces}/fh(T)w^m$

(II) The catabolic term

Starving metabolism

k(T) can, in the same way as h(T), be considered as a function of **temperature**,

$$k(T) = k1 * \exp(k2*T) \qquad (16.21)$$

As a starving fish has (a) a respiration k'(T), (b) a loss in exfoliated cells k''(T), both from epidermis and stomach and gut epithelium, and (c) a loss in urine k'''(T), k(T) can be split up into:

$$k(T) = k'(T) + k''(T) + k'''(T) \qquad (16.22)$$

where

k'(T) = "Krogh's respiration curve", see equation (16.24)

k''(T) = $b1*\exp(b3*T)w^{b4}$ ("faeces"), see equation (16.20a) and

k'''(T) = U1, (endogeneous excretion, see equation (16.23a).

From all the experimental data and references compiled in Brett & Groves (1979), the metabolism of a starving fish is adequately described as an exponential function of temperature up to a certain point when death occurs. And there is no basis for believing that k(T) will reach a maximum, and then decrease beyond this, as proposed by Ursin (1967, 1979).

In energy terms:

k(T) = k'(T) + k''(T) + k'''(T), where k'''(T) is recalculated from endogenous excretion to energy.

In nitrogen terms:

k(T) = k''(T) + k'''(T).

Feeding metabolism

This term is considered previously.

Nitrogen excretion U

U1, which represents a turnover of nitrogen, can be found from starving fish but will be an approximation to the true value of the endogenous nitrogen excretion. The true value will normally be a little lower than the value found for starving fish because these have an increased conversion of protein to fulfil their requirement for energy. See e.g. Brett & Groves (1979).

Determination of U2, which represents a nitrogen and energy loss of the assimilated food, can only be done on the basis of determinations of the total nitrogen loss on fed and starved fish, respectively. U2 is thus determined as U2 = U - U1. But in practice it will be difficult to separate U1 and U2, see Brett & Groves (1979).

The size of U can be described as:

$$\underline{\text{for } f \geq 0}: \quad U = u1 * \exp(u2*f) \exp(u3*T) w^{u4} \tag{16.23}$$

In this way

$$\underline{\text{for } f = 0}: \quad U1 = u1 * \exp(u3*T) w^{u4} \tag{16.23a}$$

and

$$\underline{\text{for } f > 0}: \quad U2 = u1 * (\exp(u3*f) - 1) \exp(u3*T) w^{u4} \tag{16.23b}$$

Oxygen consumption, a + k'(T)

The size of k'(T) can be described as:

$$k'(T) = il * \exp(i2*T) w^{i3} \tag{16.24}$$

As the total oxygen consumption of a fed or feeding fish cannot directly be separated in contributions from a and k'(T) we have:

$$\text{Total respiration} = a1 * \exp(a2*f) \exp(a3*T) w^{a4} \tag{16.25}$$

so:

$$\underline{\text{for } f = 0}: \quad k'(T) = il * \exp(i2*T) w^{i3} \approx a1 * \exp(a3*T) w^{a4} \tag{16.25a}$$

and

$$\underline{\text{for } f > 0}: \quad a1 * (\exp(a2*f) - 1) \exp(a3*T) w^{a4} \tag{16.25b}$$

16.4 APPLICATION AND EXAMPLES

The reduced metabolic growth model

In Fig. 16.2 the **spawning loss** is accounted for (i.e. true growth curve) and it is clearly showed that the concept of **asymptotic growth** (i.e. population growth curve) is not completely proved, based on weight-at-age data which are influenced by other factors.

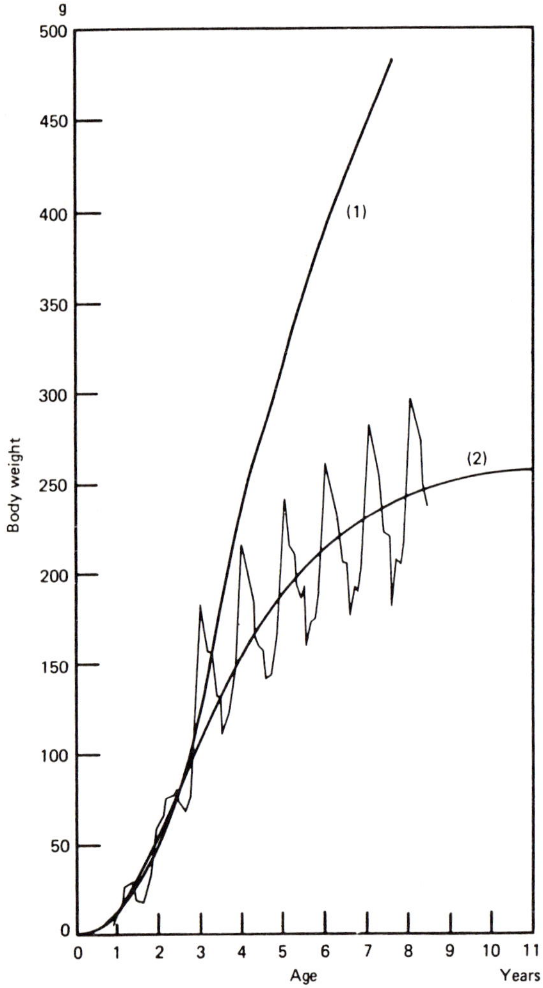

Fig. 16.2: Calculated growth course in weight of **North Sea herring**, *Clupea harengus* (1) calculated growth disregarding spawning; (2) smoothed growth using weight-at-age data at yearly intervals from seasonal observation of weight. After Ursin, 1979.

The extented metabolic growth model

From & Rasmussen (1984) made a study in aquaria of the growth of **rainbow trout**, *Salmo gairdneri.* Moist pellets (40% water) with constant composition were used as food. Examples from this study will be given here.

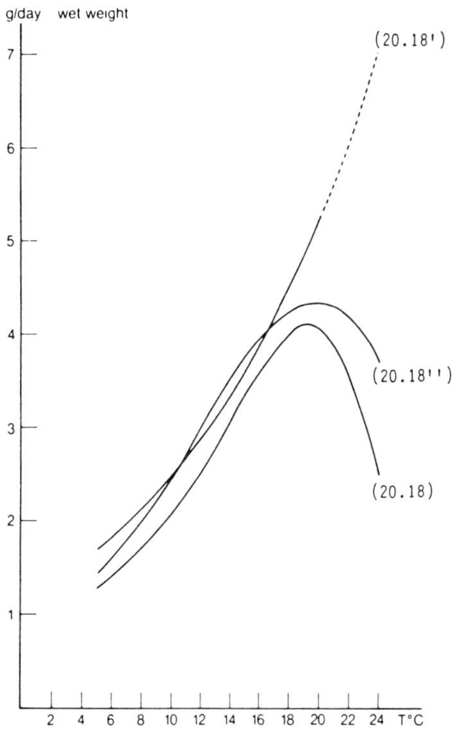

Fig. 16.3: Daily maximum food intake of moist pellets for a fish of 100 g wet weight. According to (16.18), (16.18a) and (16.18b).

(I) The anabolic term

Feeding

Fig 16.3 shows the daily **maximum food intake** (wet weight) of a 100 g fish wet weight for the models (16.18), (16.18a) and (16.18b). From Fig. 16.3 it is seen that all three models can be used to calculated the food intake up to 16-17°C. After this temperature (16.18) and (16.18ab) can be used because they take into consideration the decrease in food intake at higher temperatures.

Fig. 16.4 shows the calculated curves for h(T) from (16.18). Because different units (g wet weight, g nitrogen, g COD, Kcal) are used the h(T) curves differ. For example, converting g COD to Kcal, by g COD = 3.42 Kcal, makes the COD and calorifie curves identical.

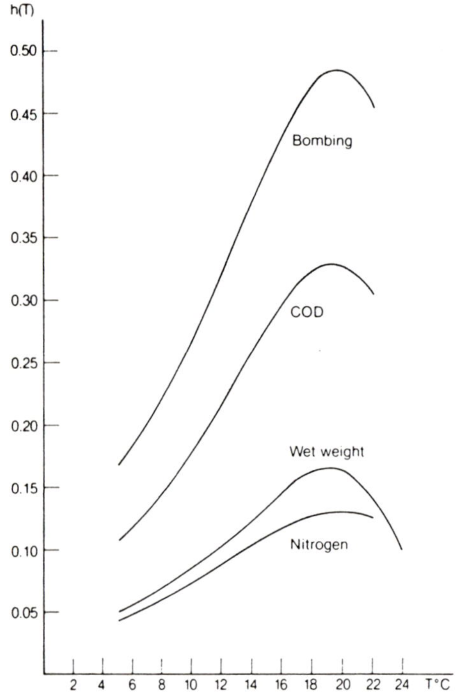

Fig. 16.4: h(T) for the different units (g wet weight, g nitrogen, g COD, and Kcal) according to (16.18).

Assimilation

Fig. 16.5 shows the calculated values of settable **faeces**, and suspended + dissolved faeces as a function of feeding level for different temperatures. The intercept for f = 0 corresponds to the non-reabsorbed metabolic residues. Fig. 16.6 shows simulated values of ß for different values of f, T, and w. The **assimilation efficiency** increases with decreasing **feeding level** at a given **temperature** and **fish weight.**

An increase in temperature at a given feeding level and fish size has negligible effect on ß. Fish weight has an effect of ß in such a way that a bigger fish assimilates more effectively than a smaller one.

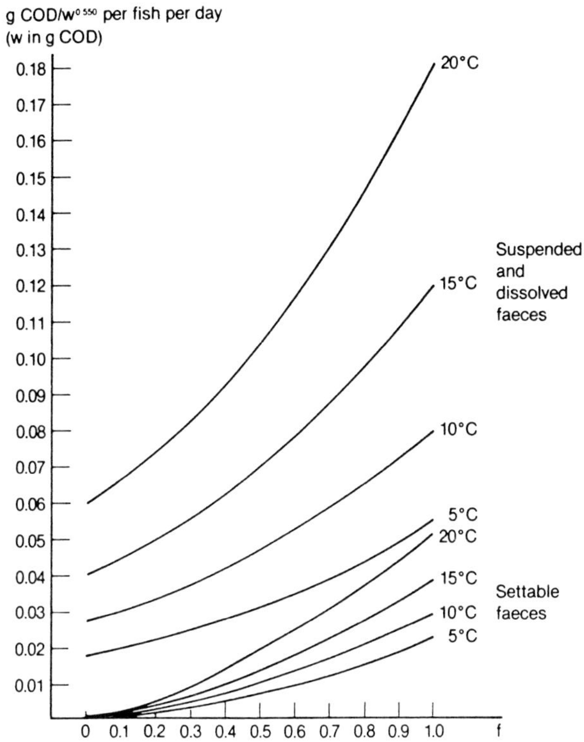

Fig. 16.5: Calculated curves for settable (16.19) and suspended + dissolved faeces (16.20) as a function of feeding level at 5, 10, 15 and 20°C.

(II) The catabolic term

Starving metabolism

Fig. 16.7 shows k(T) calculated on the basis of (16.21). Again the curves are not identical because different units are used. k(T) increases with increasing temperature.

Nitrogen excretion

In Fig. 16.8 the calculated values of NH_3-N excretion as a function of feeding level, at different temperatures are shown. The values for f = 0 corresponds to the **endogenous excretion** U1. Independent of fish size and temperature U2/U varies from 0% for f = 0 and up to 88.5% for f = 1.

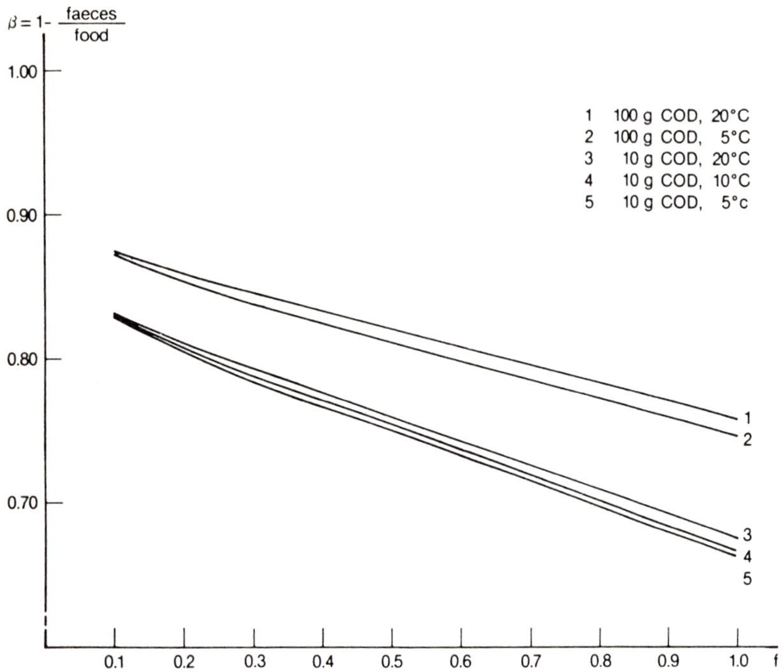

Fig. 16.6: Calculated curves of the assimilation coefficient ß for different values of f, T and w.

At a given temperature the **exogenous excretion** constitutes an increasing amount of the **nitrogen** content of the food, with increasing f, however, in such a way that a bigger fish excretes a little more than a smaller fish. As examples can be given:

Temp °C	Fish weight in g N	Feeding level	$U_2/\Delta R$% in nitrogen units
10.0	0.25 (~10 g wet)	0.1	7.1
		0.5	11.4
		1.0	22.5
	2.5 (~100 g wet)	0.1	7.4
		0.5	11.9
		1.0	23.5

Temp °C	Fish weight in g N	Feeding level	$U_2/\Delta R$% in nitrogen units
20.0	0.25 (~10 g wet)	0.1	10.0
		0.5	16.1
		1.0	35.9
	2.5 (~100 g wet)	0.1	10.4
		0.5	16.8
		1.0	37.4

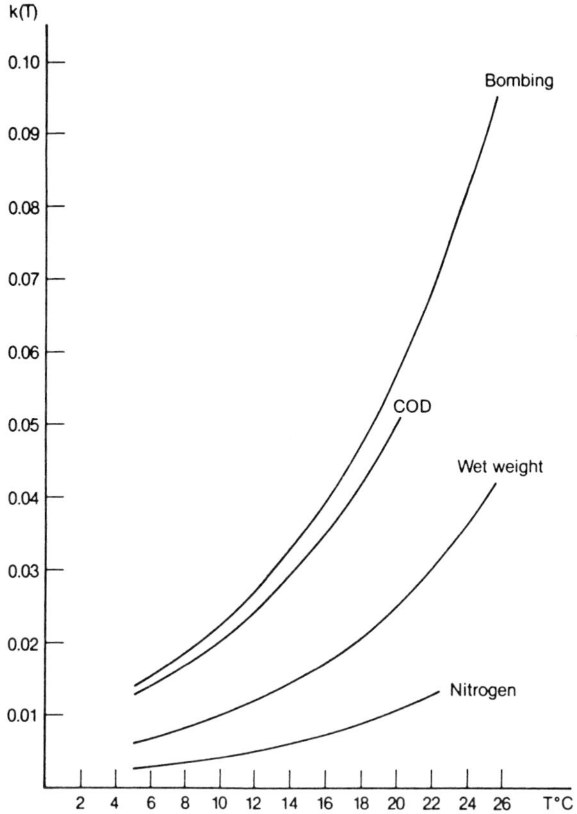

Fig. 16.7: k(T) for the different units (g wet weight, g nitrogen, g COD and Kcal) according to (16.21).

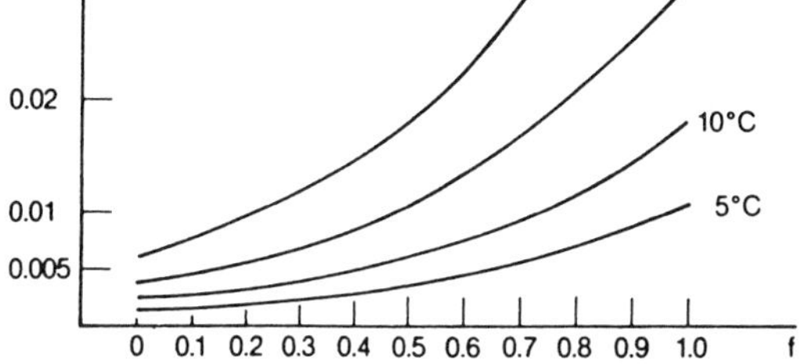

Fig. 16.8: Calculated curves of the NH_3-N excretion rates as a function of f at 5, 10, 15 and 20 °C.

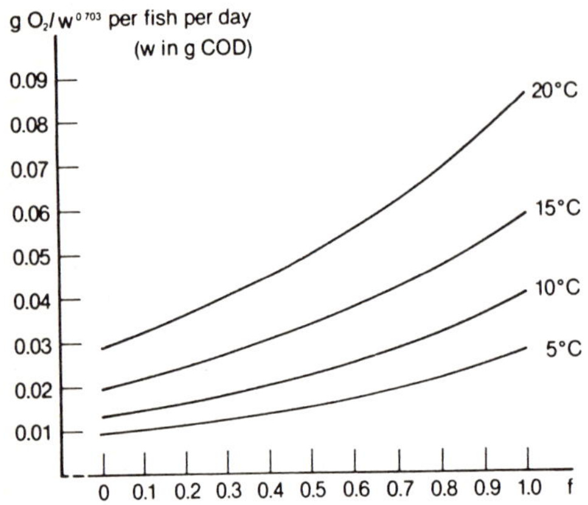

Fig. 16.9: Total respiration as a function of f at 5, 10, 15 and 20 °C.

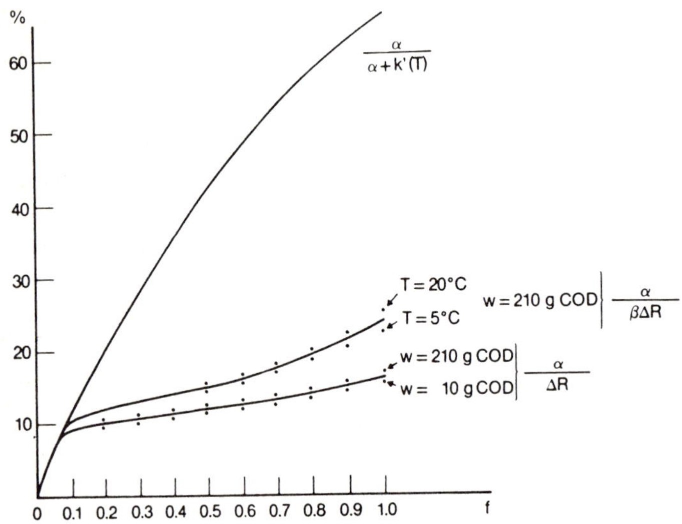

Fig. 16.10: ß in proportion of total respiration, assimilated, and ingested food.

Oxygen consumption

In Fig. 16.9, the calculated values for total respiration of fish as a function of feeding level, are shown for different temperatures. In Fig. 16.10 the calculated values of ß in proportion to total respiration, assimilated, and ingested food are shown for different values of **feeding level**,

temperature and **fish weight** (units of **COD**). Independent of fish weight and temperature ß constitutes from 0% at f = 0 to 66.3% of the total respiration at f = 1, calculated on basis of (16.25). Of course, at f = 0, ß constitutes 0% of the food, for f = 0.1 ß constitutes about 9-10% and rises to 15-16% for f = 1.0, slightly dependent on fish weight and temperature. Though the absolute values of ß for f = 0.1 is about 5% of ß for f = 1.0, ß's share of the assimilated food, ß(dR7dt) constitutes a decreasing amount with increasing fish weight but is nearly independent of temperature (for f = 0.1 about 11% of ß(dR/dt), for f = 1 about 24%).

Brett & Groves (1979) sum up the data for determinations of feeding metabolism, and conclude that the ratio of **feeding metabolic rate** to **routine metabolism** can be put equal to 1.7 ± 0.4 (standard deviation). This ratio will depend on feeding level, and a ratio of 1.7 is found in From & Rasmussen (1984) for f = 0.5, and for f = 1 the ratio is at a maximum and 2.0. Further, the feeding metabolic rate increases with increasing intake of food and increasing temperature at a given fish size.

ß determined on the basis of oxygen comsumption in connection with food intake, as most often done, is not an expression of the true "**heat increment**". Smith, Rumsey & Scott (1978a, b) are probably the only researchers in the literature who have determined the heat increment by determining the direct energy loss in connection with the physiological handling of the food, independent of the activity associated with this. However, the major part of the heat increment constitutes only a few per cent of the food, but as artificial food compositions were used, it is difficult to compare directly these results with others, see Brett & Groves (1979).

An advantage of measuring ß as oxygen consumption is, besides the problem of a correct **oxycalorific coefficient** (see Solomon & Brafield, 1972, and Elliott & Davison, 1975), that an enhanced activity in connection with an increased food intake merely results in an increased oxygen consumption, which can be measured with relatively simple technical facilities (as long as the energy consumption does not come from **anaerobic processes**) (Blazka, 1958, and Mathur, 1967).

The oxygen consumption at enhanced respiration in connection with and caused by increased food intake can be used as an indicator for ß. For example, in Brett & Groves (1979) it is claimed on account of oxygen measurements that "**apparent SDA**" constitutes 12-16% of the food, which is in accordance with the results in Fig. 16.10. Cho, Bayley & Slinger (1976) found that the fraction, for rainbow trout fed with pellets, varied from 8 to 12%, while Miura, Suzuki, Nagoshi & Yamamura (1976) found that the ratio for **biwamasu salmon,** *Oncorhyncus rhodorus* varied from 9.5-25.9%,

(18.9% in mean) and of the **assimilated food** the ratio varied from 11.9-32.3% (23.3% in mean).

(III) Simulated growth, the growth equation

Energy

The growth equation can be written down as follows on the basis of **COD** measurements:

dw/dt = food - faeces (caused by food) - feeding catabolism - exogenous excretion - starving catabolism.

For T ≤ 20.1°C the equation is:

dw/dt = (16.3) and (16.18a) - (16.20b) - (16.19) - (16.25b) - 1.8 * (16.23b) - (16.21)

$$dw/dt = f \cdot 0.086 \cdot \exp(0.0761 \cdot T) \cdot w^{0.6767} - 0.011861 \cdot [\exp(1.1057 \cdot f) - 1] \cdot \exp(0.0808 \cdot T) \cdot w^{0.5499} - 0.0162 \cdot f^{1.4259} \cdot \exp(0.0577 \cdot T) \cdot w^{0.5471} - 0.006237 \cdot [\exp(1.0885 \cdot f) - 1] \cdot \exp(0.0769 \cdot T) \cdot w^{0.7030} - 1.8 \cdot 1.2766E{-4} \cdot [\exp(2.132 \cdot f) - 1] \cdot \exp(0.1112 \cdot T) \cdot w^{0.6572} - 0.008464 \cdot \exp(0.0911 \cdot T) \cdot w^{0.7751}$$

The growth equation set up can among other things be used to simulate a course of growth where **temperature** and **feeding level** enter as variables. It can also be used to construct an **energy budget** which shows relatively how much the single terms constitute of, for example, the food. Examples are shown in Table 16.1.

In Fig. 16.11 dw/dt is shown as a function of feeding level for different temperatures for a fish of 10 and 100 g COD, respectively. dw/dR = 0 corresponds to f = **f(maintenance)**, which for a given fish weight increases a little with temperature.

When dw/dR is maximum, the fish, from purely energetical considerations, has the most effective utilization of the food, **R(opt),** e.g. (dw/dR) ~ 0.40 at 5 °C and ~ 0.33 at 20°C.

It appears from Fig. 16.11 that dw/dR for low values of f, at a given temperature, is smaller for the smallest fish. For f $\sim$ 0.75 the bigger fish have a better utilization of food than the smaller fish at a given temperature. The picture is the same for other fish sizes in such a way that the interception between the curves for the two given fish sizes is for f $\sim$ 0.55 at 5°C.

Table 16.1
Energy budget at different feeding levels

		Per cent of ΔR							
		Defecation		Excretion		Respiration			
f	ΔR gCOD	Settable faeces	Susp.+ diss. faeces	U urine	U_2 exo-gen	ß	k'(T)	k(T) biomass changes	dw/dt
T = 5°C, w = 10 g COD									
0.1	0.0596	4.80	12.37	4.59	0.9	8.73	77.64	133.50	-60.50
0.2	0.1191	6.45	13.09	2.85	1.0	9.44	38.82	66.76	3.26
0.3	0.1787	7.66	13.87	2.36	1.1	9.99	25.88	44.51	22.87
0.4	0.2382	8.66	14.71	2.20	1.3	10.59	19.41	33.38	31.36
0.5	0.2978	9.52	15.62	2.18	1.4	11.23	15.53	26.71	35.52
0.6	0.3573	10.29	16.60	2.26	1.6	11.92	12.94	22.25	37.34
0.7	0.4169	10.99	17.66	2.41	1.9	12.67	11.09	19.08	37.70
0.8	0.4765	11.63	18.81	2.61	2.2	13.48	9.70	16.69	37.19
0.9	0.5360	13.23	20.05	2.89	2.5	14.35	8.63	14.84	36.03
1.0	0.5956	12.79	21.39	3.23	2.9	15.29	7.76	13.35	34.28
T = 20°C, w = 10 g COD									
0.1	0.1865	3.64	13.27	7.62	1.5	9.04	78.57	167.20	-94.65
0.2	0.3730	4.89	14.05	4.73	1.7	9.55	39.29	83.60	-13.79
0.3	0.5595	5.81	14.89	3.92	1.9	10.11	26.19	55.74	11.55
0.4	0.7460	6.57	15.79	3.65	2.1	10.72	19.64	41.80	23.02
0.5	0.9225	7.23	16.76	3.63	2.4	11.37	15.72	33.44	28.80
0.6	1.1190	7.81	17.82	3.75	2.7	12.07	13.10	27.87	31.73
0.7	1.3055	8.34	18.95	3.99	3.1	12.82	11.22	23.89	32.90
0.8	1.4920	8.83	20.18	4.34	3.6	13.64	9.82	20.90	32.85
0.9	1.6785	9.28	21.51	4.79	4.1	14.52	8.73	18.58	32.01
1.0	1.8650	9.71	22.95	5.36	4.7	15.48	7.86	16.72	30.44

In the literature there has been discussion whether dw/dR is size-dependent, and it is often claimed that dw/dR is decreasing with increasing **fish size**. Paloheimo & Dickie (1966b) conclude that the relationship between log K = log(Δw/(TΔt)) and R is adequately described by a linear equation. This equation expresses growth as a function of rations, independent of **body weight**. Warren & Doudoroff (1971) comment on Paloheimo & Dickie (1966b) and simply state that large fish do tend to grow with less efficiency than small fish. But they give no evidence for this.

Brett & Groves (1979) claim that there is a decreasing **conversion efficiency** accompanying increasing size and say that the relationship is apparent in the studies of Kinne (1960). However, from Kinne's Table 13 it is seen that at 15 and 20°C it is the biggest fish that have the highest **gross efficiency** whereas it is the opposite at 25, 30 and 35 °C.

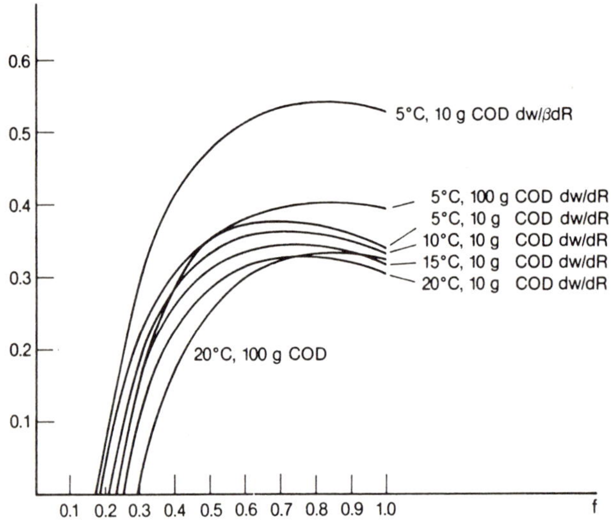

Fig. 16.11: dw/dR as a function of f for fish of 10 g COD (app. 27 g wet weight) and 100 g COD (app. 270 g wet weight) at different temperatures. Further, dw/ßdR is shown at 5°C for a fish of 10 g COD.

So it can be concluded that it is not obvious that dw/dR decreases with increasing fish size. Instead of presupposing any relationship between dw/dR and fish size, or age, it would be more profitable to carry out controlled feeding experiments to determine the parameters of the terms in a growth equation. Particularly the sizes of the **weight exponents** in the **anabolic terms** in relation to the weight exponents for the **catabolic terms.** When it is observed that the growth rate decreases as a function of **age,** e.g. Beverton & Holt (1957), it might be a true physiological fact. But it can also be an effect of **spawning**. Here a high loss of biomass has to be compensated for before additional growth can take place, e.g. Ursin (1979). The physiological growth parameters cannot be deduced from age/length-data. It might be possible that controlled aquarium experiments with, for example sexually immature fish will show a non-decreasing growth rate until maturing takes place, beyond which the observed growth rate decreases.

Nitrogen

The growth equation can be written down as follows on the basis of nitrogen measurements:

dw/dt = food - faeces - excretion (endogenous + exogenous).

For T≤ 20.1°C the equation is:

dw/dt = (16.3) and (16.18a) - (16.20b) - (16.19) - (16.23b) - (16.23a)

formel

dw/dR for an average **rainbow trout**, 3.25 g N (~ 130 g wet weight) at 5, 10, 15 and 20°C is shown in Fig. 16.12. Here the same considerations as in Fig. 16.11 about the values of dw/dR are valid.

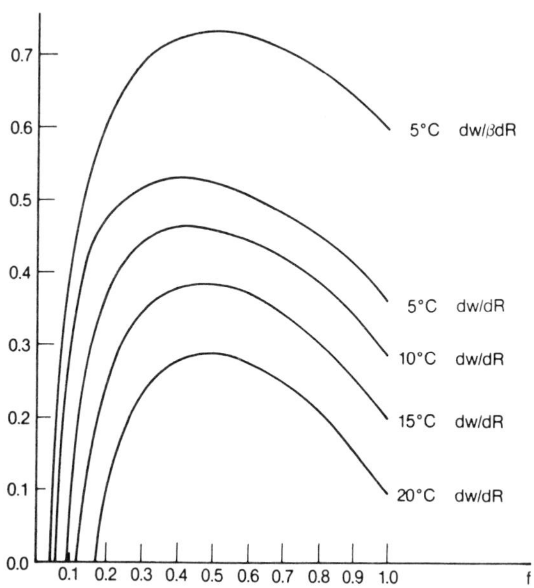

Fig. 16.12: dw/dR as a function of f for fish of 3.25 g N (app. 130 g wet weight) at different temperatures. Further, dw/ßdR is shown at 5°C for a fish of same size.

Body composition

To estimate the parameters it is only necessary to measure the fish in **energy** terms (e.g. **mg oxygen** by means of **COD**). But in growth experiments the body composition of the fish is often determined.

The fish consists of : **water** + **protein** + **nitrogen extractive** + **lipid** + **ash** + **carbohydrate**.

The terms in the relationship can be found from **proximate analysis** of fish samples.

Together carbohydrate and nitrogen extractive contribute less than about 3% of the total energy in fish **dry weight** samples (From & Rasmussen, 1984), and are normally disregarded.

Assuming that proteins consist of 16% **nitrogen**, a conversion factor of 100/16 = 6.25 is used to calculate the amount of protein in the sample from the analyzed amount of nitrogen in the sample. From equivalents of 5.65 and 9.45 cal/mg of protein and lipid respectively the energy of a sample can be calculated from the amount of protein and lipid. If the total energy content of a sample has been determined together with either protein or lipid the other can be roughly estimated from the relationship:

Total energy cal/mg dry weight (e.g. **bombing** or **COD**) = (protein * 5.65 + lipid * 9.45) cal/mg assuming that the ash does not contribute any energy. This will first take place at a ash content of 25% or more (Ostapenya, 1971). For further discussion of body composition, see e.g. Love (1970, 1980); Niimi (1972); Elliott (1976) and From & Rasmussen (1984).

As mentioned previously the only **units** in which all the quantities in the growth model can be measured are **energy** and **nitrogen**. If **wet weights** are required in a simulated course of growth it is necessary to convert the values from energy or nitrogen to values of wet weights. Doing this it must be taken into account that the conversion depends on the percent of water in the fish. An example is shown in Fig. 16.13.

Gastric evacuation

In metabolic growth modelling it is necessary to know the weight of the fish and it is important that no **stomach content** shall interfere with the true **body weight** w(t). It is necessary to know at what time a previously fed fish has evacuated its stomach content.

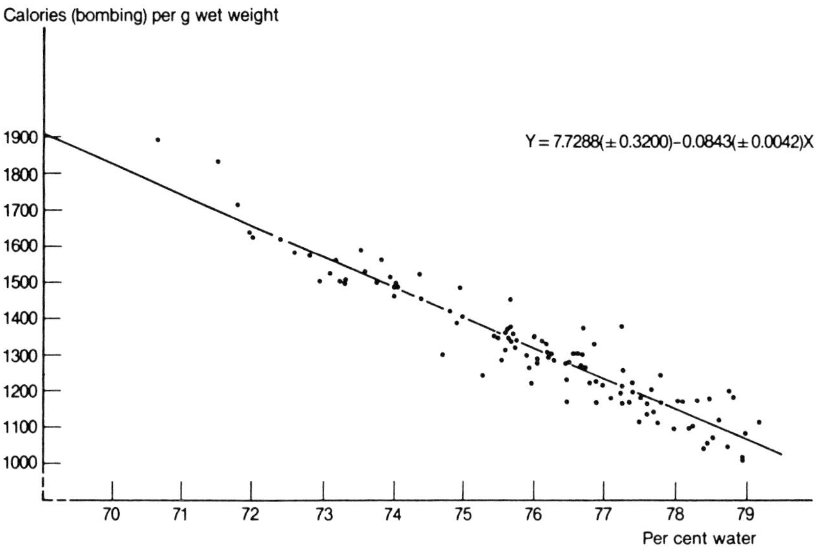

Fig. 16.13: Energy content of fish as a function of per cent water.

A reasonable **evacuation model** could express, that the rate of **stomach evacuation** at a given temperature, fish size, food composition, and particle size is a function of the amount of food present in the stomach, so that

$$dV/dt = -aV(t)^b \qquad (16.25)$$

where

dV/dt = rate of stomach evacuation

a = a constant (**instantaneous coefficient**) which might be a function of species, temperature, food type and maybe fish size, so that

b = a species specific constant

V(t) = weight (or volume) of the food in the stomach at time t.

which integrated gives:

$$V(t) = V(0) * \exp(-at) \qquad \text{for } b = 1$$

Further discussion about these matters can be found in Barrington (1957); Windell (1967, 1978); Fänge & Grove (1979); Tseitlin (1980) and Jobling (1981).

In Fig. 16.14 the calculated regressions at five temperatures are presented. The relationship between temperature 'T' and the constant 'a' can be described as:

$$a = 0.0057 * T^{0.764}$$

so that (16.26) can be expanded to:

$$V(t) = V(0) * \exp(-0.0057 * T^{0.764} * t)$$

or with actual stomach content depending on body size "incorporated":

$$\text{(actual stomach content)}(t) = \text{Ration} * \exp(-0.0057 * T^{0.764} * t)$$

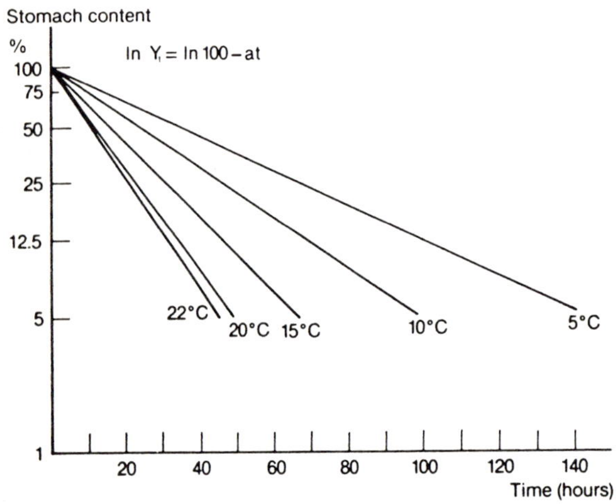

Fig. 16.14: Calculated regressions for the evacuation model at 5, 10, 15, 20 and 22°C.

As an example of the use of the evacuation model Fig. 16.15 is presented for **rainbow trout** with the same starting weight (w(0)).

From & Rasmussen (1984) show that the **evacuation time** at a given **feeding level** is independent of **fish size**. As a bigger fish in per cent body weight eats less than a smaller fish ($m < 1$) at the same feeding level it is seen that when fish eat the same amount of food in per cent of body weight then the smaller fish will have a faster gastric evacuation. It is generally accepted that the stomach volume is proportional to body weight, Kimball & Helm (1971), so it is not evident why a bigger fish eats less in per cent than a smaller fish. It is tempting to propose that a bigger fish eats less in percent because of the slower gastric evacuation. From & Rasmussen

(1984) show that a bigger fish utilizes the food better than the smaller specimen and one of the reasons for this could be the slower digestion. This explains why a bigger fish in spite of the smaller feed intake in percent of body weight can maintain an non-asymptotic course of growth.

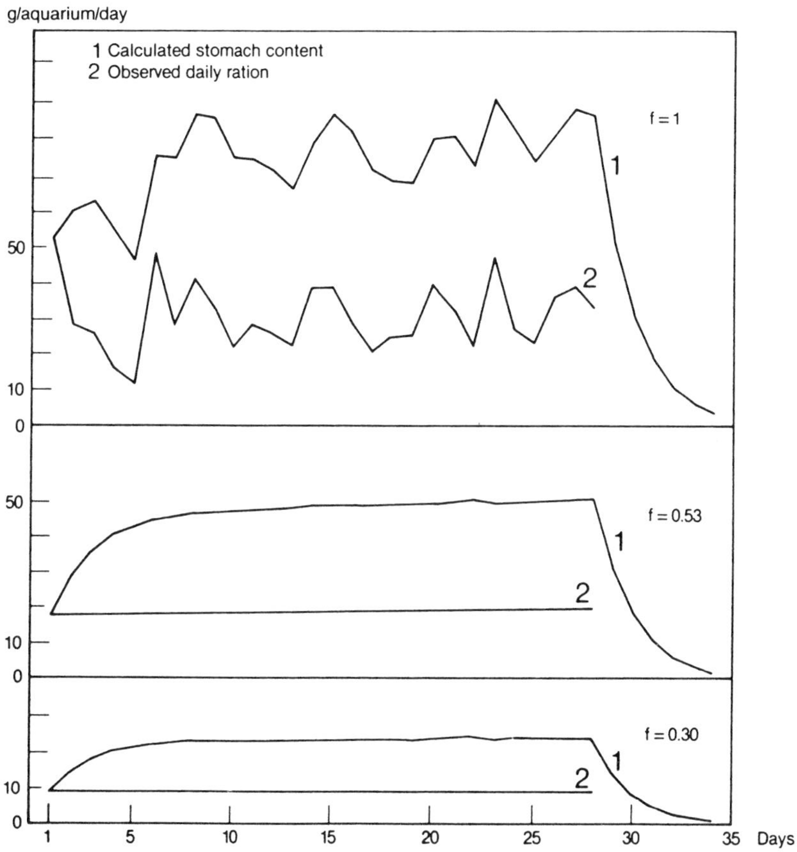

Fig. 16.15: Daily observed ration and calculated stomach content for three aquaria given three different feeding levels at 5°C.

16.5 OTHER METABOLIC GROWTH EQUATIONS

Ivlev (1939) was the first to split up the energy of the food in different terms in an energy budget. He used the following equation:

$$Q = Q' + Q_R + Q_t + Q_w + Q_v$$

where Q = heat of combustion of devoured food
Q' = heat of combustion of the growth of the organism
Q_R = heat of combustion of the excretions
Q_t = quantity of initial heat generated
Q_w = energy of **external work**
Q_v = energy of **internal work**

However, Ivlev does not (at least not in the English summary) explain the different terms in his budget. Especially his **initial** or **primary heat** has been difficult for others to understand. For example, Winberg (1960) simply denies the existence of primary heat in poikilotherms.

When Ivlev used his energy budget in practice he made the simplified assumption that the energy of the external work was approximately 20 per cent of the internal work. This assumption is not (at least not in the English summary) rendered probable in any way.

Winberg's equation

Winberg (1960) formulated a simple bioenergetic relationship implicitly incorporating temperature and fish size. This relationship has gained wide application and further improvements. Paloheimo & Dickie (1965, 1966a, b), in particular stimulated a lot of work on this subject, e.g. Kerr (1971a, b, c). The basic equation of Winberg is:

Energy of weight increase + energy of metabolism
= **physiologically useful energy** = 0.8 times energy of the **ration**,

or in letters:

$$P + T = 0.8 * R$$

The energy of metabolism T is estimated as twice (range 1.5-3.0) the energy equivalent of the **oxygen consumption** of fish at **routine level**. This idea is based on values from 5 fish species. These sparse observations led Winberg to propose the figure 2 as a universal factor which relates routine metabolism to **active metabolism**. Winberg gave no physiological explanation for why the active metabolism should be the routine metabolism multiplied by a constant. He ignored **feeding metabolism** as being physiologically distinct from active metabolism, for a fed fish has a higher metaboic rate than a fasting one, even at rest. Further, it is a great simplification to put the physiologically useful energy = 0.8 times the energy of ration, completely independent of fish species, fish size, food object, feeding level, temperature, etc.

In spite of the shortcomings of Winberg's equation Paloheimo & Dickie (1965, 1966 a, b) have used it in their three extensive papers. It is Paloheimo & Dickie's "**K-line**" that by various authors has attracted most attention. **Gross efficiency**: $K = (\Delta w/R\Delta t)$, where w is growth, R ration, and t time. So that $\log K = \log(\Delta w/R\Delta t)$ which is calculated as a function of R. This figure they call a "K-line model" and they have devoted much work to it, e.g. Warren & Davis (1967), Rafail (1968, Brett, Shelbourn & Shoop (1969), Gerking (1971), Bret & Shelbourn (1975), Ellitoo (1975a, 1979), Huisman (1976), and Staples & Nomura (1976). The only thing this term predicts is that the gross efficiency decreases with increasing ration. This is of course only true for ration sizes higher than **optimum ration** size **R(opt)**, see also Fig. 16.11.

Comparison with Davis & Warren's equation

Warren & Davis (1967) proposed an equation which resembled Ivlev's in some respects but which has terms that have been defined so as to be independent and measurable. The equation is best known in the notation used in IBP Handbook No. 3, Davis & Warren (1971):

$$C = F + U + B + R, \quad \text{cfr. section 16.2.}$$

From (16.7), (16.9) and (16.22)

$$\frac{dw}{dt} = \beta fh(T)w_t^m - k''(T)w_t^n - {}^a\beta fh(T)w_t^m - (U_2\beta fh(T)w_t^m + k'''(T)w_t^n) - k'(T)w_t^n$$

$$\Delta B = (C - F) - (R_d + R_a) - U - R_s$$

16.6 CONCLUSION AND RESEARCH NEEDS

Experiments carried out to determine the parameters in metabolic fish growth models have until now only been carried out in aquaria (Elliott, 1976; and From & Rasmussen, 1984). A question that naturally arises is whether these models can be used outside aquaria. Elliott (1975a) studied the growth rate over 12 weeks of four **brown trout, *Salmo trutta***, each placed in a rectangular trough sited in a stream. The study gave good agreement between the weights estimated on basis of aquaria experiments and the actual weights obtained at intervals of four weeks.

In future practice growth models have a great influence on **fish farming**. One of the research needs is to transfer the models so the experiments carried out in aquaria can be used in fish farming.

Further, a multitude of factors that influence growth were mentioned in section 16.1, and it was said that incorporation of all these factors in a growth model would demand an enormous amount of experiments. So, there is enough to work on, each time the chosen number of parameters have been determined, one new parameter can be added to the model. **Oxygen content** of the water, especially in fish farming is often below optimum, so incorporation of the oxygen content of the water will be important. Further, saltwater fish farming is expanding, and consequently determination of a **salinity parameter** has high priority in future research.

16.7 SYMBOL GLOSSARY

a	Instantaneous coefficient, formula (16.26)
a	a = H - k, section 16.3
a1, a2, a3, a4	Constants in the terms for respiration, e.g. Total respiration = $a1 * \exp(a2^*f) \exp(a3^*T)w^{a4}$, formula (16.25)
A	Fraction of assimilated food producing energy for catabolic processes resulting from feeding, section 16.2
b1, b2, b3, b4	Constants in the terms for faeces, e.g. suspended + dissolved faeces = $b1 * \exp(b2^*f) \exp(b3^*T)w^{b4}$, formula (16.20)
f	feeding level, section 16.2
f(maintenance)	Feeding level when dw/dt = 0, section 16.4
h(T)	Coefficient of anabolism, formula (16.3)
h1, h2, h3, h4	Constants in the terms for the coefficient of anabolism, e.g. $h(T) = h1 + h2^*T + h3^*T^2 + h4^*T^3$, formulat (16.18b)
H	Growth parameter, formula (16.10)
i1, i2, i3	Constants in the terms for respiration of starving fish, e.g. $k'(T) = i1 * \exp(i2^*T)w^{i3}$, formula (16.24)
k	Coefficient of catabolism, formula (16.6)
k'	Respiration of starving fish, formula (16.24)
k''	Loss in exfoliated cells from starving fish, section 16.3
k'''	Nitrogen excretion from starving fish, section 16.3
k1, k2	$k(T) = k1 * \exp(k2^*T)$, formula (16.21)

K	Growth parameter, formula (16.2)
l(t)	Length of fish to time t, section 16.3
m	Exponent of anabolism, formula (16.3)
n	Exponent of catabolism, formula (16.6)
q	Condition factor, section 16.3
R	Food consumed, formula (16.2)
R(opt)	The amount of food consumed which gives the best utilization of the food, section 16.4
t	Time in days, section 16.2
T	Temperature in degrees Celsius, section 16.2
u1, u2, u3, u4	Constants in the terms for nitrogen excretion, e.g. $U = u1 * \exp(u2*f)\ \exp(u3*T) w^{u4}$, formula (16.23)
U	Nitrogen excretion, section 16.3
U1	Endogenous excretion, formula (16.23a)
U2	Exogenous excretion, formula (16.23b)
V	Amount of food in the stomach, section 16.4
w(t)	Weight of fish to time t, formula (16.2)
ß	Feeding respiration, ß = A - U2, formula (16.7)
ß	Assimilation, formula (16.5)
Π	Gonadal loss, section 16.3

REFERENCES

Barrington, E.J.W., 1957. The alimentary canal and digestion. In: M.E. Brown (ed.): The physiology of fishes. Vol. 1, pp. 109-161. Academic Press.

Beamish, F.W.H., A.J. Niimi & P.F.K.P Lett, 1975. Bioenergetics of teleost fishes: Environmental influences. In: L. Bolis, H.P. Maddrell & K. Schmidt-Nielsen (eds.), Comparative Physiology - Functional Aspects of Structural Materials, pp. 187-209. North Holland Publishing Company, Amsterdam.

Bertalanffy, von I:, 1957. Quantitative laws in metabolism and growth - Quarter. Rev. Biol. 32(3): 217-231.

Beverton, R.J.H. & S.J. Holt, 1957. On the dynamics of exploited fish populations. Fisheri Invest., Lond. Ser. II Vol. XIX, 533 pp.

Blazka, P., 1958. The anaerobic metabolism of fish. Physiol. Zool. 31:

117-128.
Brafield, A.E. & D.J. Solomon, 1972. Oxy-calorific coefficients for animals respiring nitrogenous substrates. Comp. Biochem. Physiol. 43A: 837-841.
Brett, J.R., J.E. Shelbourn & C.T. Shoop, 1969. Growth rate and body composition of fingerling sockeye salmon, *Oncorhynchus nerka,* in relation to temperature and ration size. J.Fish. Res. Bd. Canada 26(9): 2353-2394.
Brett, J.R. & J.E. Shelbourn, 1975. Growth rate of young sockey salmon, *Oncorhynchus nerka,* in relation to fish size and ration level. J. Fish. Res. Bd. Canada 32: 2103-2110.
Brett, J.R. & T.D.D. Groves, 1979. Physiological energetics. In: W.S. Hoar, D.J. Randall & J.R. Brett (eds.): Fish Physiology. Vol. VIII, Bioenergetics and Growth, pp. 280-352. Academic Press.

Chapman, D.V., 1978. Production in Fish Population. In S.D. Gerking (ed.): Ecology of Freshwater Fish Production, pp. 5-25. Blackwell Scientific Populations.
Cho, C.Y., H.S. Bayley & S.J. Slinger, 1976. Energy metabolism in growing rainbow trout: Partition of dietary energy in high protein and high fat diets. In: M. Vermorel (ed.), Energy metabolism of farm animals, pp. 299-302. 7th Symp., Vichy, EAAP 19.

Davis, G.E. & C.E. Warren, 1971. Estimation of food consumption rates. In: W.E. Ricker (ed.), IBP Handbook No. 3. Methods for assessment of fish production in fresh waters, pp. 227-248. Blackwell Scientific Publications (2. ed.).

Edwards, R.W., J.W. Densem & P.A. Russell, 1979. An assessment of the importance of temperature as a factor controlling the growth of brown trout in streams. J. Anim. Ecol. 48: 501-507.
Elliott, J.M., 1975a. The growth rate of brown trout *(Salmo trutta L.)* fed on maximum rations. J. Anim. Ecol. 44: 805-821.
Elliott, J.M., 1975b. The growth rate of brown trout *(Salmo trutta L.)* fed on reduced rations. J. Anim. Ecol. 44: 832-842.
Elliott, J.M., 1976. The energetics of feeding, metabolism and growth of brown trout *(Salmo trutta L.)* in relation to body weight, water temperature and ration size. J. Anim. Ecol. 45: 923-948.
Elliott, J.M., 1979. Energetics of freshwater teleosts. Symp. Zool. Soc. Lond. 44: 29-61.
Elliott, J.M. & W. Davison, 1975. Energy equivalents of oxygen consumption in animal energetics. Oecologia (Berl.) 19: 195-201.

Fänge, R. & D. Grove, 1979. Digestion. In: W.S. Hoar, D.J. Randell & J.R. Brett (eds.), Fish Physiology. Vol. VIII. Bioenergetics and Growth, pp. 161-260. Academic Press.
Fischer, Z., 1977. Some remarks on nitrogen excretion by fish. Pol. Arch. Hydrobiol. 24(3): 355-360.
Forster, R.P. & L. Goldstein, 1969. Formation of excretory products. In: W.S. Hoar & D.J. Randall (eds.), Fish Physiology. Vol. I. Excretion, ionic regulation and metabolism, pp. 313-350. Academic Press.
From, J. & G. Rasmussen, 1984. A growth model, gastric evacuation and body composition in rainbow trout, *Salmo gairdneri* Richardson, 1836. Dana 3: 61-139.

Gerking, S.D., 1971. Influence of rate of feeding and body weight on protein metabolism of bluegill sunfish. Physiol. Zool. 44: 9-19.
Goldstein, L. & R.P. Forster, 1970. Nitrogen metabolism in fishes. In: Comparative biochemistry of nitrogen metabolism. Vol. 2, The

vertebrates, pp. 495-518.

Hemmingsen, A.M., 1960. Energy metabolism as related to body size and respiratory sufaces, and its evolution. Rept. Steno Mem. Hosp. IV: 7-58.

Huisman, E.A., 1976. Food conversion efficiency at maintenance and production levels for carp, *Cyprinus carpio L.,* and rainbow trout, *Salmo gairdneri* Richardson. Aquaculture 9: 259-273.

Ivlev, V.S., 1939. Energetitscheskij balans karpov. (Energy balance of carps). (In Russian with English summary). Zool. Zh. 18: 308-318.

Janisch, E., 1927. Das Exponentialgesetz als Grundlage einer Vergleichende Biologie. Abhandl. Z. Theorie Org. Entwicklungsmech. 2: 1-371.

Jobling, M., 1981. Mathematical models of gastric emptying and the estimation of daily rates of food consumption for fish. J. Fish. Biol. 19: 245-257.

Kerr, S.R., 1971a. Analysis of laboratory experiments on growth efficiency of fishes. J. Fish. Res. Bd. Canada 28(6): 801-808.

Kerr, S.R., 1971b. Prediction of fish growth efficiency in nature. J. Fish. Res. Bd. Canada 28(6): 809-814.

Kerr, S.R., 1971c. A simulation model of lake trout growth. J. Fish. Res. Bd. Canada 28(6): 815-819.

KImball, D.C. & W.T. Helm, 1971. A method of estimating fish stomach capacity. Trans. Amer. Fish. Soc. 100(3): 572-575.

Kinne, O., 1960. Growth, food intake, and food conversion in a eutyplastic fish exposed to different temperatures and salinities. Physiol. Zool. 33: 288-317.

Krokhin, E.M., 1959. Determination of the daily food ration of young sockeye and three-spined stickleback by the respiration method. Fish. Res. Bd. Canada Translation Series 209, 14 pp.

Love, R.M., 1970. The chemical biology of fishes. Academic Press, 547 pp.

Love, R.M., 1980. The chemical biology of fishes. Vol. 2. Advances 1968-1977. Academic Press, 943 pp.

Mathur, G.B., 1967. Anaerobic respiration in a cyprinoid fish *Rasbora daniconius* (Ham.). Nature 214: 318-319.

Miura, T., N. Suzuki, M. Nagoshi & K. Yamamura, 1976. The rate of production and food consumption of the biwamasu, *Oncorhynchus rhodurus,* population in lake Biwa. Res. Popul. Ecol. 17: 135-154.

Niimi, A.J., 1972. Bioenergetics and growth of largemouth bass *(Micropterus salmoides)* with feeding in relation to body weight and temperature. Ph. D. Thesis University of Guelph, 102 pp.

Ostapenya, A.P., 1971. Biomass and how to express it. In: G.G. Winberg (ed.): Methods for estimation of production of aquatic animals, pp. 11-31. Academic Press.

Paloheimo, J.E. & L.M. Dickie, 1965. Food and growth of fishes. I. A growth curve derived from experimental data. J. Fish. Res. Bd. Canada 22(2): 521-542.

Paloheimo, J.E. & L.M. Dickie, 1966a. Food and growth of fishes. I. Effects of food and temperature on the relation between metabolism and body weight. J. Fish. Res. Bd. Canada 23(6): 869-908.

Paloheimo, J.E. & L.M. Dickie, 1966b. Food and growth of fishes. I. Relations among food, body size, and growth efficiency. J. Fish. Res. Bd. Canada

23(8): 1209-1248.
Parker, R.R. & P.A. Larkin, 1959. A concept of growth in fishes. J. Fish. Res. Bd. Canada 16(5): 721-745.
Pütter, A., 1920. Studien über physiologische Ähnlichkeit VI: Wachstumsähnlichkeiten. Pflügers Arch. Gesamte Physiol. Menschen Tiere 179-180: 298-340.

Rafail, S.Z., 1968. A statistical analysis of ration and growth relationship of plaice *(Pleuronectes platessa)* J. Fish. Res. Bd. Canada 25(4): 717-732.
Rasmussen, G. 1983. Recent investigations on the population dynamics of eels *(Anguilla anguilla L.)* in some Danish streams. Proc. 3rd Brit., Freshw. Fish. Conf.: 71-77.
Rasmussen, G. & B. Therkildsen, 1979. Food, growth and production of *Anguilla anguilla L.* in a small Danish stream. Rapp. P.-v. Reun. Cons. int. Explor. Mer. 174: 32-40.
Ricker, W.E., 1975. Computation and interpretation of biological statistics of fish populations. Bull. Fish. Res. Bd. Canada Bulletin 191, 382 pp.
Ricker, W.E., 1979. Growth rates and models. In: W.S. Hoar, D.J. Randall & J.R. Brett (eds.): Fish Physiology. Vol. VIII. Bioenergetics and Growth, pp. 678-743. Academic Press.

Smith, R.R., G.L. Rumsey & M.L. Scott, 1978a. Net energy maintenance requirements of salmoids as measured by direct calorimetry: Effects of body size and environmental temperature. The Jorunal of Nutrition 108(6): 1017-1024.
Smith, R.R., G.L. Rumsey & M.L. Scott, 1978b. Heat increment associated with dietary protein, fat, carbohydrate and complete diets of salmonids: Comparative energetic efficiency. The Journal of Nutrition 108(6): 1025-1032.
Solomon, D.J. & A.E. Brafield, 1972. The energetics of feeding metabolism and growth of perch *(Perca fluviatilis L.)* J. Anim. Ecol. 41: 699-718.
Sparre, P., 1979. Some necessary adjustments for using the common methods in eel assessment. Rapp. P.-v. Reun. Cons. int. Explor. Mer.: 41-44.
Sperber, O., J. From & P. Sparre, 1977. A method to estimate the growth rate of fishes, as a function of temperature and feeding level, applied to rainbow trout. Medd. Danm. Fisk.- og Havunders. N.S. 7: 275-317.
Staples, D.J. & M. Nomura, 1976. Influence of body size and food ration on the energy budget of rainbow trout *Salmo gairdneri* Richardson. J. Fish Biol. 9: 29-43.
Stauffer, G.D., 1973. A growth model for salmonids reared in hatchery environments. Ph. D. Thesis, Univ. of Washington, Seattle, 213 pp.
Stewart, D.J., D. Weininger, D.V. Rottiers & T.A. Edsall, 1983. An energetic model for lake trout, *Salvelinus namaycush* . Application to the lake Michigan population. Can. J. Fish. Aquat. Sci. 40(6): 681-698.

Tseitlin, V.B., 1980. Duration of gastric digestion in fishes. Mar. Ecol. Prog. Ser. 2: 277-280.

Ursin, E., 1967. A mathematical model of some aspects of fish growth, respiration, and mortality. J. Fish Res. Bd. Canada 24(11): 2355-2453.
Ursin, E., 1979. Principles of growth in fishes. Symp. Zool. Soc. Lond. 44: 63-87.

Warren, C.E. & G.E. Davis, 1967. Laboratory studies on the feeding, bioenergetics, and growth of fish. In: S.D. Gerking (ed.), The biological basis of freshwater fish production, pp. 175-214. Blackwell Scientific Publications.
Warren, C.E. & P. Doudoroff, 1971. Bioenergetics and growth. In: Biology and

Water pollution control, pp. 135-167. Sauners.
Winberg, G.G., 1960. Rate of metabolism and food requirements of fishes. Fish. Res. Bd. Canada. Translation Series 194, 202 pp.
Winberg, G.G., 1961. New information on metabolic rate in fishes. Fish. Res. Bd. Canada Translation Series 362, 11pp.
Windell, J.T., 1967. Rates of digestion in fishes. In: S.D. Gerking (ed.), The biological basis of freshwater fish production, pp. 151-173. Blackwell Scientific Publications.
Windell, J.T., 1978. Digestion and daily ration of fishes. In: S.D. Gerking (ed.), Ecology of freshwater fish production, pp. 159-183. Blackwell Scientific Publications.

CHAPTER 17

SEDIMENT-WATER EXCHANGE MODELS

by
Lars Kamp-Nielsen

17.1 INTRODUCTION

Sediments consist of particulate material either brought to the water body from outside by runoff from the drainage basin, or directly be wet or dry deposition. This part of the sediments are the allochtonous part. However, some of the dissolved elements undergo a transformation in the water body by biological and chemical processes, e.g. carbon dioxide is fixed by photosynthetic organisms, which partly sediments as dead or living particulate carbon. And due to increased pH as a result of photosynthesis the solubility product of calcium carbonate can be exceeded and calcite precipitation may occur. This part of the sediments are the endogenic part. The deposited material is subject to further diagenetical processes in the sediment environment and a third authigenic fraction is generated.

On a geological time scale sediments are rather young structures, but they play an important role in the global and local cycling of elements like nutrients, organic matter and metals. All of which are important for the productivity in both the terrestrial and the aquatic ecosystems. For some conservative elements the sediments can be considered almost as a permanent sink, but for other elements the sediments act only as a temporary sink which is highly integrated in the overall metabolism of the waterbody.

After settling the sedimented material undergoes a certain compaction and a very "concentrated" environment is generated with slow transports of dissolved gases and ions. High concentrations develop and stimulate precipitation reactions and together with a high concentration of surfaces and strong gradients at the sediment-water interface the sediment becomes the habitat for an intensive and diverse microbial metabolism dominated by anaerobic processes.

Both the role as a more or less permanent sink, and the role as a quantitative important compartment integrated in the overall aquatic metabolism justifies the effort devoted to sediment research. And the very complex interactions of physical, chemical and biological processes justify the use of models in the study of the sediment-water interactions.

17.2 ESSENTIAL SUBMODELS

In the following a separation of the various processes involved in the sediment-water interactions will be made by defining a set of submodels each of which are confined in time and space and serve as linkages between compartments in the considered ecosystem.

The following are are considered as essential submodels:

1. Sedimentation understood as the transport of particulate matter to the sediment from the water phase, including resuspension and horizontal transport.
2. Sorption equilibria at the sediment-water interface.
3. Physical, chemical and biological processes occuring in the sediment e.g. diffusion, bioturbation, sorption equilibria, mineralization, biological oxidation/reduction.

Empirical, aggregated models

On an annual scale, sediments almost exclusively act as a sink for elements transported to the water body and the net rate for retention can be approached empirically be means of simple mass balances:

$$V \frac{dC}{dt} = (1 - f)Q\ C_i\ -\ Q\ C \quad \text{(Piontelli \& Tonolli, 1964)} \qquad (17.1)$$

with the time-dependent solution:

$$C = (1 - f)C_i\ (1 - e^{-t/\beta}) + C_0\ e^{-t/\beta} \qquad (17.2)$$

where V is the volume of the water body, C the concentration of the element, C_i the concentration of the element in the inflow, C_0 the initial concentration of the element, Q is the volume rate of inflow (and outflow), f is the fraction of incoming element deposited in the sediment, and ß is the hydraulic residence time (=V/Q). This model suggests that net sedimentation is a function of the incoming mass of the element. The phenomenology behind this concept is an element load as particles sink to the bottom after arrival to the water body. The concept might be valid for true conservative substances like many minerals. But most elements are a subject to chemical-biological transformations in the water and a more general model can be introduced with ¥ - a sedimentary loss coefficient relating sedimentation to the element concentration in the water phase:

$$V \frac{dC}{dt} = (1 - f_0)Q\ C_i\ -\ ¥\ V\ C \qquad (17.3)$$

with the time-dependent solution:

$$C = \frac{(1 - f_0)\ \frac{Q\ C_i}{V}}{¥}\ [1 - (1 - \frac{¥\ C_0}{[1 - f_0)\ \frac{Q\ C_i}{V}})\ e^{-¥ß}] \qquad (17.4)$$

where f_0 is the fraction of the incoming element lost through the outflow: (17.4) has the steady state solution:

$$C = \frac{(1 - f_0)\ Q\ C_i}{V\ ¥} \qquad (17.5)$$

or:

$$C = \frac{L}{¥\ z} \qquad (17.6)$$

where L is the specific areal loading (e.g. g $m^{-2}yr^{-1}$) and z is the mean depth of the water body. This model was applied for phosphorus for the first time by Vollenweider (1964) on swiss lakes and an almost constant L/C ratio was reported suggesting that ¥ was inversely related to lake depth. This is not necessarily true and the model was revised (Vollenweider, 1969) assuming that outflow and inlake concentrations were identical:

$$V \frac{dC}{dt} = Q\ C_i\ -\ Q\ C\ -\ ¥\ C\ V \qquad (17.7)$$

with the time dependent solution:

$$C = \frac{L}{¥z\ +\ z/ß}\ [1\ -\ e^{-(1/ß\ +\ ¥)t}\ ]\ +\ C_0\ *\ e^{-(1/ß\ +\ ¥)t} \qquad (17.8)$$

and the steady state solution:

$$C = \frac{L}{z(\ 1/ß\ +\ ¥)} \qquad (17.9)$$

Instead of using ¥ an apparent settling rate V_s can be introduced and (17.7) can be written:

$$V \frac{dC}{dt} = Q\ C_i\ -\ V_s\ C\ A\ -\ Q\ C \qquad (17.10)$$

with the steady state solution:

$$C = \frac{L}{V_s + z/ß} \qquad (17.11)$$

The difference between (17.9) and (17.11) is the assumption of a depth-dependent settling velocity in (17.9) and a constant rate in (17.11).

A third model proposed a constant fraction of the incoming phosphorus being retained in the lake (Sillon and Rigler, 1974):

$$R = \frac{Q\ C_i - Q\ C_0}{Q\ C_i} \qquad (17.12)$$

when C is the outflow concentration. If C_0 is set equal to C, (17.12) can be transformed into:

$$C = \frac{L\ ß}{z}(1 - R) \qquad (17.13)$$

It should be stressed that the three various models for net sedimentation rates are interrelated by:

$$R = \frac{1}{1 + (1\ /\ ß\ ¥)} \qquad (17.14)$$

and

$$R = \frac{V_s}{V_s + (z/ß)} \qquad (17.15)$$

Consequently a "selection of a "best" model should likely be termed an irresolvable issue on theoretical grounds" (Rechhow, 1979).

Several attempts to relate these net sedimentation rates to morphometric and hydraulic characteristics by empirical extensions of the one-box models have been made. Based on cross-sectional analysis from many lakes in various regions a number of phosphorus models have been developed:

$$¥ = 10/z = \ln 5.5 - 0.85 \ln z \quad \text{(Vollenweider, 1975)} \qquad (17.16)$$

$$R = 0.426 \exp(-0.271\ z/ß) + 0.574 \exp(-0.00949\ z/ß) \qquad (17.17)$$
(Kirchner & Dillon, 1975)

$$R = \frac{16}{16 + z/t} \quad \text{(Chapra, 1975)} \qquad (17.18)$$

$$R = \frac{1}{1 + 1.12\ (1/ß)^{0.49}} = \frac{1}{1 + \sqrt{1/ß}} \qquad (17.19)$$

(Larsen & Mercier, 1975; Vollenweider, 1976)

$$¥ = 0.65 \quad \text{(Jones \& Bachmann, 1976)} \qquad (17.20)$$

Since R is negatively correlated to flushing rate $ß^{-1}$ and positively correlated with ¥, R is negatively correlated to ¥. This puzzling consequance led to relations between sedimentation coefficients and loading:

$$¥ = 0.129\ (L/z)^{0.549} \quad \text{(Canfield \& Bachmann, 1981)} \qquad (17.21)$$

which can be transformed to:

$$R = 0.129\ ß^{0.451}\ P_i^{0.549} / (1 + 0.129\ ß^{0.451}\ P_i^{0.549}) \qquad (17.22)$$

where P_i is the average inlet concentration.

A similar result was obtained for Finnish lakes with a simple model by Frisk et al. (1981):

$$R = P_i\ ß / (K_R + P_i\ ß) \qquad (17.23)$$

where K_R has a value of 30 g P m^{-3}.

The comprehensive OECD study (OECD, 1982) ended up with a regression analysis showing that inlake phosphorus concentration:

$$P = 1.55 * P_i\ (1 + \sqrt{ß})^{0.82} \qquad (17.24)$$

and since $R = 1 - P/P_i$:

$$R = 1 - 1.55 * P_i^{-0.18} / (1 + \sqrt{ß})^{0.82} \qquad (17.25)$$

As seen from (17.25) R is increasing with P_i, but this is not always true.

As shown above a variety of empirical models predicting the net sedimentation of phosphorus has been developed over past years. They are based on multi-lake data and show an inverse relation between phosphorus retention and flushing rate and a retention increasing with increased inflow phosphorus concentrations, but for individual lakes very often the opposite is observed (Gächter et al., 1983; Nürnberg, 1984). When longitudinal forcasts for an individual lake are made, one should be very careful. Based on 20 individual cases from European lakes, where phosphorus load actually has been reduced it was observed that the R-value very often became negative and in all cases decreased after reduced loading, but after some time it increased again and stabilized at a value lower than before phosphorus reduction:

$$\frac{P_{post}}{P_{pre}} = \left(\frac{P_{i,post}}{P_{i,pre}}\right)^{0.65} \qquad (17.26)$$

where P_{pre} and P_{post} are steady state concentrations before and after restoration. An exponent which is considerably lower than the exponent of 0.82 found in the OECD-study (Sas 1988).

These empirically derived one-box models are highly aggregated and consider the whole water body, including the sediment, as a black box, where net sedimentation is a one-directional flux like inflow and outflow. However, very little information on the behaviour of the ecosystem is gained from these types of models, and the relation between constants are often very speculative and do not have very much justification in reality. But if the models are carefully used within their limits, and if they are improved with uncertainty analysis, they can be helpful in water management.

If more information on specific processes and steering variables are wanted, theoretical models have to be used. These models are much less aggregated in space and time and often use a time resolution of one day, where empirical models use a year.

Theoretical models

Sedimentation

The first step in the sediment-water exchange process is the sedimentation whereby particulate material is transported from the water body to the sediment surface. Theoretically any small particle (< 0.2 mm) will sink through a stagnant water column with the velocity:

$$V = 1/18\ d^2\ g\ (r_A - r_W)\ n^{-1} \quad \text{(Stoke's Law)} \qquad (17.27)$$

where d is the diameter of the particle, g is the gravitational force, r_A the density of the particle, r_W the density of the water, and n is the viscosity of the water. Often (17.27) is modified by application of a form resistance coefficient ($f_r = V/V_r$):

$$V = 1/18\ d^2\ g\ (r_A - r_W)\ (n\ f_r)^{-1} \qquad (17.28)$$

f_r has a value of up to 5 (Reynolds, 1984) for planktonic algae.

From (17.28) is can be seen that V, d, r_A and f_r are all factors which can be regulated by the phytoplankton themselves. d is varying from 2 to about 500 μm and due to the second power dependence it is by far the strongest regulator of sinking rates when various phytoplankton species are compared. r_A varies from 1,000 - 1,263 and diatoms are the species with the highest densities. Often the individual species are able to regulate their density by lipid accumulation, ionic regulation, mucilage excretion, forming of gas vacuoles, etc. (Reynolds, 1984). Factors which directly regulate the sinking rate on a short term basis are often related to nutritional state. It has been reported that within the same population the sinking rate can change with a factor of five during a changing nutrient level (Kamp-Nielsen, 1980b, Smayda, 1970, 1974).

However, the fastest rates which can be obtained for planktonic algae are less than 1 mm s^{-1} and at the same time even light winds (3-4 m s^{-1}) are able to develop turbulence exceeding the intrinsic movements of algae by a factor of 10. But at the lower boundary of a mixed layer the turbulent forces are decreased and the actual loss of particles from a mixed layer are more a "disentrainement" from the mixed-layer motion. Thus the critical factor is not the intensity of turbulence, but rather the absolute depth of the mixed layer (Reynolds 1984). The exponential rate of sinking loss per unit time from a continuously mixed layer can be formulated:

$$K_s = s\, h_m^{-1} \qquad (17.29)$$

where K_s is a loss rate (d^{-1}), s the sinking velocity (m d^{-1}) and h_m the depth of the mixing layer (m).

Once settled at the sediment surface a particle does not necessarily stay there. Waves and currents create sheer stress acting upon the sediments and cause resuspension, transport and resettling at places with less stress. To include resuspension and horizontal transport of sediments simple empirical models can be used to describe the areas, where erosion, transportation and accumulation, respectively, dominate. The distribution is affected by the potential, effective fetch and water depth as shown by Håkonson and Jansson (1983) (Fig. 17.1). Since the various areas are characterized also by water content in the sediment surface layer (0-1 cm) we get:

Erosion areas:	$W < 50\%$
Transportation areas:	$W = 50\text{-}75\%$
Accumulation areas:	$W > 75\%$

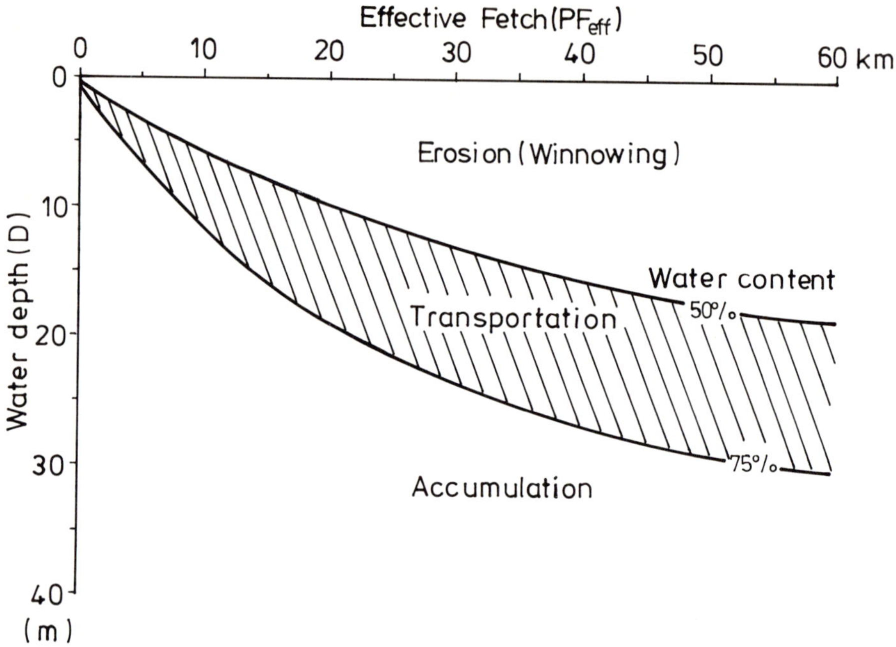

Fig. 17.1: The ETA-diagram (erosion-transportation-accumulation) for superficial sediments (0-1 cm). From Håkanson & Jansson (1983).

A mapping of the sediments according to water content can be used to identify the various areas. The distribution of sediments is only valid on a long term scale (years); if a daily or seasonal time resolution is necessary, the actual wind direction and speed must be used as forcing functions or more refined hydrodynamic approaches. The empirical approaches should be carefully used since local effects can modify the diagram, e.g. river inlets creating currents and plumes, upwelling areas, vegetation covered areas and strong bottom slopes.

Sorption Equilibria at the Sediment-water Interface

Dissolved elements will establish a sorption equilibrium with the particles at the sediment surface. If the physico-chemical environment (pH, E_h, temperature and concentration levels) is different from that in the overlying water the equilibrium changes, and dissolved elements can be sorbed or desorbed. The sorption equilibria are most conveniently described by conventional sorption isotherms, e.g.:

$$S = S_{max} C / (1 - k C) \quad \text{(Langmuir equation)} \tag{17.30}$$

where S is the sorbed amount at the equilibrium concentration C, S_{max} is the sorption capacity and k is a constant relating to bonding energy. (17.30) has the advantage that a sorption maximum can be calculated. The equation assumes that the sorption energy does not vary with surface saturation. This is considered in:

$$S = k\, C^{1/n} \quad \text{(Freundlich equation)} \tag{17.31}$$

where S is the sorbed amount of equilibrium concentration C and k and n are constants. The Freundlich isotherm can be modified by a sorption energy linearily decreasing with increased surface saturation:

$$S = S_{max} \; RT/\, b \ln (AC) \quad \text{(Temkin equation)} \tag{17.32}$$

where R is the gas constant, T is the absolute temperature, and A and b are constants.

Selection of a "best" model is often done by 'goodness of fit' to experimental data, but in some cases hysteresis effects are found. The irreversability is due to chemical reactions occuring simultaneously with the sorption reactions, and the process is termed chemosorption. If such processes are expected, it becomes questionable wheter the sorption can be modelled as a time independent equilibrium, and it rather should be modelled with a differential equation.

In very turbulent systems with extensive contact between water and sediment with a high sorption capacity, the mass transfer across the sediment-water interface can be assumed to be controlled mainly by diffusion-advection and sorption equilibria. The large, shallow L. Balaton is such a case and the sediment water exchange has been modelled by Van Raaphorst and Brinkman (1985): For the mass transfer of phosphorus a model was formulated:

$$\frac{\partial C}{\partial t} = D \frac{\partial^2 C}{\partial z^2} - U \frac{\partial C}{\partial z} - K(C - C_e) \tag{17.33}$$

where C is the phosphate concentration in the pore water, C_e the equilibrium concentration, K the sorption-desorption rate, D the effective diffusion coefficient, U the advective transport of pore water and z the depth in the sediment.

(17.34) has the steady state solution:

$$(C - C_e) = (C_0 - C_e) \exp (U - (U^2 + 4 D K)^{0.5}) (2D)^{-1} z \qquad (17.35)$$

and if the seepage is neglected ($U > 10^{-7}$ m s^{-1}) (17.35) is reduced to:

$$(C - C_e) = (C_0 - C_e) \exp (- (K/D)^{0.5}) z \qquad (17.36)$$

and if a film model for diffusive transport is assumed, and the flux at z = 0 is calculated, a mass transfer K_L can be derived:

$$K_L = (K D)^{0.5} \qquad (17.37)$$

However, in most cases a metabolization takes place in the sediment, and a mobilization term has to be added to (17.33). For nutrients and pollutants transported to the sediments incorporated in organic matter, this process is described as a mineralization, and is often approached by a 1. order temperature dependent process:

$$\frac{dC}{dt} = k \; C \; f(t) \qquad (17.38)$$

where k is a mineralization constant and f(t) is a temperature function e.g.:

$$f(t) = k_{20} * V^{(t-20)} \qquad (17.39)$$

where k_{20} is the rate at 20°C, V is a constant, and t is the temperature.

Or if an optimum temperature is considered:

$$f(t) = k_{opt} \exp a(t - t_{opt}) * \left(\frac{t_{max} - t}{t_{max} - t_{opt}} \right) a(t_{max} - t_{opt}) \qquad (17.40)$$

where k_{opt} is the optimum rate, a is a constant, t_{max} is a maximum temperature, and t_{opt} is the optimum temperature (Lassiter & Kearns, 1974). Many other versions of non-symmetric temperature dependence models exist, cf Jørgensen (1986).

The mineralization of organic material in the sediment is a complex mixture of many different components being metabolized by a variety of organisms. Consequently, a distribution of decay rates and substrates should be applied rather than average values. This involves the identification of a tremendous amount of substrates and the decay rates for also a tremendous number of organisms in various life stages, but such a highly reductionistic approach does not to be fruitful. Overall decay rates can be determined from turnover of sedimentary organic matter, calculated from the depth distribution of material and sediment age, and a time resolution can be achieved by looking into seasonal variations in oxygen uptake or simply distributing annual decay rates by means of seasonality in temperature.

Since sediment depth is a function of time, the variability in decay rates can be approached by use of partial differential equations:

$$\partial C/\partial t \;=\; - K(\beta)\, g(T,OX)C \;-\; \partial Z/\partial t\; \partial C/\partial z \qquad (17.41)$$

where

$$K(\beta) \;=\; \int_0^\infty k\, C_0(k) \exp(-k\beta) \;-\; dk/ \int_0^\infty C_0(k) \exp(-k\beta)\, dk \qquad (17.42)$$

where: C is a degradable part of the organic material or organically bound element, $K(\beta)$ is the decay constant for the total distribution of the element, g(T,OX) is the influence of temperature T and oxygen OX on "biological" age relative to chronological age; Z is the displacement of sediment surface relative to t=0, z is the sediment depth expressed as a depth coordinate; $C_0(k)$ is distribution of rate constants or various fractions of the element C, when the sedimenting element arrives at the sediment surface. The model contains further expressions allowing the decay (and other processes) to switch between aerobic and anaerobic processes according to oxygen level (Jørgensen et al., 1982). The background for applying this model was to investigate the relative importance of increased diffusion range, sediment "dilution" and distribution of decay rates during a recovery after loading reduction.

A more simple division in three compartments was made in the phosphorus model applied to L. Glumsø (Jørgensen, 1975). An easy degradable part of fresh sedimented material, a more slowly degradable part below the surface sediment and a non-exchangeable part extrapolated from the phosphorus concentrations in the deeper parts of the sediment.

Other microbial processes like nitrification-denitrification, sulphate reduction and methane formation are kinetically described as first-order processes, eventually as a Michaelis-Menten-like substrate dependence with a rather steady supply of substrate. But often rapid redox changes occur, and the switch from aerobic to anaerobic processes, and vice versa, do not occur instantaneously. First the population of bacteria has to develop, and if a correct description has to be used, the bacterial production has to be introduced as a state variable:

$$\frac{dS}{dt} = -\frac{1}{Y}\frac{dX}{dt} = -\frac{1}{Y}\,\mu\, S \frac{S}{+ K_s}\, X \qquad (17.43)$$

where S is the substrate concentration, Y is the average growth yield, X is the concentration of microorganisms, μ is the maximum growth rate, and K_s is a half saturation constant. But since the microbial growth rate is generally higher than the decay rates, a compensation for microbial growth or adaption of a population can be taken into account by using a high temperature constant for the total process of growth and substrate utilization (Van Straaten, 1986).

Transport in the Sediment

Transport in the sediment is generally slow compared with the vertical and horizontal transport in the overlying water, but in principle the same types of diffusive and advective transport occur. Diffusive processes are driven by concentrations gradients, and delayed by the diffusion resistance due to high concentration of particles increasing the distance for a dissolved ion, or molecule, to travel. This modification is most conveniently applied by a correction factor related to porosity of the sediment:

$$\frac{\partial C}{\partial t} = D\,\frac{\partial^2 C}{\partial x^2}\,k\,\pi \qquad (17.44)$$

where C is the concentration of an element, D is the diffusion coefficient, $\partial C/\partial x$ is the concentration gradient of the element, k is a diffusion resistance factor with a value of approximately 0.3 (Frissel & Reininger, 1974), and π is the porosity. Under certain boundary conditions: $C_{(0,t)}/C_0$ and that C_0 is constant over a certain time, and the approximate solution is:

$$C_{(x,t)} / C_0 = \text{erfc}\ x/2\ \sqrt{DT}$$

where erfc is the complementary error function.

If diffusion in all three dimensions has to be described we get:

$$\frac{\partial C}{\partial t} = D\left(\frac{\partial^2 C}{\partial x^2}\quad\frac{\partial^2 C}{\partial y^2}\quad\frac{\partial^2 C}{\partial z^2}\right) \qquad (17.45)$$

where x, y and z are space coordinates.

Description of sediment diagenesis by combination of a diffusion term and various reaction terms has been applied in the classical study by Berner (1974, 1977) on nitrogen, sulphur and other compounds. Vanderborght &

Billen (1974) also used diffusion-reaction models in their treatment of nitrogen distribution in marine sediments.

But often both advective transport and turbulent diffusion participate in the transport of dissolved elements in the sediment, and an advection term is added to the diffusion equation:

$$\frac{\partial C}{\partial t} = D \frac{\partial^2 C}{\partial x^2} - U \frac{\partial C}{\partial t} \quad (17.46)$$

where U is the velocity of the pore water. If also turbulent diffusion is expected (17.46) is changed to:

$$\frac{\partial C}{\partial t} = D \frac{\partial^2 C}{\partial x^2} - (U + U_t) \frac{\partial C}{\partial t} \quad (17.47)$$

where U_t is the turbulent velocity fluctuation.

Advection is important in lakes where water balances have shown that either seepage or drainage through the lake bottom occur, and to get a resolution in time the velocity can be determined by Darcy's law:

$$Q = K \frac{dh}{dx} \quad (17.48)$$

where Q is the water flow, K is the average hydraulic conductivity, and dh/dx is the hydraulic gradient.

The turbulent diffusion and advection created by the pumping activity of benthic invertebrates is identified when the mass transfer of dissolved elements is higher than can be described by a seepage-molecular diffusion model alone. Often apparent diffusion coefficients are two orders of magnitude higher than the molecular diffusion found, and the differences have been attributed to bioturbation and/or gasconvection (Skogheim & Erlandsen, 1979).

17.3 APPLICATION OF SEDIMENT MODELS

The number of unit processes necessary for a total sediment-water exchange model depends on the resolution in time, space and the compartments wanted. The development in sediment phosphorus models from very simple black-box models towards complex multilayer models is shown in Fig. 17.2.

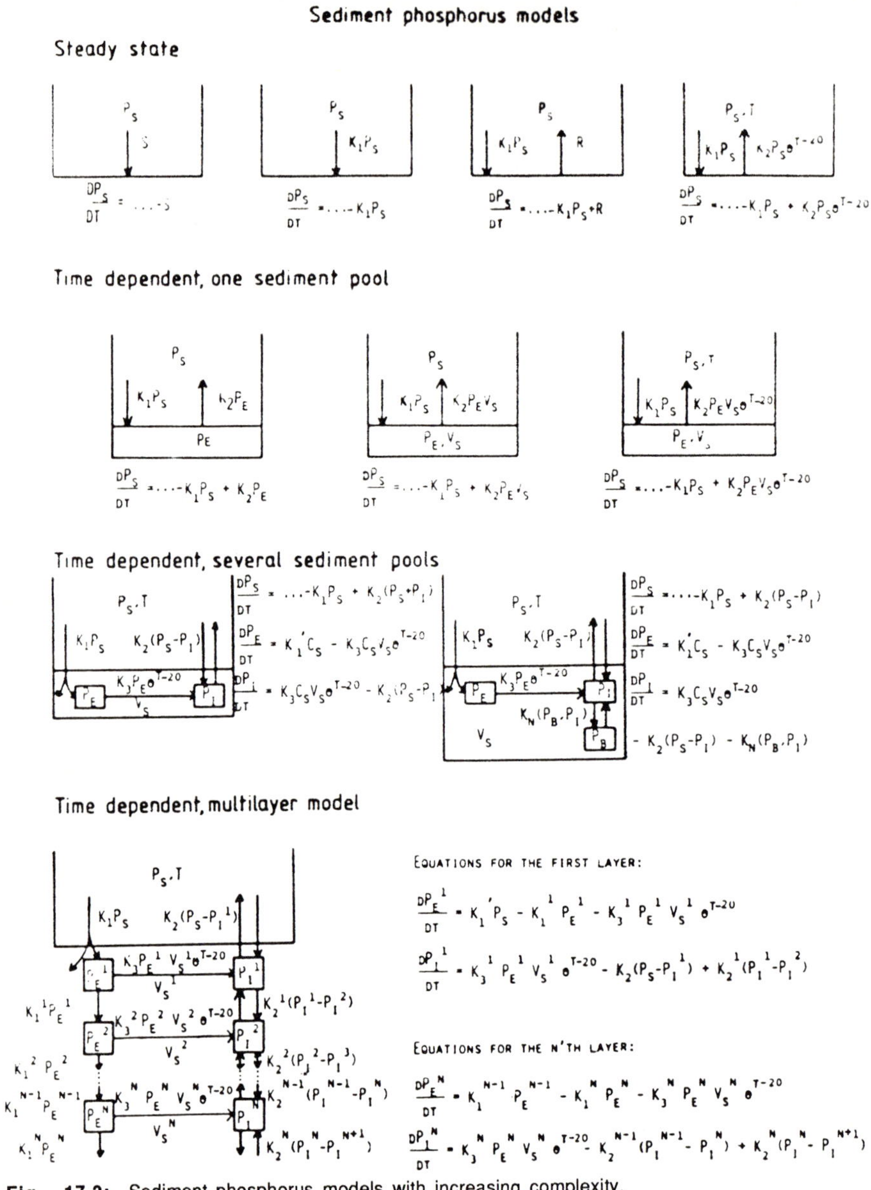

Fig. 17.2: Sediment phosphorus models with increasing complexity. P_S = dissolved phosphate in lake water, P_E = concentration of exchangeable, sedimentary phosphorus, P_I = concentration of interstitial phosphate, P_B = sorbed, sedimentary phosphate. C_S = concentration of exchangeable phosphorus in active layer, V_S = volume of active layer, K_1 = sedimentation rate constant, K_2 = diffusion rate constant, K_3 = mineralization rate constant, 0 = temperature coefficient, T = temperature, S = sedimentation rate, R = release rate. Superscripts indicate number of sediment layers. (From Kamp-Nielsen, 1983).

In cases where only a whole lake mass balance is available, simple models treating the sediment as a black-box must be used. If sediment concentrations of the element are measured, a one sediment pool model can be brought into use, but if a time resolution lesser than a year, e.g. a week or a day, both mass balances and inlake concentrations of the element estimated with this frequency is needed for calibration of sedimenation and release rates, unless these rates are measured directly. And if even more complex models including sorption equilibria, advective and diffusive transport and age-distributed mineralization rates are incorporated in the model, an extensive experimental and field measurement programme is needed.

In most cases the available data determine the constraints on the complexity to be used, since "modelling moral" prohibits the introduction of more state variables than measured. In cases where the necessary data are available, it is possible to optimize the complexity by testing a whole ecosystem model against a set of sediment-water exchange models with increasing complexity. Such a study was performed on L. Glumsø by the author (Kamp-Nielsen, 1980a).

The Glumsø-model has 17 state variables, 6 of them describing the independent phosphorus cycle (Table 17.1):

Table 17.1

Variable	Compartment	Unit
PS	Dissolved phosphate	g m^{-3}
PC	Phosphorus in algae	g m^{-3}
PD	Detrital phosphorus	g m^{-3}
PB	Easily degradable sediment phosphorus	g m^{-3}
PE	Total exchangeable sediment phosphorus	g m^{-3}
PI	Dissolved, interstitial phosphate	g m^{-3}

In the original version of the Glumsø model (Jørgensen et al., 1978) the 3 sediment phosphorus compartments (17.49 - 17.51) were described:

$$dPB/dt = QSED/AB - QBIO - QDSORP \qquad (17.49)$$

where QSED is the sedimentation of non-exchangeable phosphorus, AB is the proportion of sedimentation which is easily degradable, QBIO is the mineralization of the easily degradable part, and QDSORP is the desorption of phosphate from the uppermost sediment layer.

$$dPI/dt = (AE/AI)\ KE * PE - QDIFF/AI \quad (17.50)$$

where AE and AI are volume ratios for converting sediment variables to water variables, KE is the decomposition rate constant for PE and QDIFF is the diffusion rate for interstial phosphate.

$$dPE/dt = KF * (SA * PC - QSED + SD * PD)/AE - KE * PE \quad (17.51)$$

where KF is the proportion of the sedimented phosphorus, which is exchangeable, SA is the settling velocity of algae, SD is the settling velocity of detrital phosphorus.

This rather complicated model was calibrated and validated on data from 1974-1975 and data from 1981-1983, respectively. To test the sensitivity of the total model against the complexity of sediment submodels, a range from more simple to more complex models was applied.

Model I: A constant release from the sediment:

$$dPS/dt = \ldots\ldots. + 3.1 * 10^{-3} \ldots\ldots. \quad (17.52)$$

a release calculated from the average measured release at 7°C in the laboratory.

Model II: A constant fraction of the sedimented phosphorus is released (as described by Sonzogni et al., 1976):

$$dPS/dt = \ldots\ldots\ldots + KF(SA * PC + SD * PD) \ldots\ldots \quad (17.53)$$

Model III: From treating the sediment as a black-box we move into the category with true sediment compartments. A pool of sediment phosphorus is built up from sedimentation, but is degraded by a first order mineralization:

$$dPS/dt = \ldots\ldots. + AE * KE * PE \ldots\ldots\ldots\ldots. \quad (17.54)$$

$$dPE/dt = KF(SA * PC - QSED + SD * PD) / AE - KE * PE \quad (17.55)$$

Model IV: As model III, but with a temperature dependent mineralization:

$$KE = KE_{20}\ \pi^{(T-20)} \quad (17.56)$$

where π is a temperature coefficient, T is the temperature (°C) and KE_{20} is the rate at 20°C.

Model V: Transport limitation is introduced by considering both a particulate and a dissolved sediment fraction:

$$dPS/dt = ... + QDIFF/AI .... \quad (17.57)$$

$$dPI/dt = (AE/AI)(KE * PE) - QDIFF/AI \quad (17.58)$$

$$dPE/dt = KF(SA * PC - QSED + SD * PD) / AE - (KE * PE) \quad (17.59)$$

Model VI: As model V, but with an addition of a special, active surface layer and a sorption equilibrium established at the sediment surface:

$$dPS/dt = ... + QDIFF/AI + AB(QBIO + QDSORP) .... \quad (17.60)$$

$$dPE/dt = KF(SA * PC - QSED + SD * PD) / AE - (KE * PE) \quad (17.61)$$

$$dPI/dt = (AE/AI)(KE * PE) - QDIFF/AI \quad (17.62)$$

Model VII: As model VI, but with a temperature dependent mineralization as (17.56). This model is the original as described in (17.49) to (17.51).

Model VIII: A model VII, but QDSORP is only working when oxygen is below a certain threshold value at the sediment surface (0.5 mg O_2/l). A table function describes oxygen concentrations and together with sedimentation rates and temperature the threshold value is switched on and off.

The various model versions simulated the response to a 90% reduction in phosphorus load from waste water corresponding to about 82% reduction in total phosphorus load. The model computes the annual average concentration of total phosphorus during the 10 years following the load reduction. Since the hydraulic residence time in L. Glumsø varies between 2 and 8 months due to a variation in runoff from year to year, the simulations are run with an average residence time and with ± 75%.

The simulations shown in Fig. 17.3 demonstrate a significant influence of hydraulic loading as expected in a lake with a relatively short residence time. It further demonstrates the effects on the transient during the recovery and the magnitude and stability of the new steady state. The simple submodels (model I and II) and the complex model VIII with aerobic/anaerobic shifts produce labile steady states when hydraulic residence time is increased from 5 to 8 months. In the other cases stable steady states were achieved after 3-4 years.

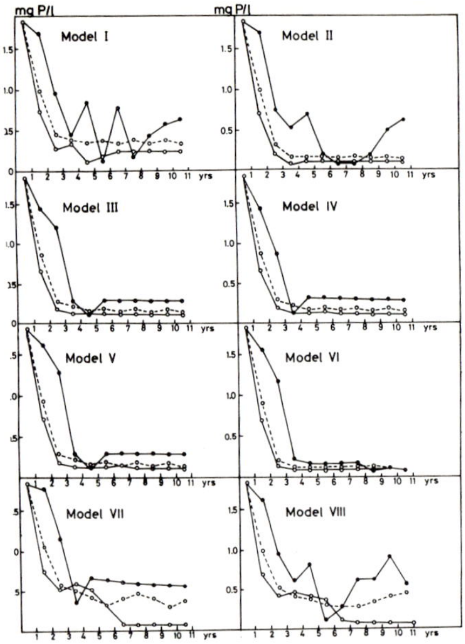

Fig. 17.3: Time-response for total phosphorus after reduction in loading described by various model versions (model I-VIII). The broken line is the time response at the actual hydraulic conditions. The full line with black dots is at 25% of the actual hydraulic loading and the full line with open circles is at 175% of the actual hydraulic loading (From Kamp-Nielsen 1980a).

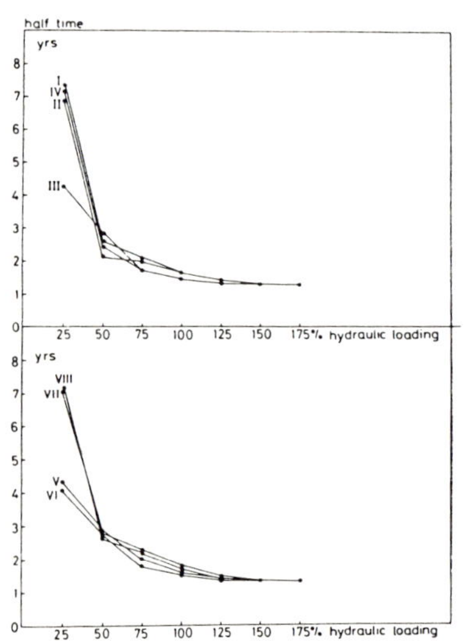

Fig. 17.4: The influence of hydraulic loading on the half time for the initial concentration of total phosphorus described by various model versions (Model I-VIII). (From Kamp-Nielsen 1980a).

Fig. 17.4 shows the response times of the various models versus hydraulic residence time, and it demonstrates that at residence times of one year or shorter, the transients are almost independent of both residence time and model type. This indicates that at short residence times the importance of the sediment pool is small, and that the recovery can be simply explained by a dilution model as shown for L. Norrviken (Ahlgren, 1977).

For management purposes and also for the validation procedure both the correct transient description and the magnitude of the new steady state are important, and in Fig. 17.5 the new steady states 6-10 years after the loading reduction are shown for the various model versions as a function of hydraulic loading. The total phosphorus concentration in the new steady states vary from 0.08 to 0.8 mg P/l. With all submodels except model VI steady state concentrations are reduced with increasing hydraulic loading, but the influence of the model version is reduced when the retention time falls below 4 months, except for the very simple rate expression in model I.

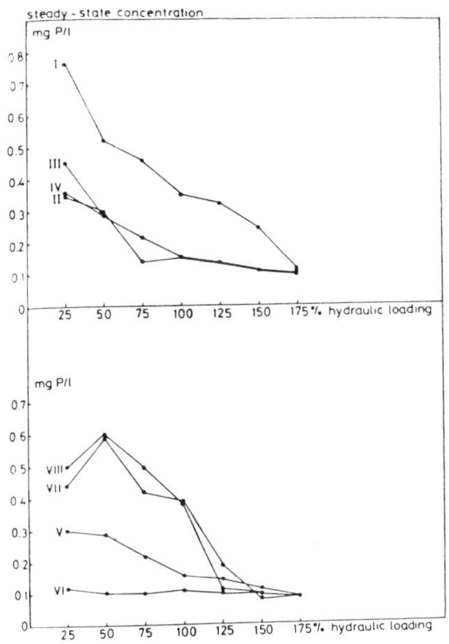

Fig. 17.5: The influence of hydraulic loading on the steady state concentration of total phosphorus after reduction in loading described by various model versions (model I-VIII). The average concentations, in the period 6-11 years after loading reduction, are considered as new steady state concentations. (From Kamp-Nielsen, 1980a).

At a retention time of one year and more, the simulations show that the steady state concentration is very sensitive to choice of model. Only model

VIII gives the same concentration as model VII, which is the version actually validated on L. Glumsø (Jørgensen et al., 1987a) and on several other lakes (Jørgensen et al., 1987b). On the other hand introducing aerobic/anaerobic shifts does not affect the response significantly. The lack of influence of these shifts is explained by the reduced influence of the shifts at higher temperatures, where both the temperature itself and high sedimentation rates and low oxygen concentrations at the sediment surface make the model shift to anaerobic metabolism (Kamp-Nielsen, 1975b).

From Fig. 17.5 it can be concluded that the response for models with several sediment pools and with a temperature dependent mineralization rate is significantly different from the simpler versions (model I-VI). The explanation lies in the magnitude of the exchangeable phosphorus pool PE. In lakes this pool varies from 0.2 to 4.1 g P/m^2 (Jacobsen, 1974; Kamp-Nielsen, 1974, 1977), and the annual gross release from these pools are in the range 0.2 to 2 g P/m^2/year (Kamp-Nielsen, 1974, 1975b, 1977). These rates and pools correspond to turnover rates of about one year for the exchangeable pool and is supported by the annual amplitude of 3 g P/m^2. This conclusion is further supported from the results of the EMIP-study (Sas, 1988) on 18 European lakes, where internal loading or net release from the sediments only occurred a few years after reduced loading. The only exception was L. Søbygaard, where a tremendous pool of iron-bound phosphorus of about 100 g P/m^2 has maintained an internal loading of 4-7 g P/m^2/year for several years after load reduction (Jeppesen et al., 1986).

17.4 PARAMETER ESTIMATION IN SEDIMENT SUBMODELS

For simple steady state models with low time and space resolution net retention rates can be estimated from proper mass balance studies. If one pool in the sediment is used, both gross sediment rate, release rate and net sedimentation rates have to be estimated. A calibration procedure by trial and error as done by Lorentsen et al. (1976), can be used, and much information on sensitivity to initial pool size and to rate constants can be derived, but fundamentally much information is also hidden, since the three process rates are interrelated.

If high resolution in time and space is wanted, either laboratory experiments or field measurements and, best, a combination of these, have to be performed. Gross sedimentation rates can be measured in the field by sediment traps carefully designed. In very shallow water bodies resuspension might introduce significant errors.

Sedimentation rate constants can be estimated by linear regression between trap catches and average concentrations of the considered element in the mixed layer above the trap. Short term exposures are recommended, since sedimentation rates may vary considerably in time (Kamp-Nielsen, 1980b). Corrections for resuspension can be made by comparison of element concentrations in seston, trap catch and the upper part of the sediment, or empirical relations between sedimentation rate constants and the magnitude and duration of the sheer forces.

The size of the exchangeable sediment pool of the element can be estimated, assuming a first order process and performing a non-transport limited exhaustion experiment (Jakobsen, 1974); or by a differential extraction procedure where the solvent can be assumed to extract the mobile pool. For sediment phosphorus several schemes exist (Psenner et al., 1984; Bostrøm et al., 1986), but only a few of the extracted pools have been shown to identify the phosphorus actually released in nature. Loosely adsorbed phosphorus, reductable iron phosphates and a certain fraction of the organically bound phosphorus are generally believed to be easily exchangeable, whereas calcium bound phosphorus is believed not to be exchangeable, since after some time it is converted into apatite.

Mineralization rate constants and sorption equilibria constants can be determined experimentally without or with addition of poisons such as mercury, or antibiotics suppressing the mineralization (Kamp-Nielsen, 1981). These experiments can be performed at various temperatures, pH and redox conditions in the laboratory. To exclude transport limitations the experiments have to be carried out with completely stirred samples either representing the whole assumed active layer of the sediment (down to 5-10 cm sediment depth) or by integration of samples from smaller sediment layers.

Transport processes in the sediment are also best determined experimentally in the laboratory. Apparent diffusion can be determined by using a conservative tracer, e.g. lithium ions, applied to the well mixed water volume above the sediment core. Either by splitting a series of identical cores with various time intervals and pressing or centrifuging the interstitial water off or by taking samples from one core with a filter mounted syringe, the variation with time and depth in the sediment can be estimated. By solving (17.43) with respect to either time or depth, an apparent diffusion coefficient can be estimated for the various sediment depths and by relating the distribution to variations in porosity, a diffusion correction factor can be found. Since diffusion also depends on size of the ions, a correction also has to be made for the difference in molecular diffusion of the conservative element and the considered ion. The influence of bioturbation can be investigated by determing the response in apparent diffusion to introduction of increased number of benthic organisms, or estimates can be made on their activity measured in the laboratory with isolated individuals.

17.5 SELECTION OF SUBMODELS

Table 17.2
Characteristics of various sediment phosphorus models. (From Kamp-Nielsen, 1983).

Type	Number of sediment pools	Characterization of model	References
Steady state	0	Constant fraction of P-loading is sedimented	Piontelli & Tonolli, 1964
	0	Constant fraction of lake-P is sedimented	Vollenweider, 1969
	0	Constant settling velocity	
	0	Constant settling velocity corrected for stratification period	Sonzogni et al., 1976
	0	Net sedimentation related to depth and hydraulic retention time	Dillon & Rigler 1974, 1975
Empirical, predictive	0	Constant settling velocity	Vollenweider, 1975
	0	Retention a function of depth and hydraulic retention time	Kirchner & Dillon, 1975
	0	Constant settling velocity	Chapra, 1975, 1977
	0	Net sedimentation related to depth and hydraulic retention time	Larsen & Mercier, 1975
	0	Constant fraction of lake-P is sedimented	Jones & Bachmann, 1976
	0	Retention a function of hydraulic residence time and depth. Separate models for 3 lake classes	Reckhow, 1977
Empirical, predictive	0	Retention a function of hydraulic residence time and depth	Walker, 1977
Dynamic	0	Constant release rate	Straskraba & Dvorakova, 1977
	0	Settling rate controlled by stratification and flocculation	Snodgrass & O'Melia, 1975
	0	Release rate controlled by concentration gradient	Tessenow, 1972
	0	Release rate controlled by temperature and sedimentation	Nyholm, 1978
	0	Release rate controlled by concentration gradient and sorption on particles	Imboden & Lerman, 1978
	0	Release rate controlled by diffusion and redox conditions	Sjøberg et al 1972, Kamp-Nielsen, 1974, Imboden & Gächter, 1977

Table 17.2 (Continued)

Type	Number of sediment pools	Characterization of model	References
Dynamic	1	Release rate controlled by sedimentation and diffusion	DiGiano & Snow, 1977
	1	1. order mineralization of sediment-P in active layer	Chen & Orlob, 1975, Dahl-Madsen & Nielsenl, 1973, Lorentzen et al., 1976, Hornberger & Spear, 1980
	2	1. order mineralization followed by diffusion of interstitial phosphate	Jørgensen et al., 1978
	3	1. order mineralization from a special active surface sediment and conventional mineralization and diffusion	Jørgensen et al., 1978
	3	1. order mineralization followed by sorption and diffusion changing with oxygen conditions	Kamp-Nielsen, 1976, 1981
	Several	1. order mineralizatio from organic phosphorus pools follwed by diffusion	Youngberg, 1977, Schnellenberg et al., 1975, Bloomfield et al., 1973
	76	1. order mineralization followed by diffusion between several sediment layers	Lung et al., 1976
	250	1. order mineralization followed by sorption and diffusion coupled with an oxygen model	Kamp-Nielsen, 1978, Mejer, 1978

A variety of sediment submodels with a wide span in complexity have been applied in the modelling of aquatic ecosystems. Examples of sediment phosphorus submodels used in lake models have been compiled by Kamp-Nielsen (1983) in Table 17.2.

With the background in sediment processes in comparison with the processes in the water above, some guidelines can be set up for the selection of complexity of sediment submodels. A useful comparison between rate constants for the behaviour of an element in a lake has been set up by Imboden & Lerman (1978) (Table 17.3).

As seen from the table, dilution and decay rates are faster than sediment "dilution" and diffusional fluxes. This means that decomposition at the sediment surface is important, but transport-limited decay in the deeper parts of the sediment is less important. And if the range of diffusion is increased (when moving deeper in the sediment), or if the gradients are becoming smaller with increasing sediment depth, the participation of the deeper situated pools are strongly reduced.

Table 17.3
Rate constants (reciprocal residence time) for a dissolved species in a lake, for diffusional flux at the sediment-water interface (l/t_d), sedimentation and growth of pore water volume (l/t_u), water outflow (l/t_w) and first-order chemical removal or decay (l/t_r). All values have the dimension $year^{-1}$.

l/t_d	l/t_u	l/t_w	l/t_r
0 - 0.1	0 - 0.05	0.1 - 1	0 - 35

Consequently, in lakes with short residence times the sediment fluxes are not important, if both water renewal and sediment-water exchange rates are evenly distributed over the year, but if exchange rates both to and from the sediment are fast in the summer and water renewal is slow at that time, the sediment can cause a significant delay in response to reduced loading. In such cases a high resolution in time and in compartments are necessary for the understanding of the ecosystem behaviour.

In lakes with residence times of several years, the sediment will not delay the response, since hydraulic residence time, and the "dilution" effect will be the most important, but variations from year to year in loading, temperature and residence time may create various seasonal patterns in sediment water exchange rates. If these fluctuations are important for the lake behaviour, complex models must be used.

In selection of the appropirate complexity for sediment submodel one has clearly to define what resolution in time and space is needed. If the question to be answered concerns the whole water body as an ecosystem, the turnover time for the element in the sediment has to be roughly estimated and compared to that in the overlying water, and if it is considerably slower, both pools and rates in the sediment have to be considered in a sediment submodel. The localization of the pool assumed to participate in the exchange can be confined from the distribution in the sediment. If a strong gradient is found, the pool confines itself to the surface sediments, and the transport processes are not so important. But if the gradient is weak due to slow decay or mechanical disturbances (bioturbation and sheer induced mixing), the diffusion description becomes important. Finally, it should be stressed that sediment cores offer an opportunity to test various submodels as far as processes in the sediment are concerned.

REFERENCES

Ahlgren, I., 1977. Role of sediments in the process of recovery of a eutorphicated lake. Proc. Int. Symp. Interactions between sediments and freshwater. Junk & Pudoc. Haag. 372-378.

Berner, R.A., 1974. Kinetic models for the early diagenesis of nitrogen, sulphur, phosphorus and silicone in anoxic marine sediments. In: The Sea. Vol. 5. E.D. Goldberg (ed.). 427-450.
Berner, R.A., 1977. Stochiometric models for nutrient regeneration in anoxic sediments. Limnol. Oceanogr. 22: 781-786.
Bloomfield, J.A., Park, R.A., Scavia, D. and Zahorcak, C.S., 1973. Aquatic modelling in the EDFB, US-IBP. In: Modelling the eutrophication process. E.J. Middlebrooks, D.H. Falkenberg and T.W. Maloney (eds.). Utah Water Research Lab. Utah State University. 139-159.
Bostrøm, B., Jansson, M., and Forsberg, C., 1982. Phosphorus release from sediments. Arch. Hydrobiol. Beih. Ergebn. Limnol. 18: 5-59.

Canfield, D.E. and Bachmann, R.W., 1981. Prediction of total phosphorus concentrations, chlorophyll a and Secchi depths in natural and artificial lakes. Can. J. Fish. Aquat. Sci. 38: 414-423.
Chapra, S.C., 1975. Comment on "An empirical method of estimating the retention of phosphorus in lakes" by W.B. Kirchner and P.J. Dillon. Water Resources Res., 11: 1033-1034.
Chapra, S.C., 1977. Total phosphorus model for the Great Lakes. J. Environ. Eng. Div., ASCE 103 (EE2): 147-161.
Chen, C.W., 1970. Concepts and utilities of ecological models. J. Sanit. Eng. Div. ASCE 96 No. A5.
Chen, C.W., and Orlob, G.T., 1975. Ecological simulation for aquatic environments. In: System and simulation in ecology. B.C. Patten (ed.) Academic Press. New York. Vol. III: 475-587.

Dahl-Madsen, K.I., and Strange Nielsen, K., 1973. Eutrophication models for ponds. Vand 5f: 24-31. (In Danish).
DiGiano, F.A. and Snow, P.D., 1977. Consideration of phosphorus release from sediments in a lake model. Proc. Int. Symp. Interactions between sediments and freshwater. Junk & Pudoc. Haag. 318-324.
DiGiano, F.A. and Rigler, F.H., 1974. A test of a simple nutrient budget model predicting the phosphorus concentration in lake water. J. Fish. Res. Bd. Can. 32: 1771-1778.
DiGiano, F.A. and Rigler, F.H., 1975. A simple method for predicting the capacity of a lake for development based on lake trophic status. J. Fish. Res. Bd. Can. 32: 1519-1531.

Frisk, T., Niemi, J.S. and Kinnunen, K.A.I., 1981. Comparison of statistical phosphorus retention models. Ecol. Modelling. 12: 11-17.
Frissel, M.J. and Reininger, P., 1974. In: Simulation of accumulation and leaching in soils. Wageningen. The Netherland. 88 pp.

Gächter, R., Imboden, D., Bührer, H. and Stadelmann, P., 1983. Mögliche Massnahmen zur Restaurierung des Sempachersees. Schweiz. Z. Hydrol. 45: 246-266.

Hornberger, G.M. and Spear, R.C., 1980. Eutrophication in Peel Inlet. I. The problem-defining behavior and a mathematical model for the phosphorus scenario. Water Res. 14: 29-42.
Håkonson, L. and Jansson, M., 1983. Principles of Lake sedimentology. Springer Verlag. Berlin.

Imboden, D.M. and Gächter, R., 1977. A Dynamic lake model for trophic state prediction. J. Ecol. Modelling. 4: 77-98.
Imboden, D.M. and Lerman, A., 1978. Chemical models of lakes. In: Lakes: Chemistry, Geology, Physics. A. Lerman (ed.) Springer Verlag. New York: 341-356.

Jacobsen, O.S., 1974. Release of mobile phosphorus from Danish lake sediments. Proc. 4th Nordic Sed. Symp. SNV PM-series 694. 144-159.

Jeppesen, E., Kristiansen, A., Søndergaard, O. and Sortkjær, O., 1986. Lavvandede, hurtigt gennemstrømmede søers reaktion på nedsat forforbelastning. Yearbook Freshwater Laboratory. Ministry of Environment. Denmark (In Danish).

Jones, J.R. and Bachmann, R.W., 1976. Predictions of phosphorus and chlorophyll levels in lakes. J. Water Poll. Contr. Fed. 48: 2176-2181.

Jørgensen, S.E., 1976. A eutrophication model for a lake. Ecol. Modelling. 2: 147-165.

Jørgensen, S.E.(ed.), 1986. Fundamentals of Ecological Modelling. Developments in Environmental Modelling. 9. Elsevier. Amsterdam.

Jørgensen, S.E., Mejer, H. and Friis, M., 1978. Examination of a lake model. Ecol. Modelling. 4: 253-278.

Jørgensen, S.E., Kamp-Nielsen, L. and Mejer, H., 1982. Comparison of a simple and a complex sediment phosphorus model. Ibid. 16: 99-124.

Jørgensen, S.E., Kamp-Nielsen, L., Christensen, T., Windolf-Nielsen, J. and Westergaard, B., 1986. Validation of a prognosis based upon a eutrophication modelling, Ibid. 32: 165-82.

Jørgensen, S.E., Kamp-Nielsen, L. and Jørgensen, L.A., 1986. Examination of the generality of eutrophication models. Ibid. 32: 251-266.

Kamp-Nielsen, L., 1974. Mud-water exchange of phosphate and other ions in undisturbed sediment cores and factors affecting the exchange rates. Arch. Hydrobiol. 73: 218-237.

Kamp-Nielsen, L., 1975a. Seasonal variation in sediment-water exchange of nutrient ions in L. Esrom. Verh. Internat. Verein. Limnol. 19: 1057-1065.

Kamp-Nielsen, L., 1975b. A kinetic approach to the aerobic sediment-water exchange of phosphorus in L. Esrom. Ecol. Modelling. 1: 153-160.

Kamp-Nielsen, L., 1977a. Modelling the temporal variation in sedimentary phosphorus fractions. Proc. Int. Symp. Interactions between sediments and freshwater. Junk & Pudoc. Haag: 277-186.

Kamp-Nielsen, L., 1978. Modelling the vertical gradients in sedimentary phosphorus fractions. Verh. Internat. Verein. Limnol. 20: 720-727.

Kamp-Nielsen, L., 1980a. The influence of changed phosphorus loading to hypertrophic L. Glumsø. Proc. SIL Workshop on hypertrophic ecosystems. Växsjö, Sweden. Developments in Hydrobiology. 2: 29-36.

Kamp-Nielsen, L., 1980b. Intensive measurements of sedimentation in L. Esrom and L. Glumsø. Proc. 8th Nordic Symp. Sed. Salten Skov, Denmark. 44-56.

Kamp-Nielsen, L., 1983. Sediment-water exchange models. Developments in Environmental Modelling 4A. Elsevier. Amsterdam. 387-416.

Kamp-Nielsen, L., Jørgensen, S.E. and Jørgensen, L.A., 1981. A sediment-water exchange model for lakes in the Upper Nile Basin. In : Progress in Ecological Engineering and Management by Mathematical Modelling. D.M. Dubois (ed.) Editions Cebedoc. Liege, Belgium. 557-582.

Kirchner, W.B. and Dillon, P.J., 1975. An empirical method of estimating the retention of phosphorus in lakes. Water Resources Res. 11: 182-183.

Larsen, D.P. and Mercier, H.T., 1975. Lake phosphorus loading graphs: an alternative. National Eutrophication Survey. Working paper No. 174. EPA Corvallis. Oregon.

Lassiter, R.R. and Kearns, D.K., 1975. Phytoplankton population changes and nutrient fluctuations in a simple aquatic ecosystem model. Modelling the eutrophication Process. Ann. Arbor Science. 131-138.

Lorentzen, M.W., Smith, D.J. and Kimmel, L.V., 1976. A long-term phosphorus model for lakes: application to L. Washington. In: Modelling biochemical

processes in aquatic ecosystems. R.P. Canale (ed.). Ann Arbor Science. Ann Arbor. 75-92.
Lung, W.S., Canale, R.P. and Freedman, P.L., 1976. Phosphorus model for eutrophic lakes. Water Res. 10: 1101-1115.

Mejer, H., 1978. Sediment phosphorus model. Proc. 6th Nordic Sed. Symp. Hurdal, Norway. 137-145.

Nyholm, N., 1978. A simulation model for phytoplankton growth and nutrient cycling in eutrophic, shallow lakes. Ecol. Modelling. 4: 279-311.
Nürnberg, G.K., 1984. The prediction of internal loading in lakes with anaerobic hypolimnia. Limnol. Oceanogr. 29: 111-124.

OECD, 1982. Eutrophication of waters. Monitoring, assesment and control. OECD. Paris.

Piontelli, R. and Tonolli, V., 1964. The time of retention of lacustrine waters in relation to the phenomenon of enrichment ín introduced substances, with special reference to the L. Maggiore. Mem. Ist. Ital. Idrobiol. 17: 247-266.
Psenner, V.R., Pucsko, R. und Sager, M., 1984. Die Fraktionierung organischer und anorganischer Phosphorverbindungen von Sedimenten. Arch. Hydrobiol. Suppl. 70: 111-155.

Reckhow, K.H., 1977. Phosphorus models for lake management. Ph. D. dissertation, Harvard University.
Reckhow, K.H., 1979. Empirical lake models for phosphorus: development, applications, limitations and uncertainty. In: Perspectives on lake ecosystem modelling. D. Scavia and A. Robertson (eds.). Ann Arbor Science. Ann Arbor: 193-222.
Reynolds, C.S., 1984. The ecology of freshwater phytoplankton. Cambridge University Press. Cambridge.

Sas, H., 1988. Lake restoration by nutrient control - expectations, experiences and extrapolations. In press.
Schnellenberg, G., Hoeg, S., Kozerski, H.-P., and Cramer, F., 1975. Biokypernetisches Modell der Bodenette südlich des Darss des Zingst. Ber. Forsh. - Stelle Umweltgest. Akad. Wiss. DDR, Berlin. Cited in: H.-P. Kozerski, 1979. Neuere Arbeiten zur mathematischen Modellierung des Nährstoffaustausches zwischen Sediment und Freiwasser. Acta Hydrophysica. 24: 5-16.
Sjøberg, S., Wulff, F. and Wahlström, P., 1972. Computer simulation of hydrochemical and biological processes in the Baltic. Contrib. Askö Labor. Stockholm. 1: 1-91.
Skogheim, O. and Erlandsen, A.H., 1979. Transport mechanisms of solutes in lake sediments. Proc. 7th Nordic Symp. Sed. Aneboda, Sweden. 124-134.
Smayda, T.J., 1970. The suspension and sinking of phytoplankton in the sea. Oceanogr. Mar. Biol. Ann. Rev. 8: 353-414.
Smayda, T.J., 1974. Some experiments on sinking characteristics of two freshwater diatoms. Limnol. Oceanogr. 19: 628-635.
Snodgrass, W.J. and O'Melia, C.R., 1975. Predictive model for phosphorus in lakes. Environ. Sci. Technol. 9: 937-945.
Sonzogni, W.C., Uttormark, P.C. and Lee, G.F., 1976. A phosphorus residence time model: theory and application. Water Res. 10: 429-435.
Straskraba, M. and Dvorakova, M., 1977. Probleme der Simulation von Talsperrenökosystemen. Acta Hydrochim. et Hydrobiol. 5: 211-230.

Tessenow, U., 1972. Lösungs-, Diffusions- und Sorptionsprozesse in der

Obersicht von Seesedimenten. I. Ein Langzeitexperiment unter anaeroben und aeroben Bedingungen in Fliessgleichgewicht. Arch. Hydrobiol. Suppl. 38: 353-398.

Van Raaphorst, W. and Brinkman, A.G., 1985. The calculation of transport coefficients of phosphate and calcium fluxes across the sediment-water interface from experiments with undisturbed sediment cores. Wat. Sci. Tech., 17: 941-951.

Van Straten, G., 1986. Hypothesis testing and parameter uncertainty analysis in simple phytoplankton-P models. In: Modelling and managing shallow lake eutrophication. Springer Verlag. Berlin.

Vanderborght, J.P. and Billen, G., 1975. Vertical distribution of nitrate concentration in interstitial water of marine sediments with nitrification and denitrification. Limnol. Oceanogr. 20: 953-961.

Vollenweider, R.A., 1964. The correlation between area of inflow and lake budget. Proc. Ger. Limnol. Conf. Lunz. Germany.

Vollenweider, R.A., 1969. Possibilities and limits of elementary models concerning the budget of substances in lakes. Arch. Hydrobiol. 66: 1-36.

Vollenweider, R.A., 1975. Input-output models with special reference to the phosphorus loading concept in limnology. Schweiz. Z. Hydrol. 37: 53-84.

Vollenweider, R.A., 1976. Advances in defining critical loading levels for phosphorus in lake eutrophication. Mem. Ist. Ital. Idrobiol. 33: 53-83.

Walker, W.W., 1977. Some analytical methods applied to lake water quality problems. Ph. D. dissertation, Harvard University.

Yongberg, B.A., 1977. Application of the model CLEANER to a stratified reservoir system. Ph. D. dissertation. Rensselear Polytechnic Institute. New York.

APPENDIX

Table 1
Volumes and densities of the unit world

Compartment	Volume (m^3)	Density (kg/m^3)	Organic carbon content (fraction)
Air	$6.0 * 10^9$ (6000-m high)	1.21	-
Water	$7.0 * 10^6$ (10-m deep)	1000	-
Soil	$4.5 * 10^4$ (15-cm deep)	1500	-
Sediment	$2.1 * 10^4$ (3-cm deep)	1500	0.02
Suspended sediment	35.0 (5-ppm vol)	1500	0.04
Biota	7.0 (1-ppm vol)	1000	-

Table 2
Definition of fugacity capacities

Compartment		Definition of Z (mol/m^3Pa)
Air (subscript A)	$1/RT$	R = 8.314 Pa m^3/mol K T = Temperature (K)
Water (subscript W)	$1/H$ or C^S/P^S	C^S = aqueous solubility (mol/m^3) P^S = vapor pressure (Pa) H = Henry's law constant (Pa m^3/mol)
Solid Sorbent (e.g. soil, sediment, particles) (subscript S)	$K_{sw}p_s/H$	K_{sw} = partition coefficient (l/kg) p_s = density (kg/l)
Biota (subscript F)	$K_{fw}p_f/H$	K_{fw} = bioconcentration factor (l/kg) p_f = density (kg/l)

TABLE 3
Activity coefficient f at different ionic strengths

I	$\frac{\sqrt{I}}{1+\sqrt{I}}$	f for Z = 1	f for Z = 2	f for Z = 3
0	0	1.00	1.00	1.00
0.001	0.03	0.97	0.87	0.73
0.002	0.04	0.95	0.82	0.64
0.005	0.07	0.93	0.74	0.51
0.01	0.09	0.90	0.66	0.40
0.02	0.12	0.87	0.57	0.28
0.05	0.18	0.81	0.43	0.15
0.1	0.24	0.76	0.33	0.10
0.2	0.31	0.70	-	-
0.5	0.41	0.62	-	-

I = ionic strength, Z = charge, f = activity coefficient

LIST OF AUTHORS

Collins, Carol D., NY State Science Service, Albany, NY 12230.

From, Jon, Freshwater Laboratory (DMU), Fishery Department, DK-8600 Silkeborg, Denmark.

Gromiec, M.J., Institute of Meteorology and Water Management, 01-673 Warszawa, ul. Podlesna 61.

Hahn, Hermann H., Institut für Siedlungswasserwirtschaft der Universität Karlsruhe, D-7500 Karlsruhe 1.

Jensen, Arne, The Royal Danish School of Pharmacy, Dept. of Chemistry AD, 2 Universitetsparken, DK-2100 Copenhagen, Denmark.

Jørgensen, Sven Erik, The Royal Danish School of Pharmacy, Dept. of Chemistry AD, 2 Universitetsparken, DK-2100 Copenhagen, Denmark.

Kamp-Nielsen, Lars, Freshwater-Biological Laboratory, University of Copenhagen, DK-3400 Hillerød, Denmark.

Legovic, Tarzan, Institute of Ecology, University of Georgia, Athens, GA 30602.

Marjanovic, N., University of California, Davis. Department of Civil Engineering.

Ohgaki, Shinichiro, Department of Urban Engineering, University of Tokyo.

Orlob, Gerald T., University of California, Davis. Department of Civil Engineering.

Park, Richard A., Holcomb Research Institute, Butler University, Indianapolis, IN 46208.

Rasmussen, Gorm, Freshwater Laboratory (DMU), Fishery Department, DK-8600 Silkeborg, Denmark.

Tassin, Bruno, Centre d'Enseignement et de Recherche pour la Gestion des Ressources Naturelles et l'Environment, Ecole Nationale des Ponts et Chaussées, La Courtine, Boite 105, 93 194 Noisy le Grand, France.

Thévenot, Daniel R., Laboratoire de Bioélectrochimie et d'Analyse du Milieu, U.F.R. de Sciences et de Technologie, Université Paris Val de Marne, Avenue du Général de Gaulle, 94 010 Créteil Cedex, France.

Wantawin, Chalermraj, Environmental Engineering Division, Asian Institute of Technology.

Wong, George T.F., Department of Oceanography, Old Dominion University, Norfolk, VA 23508.

INDEX